Environmental Impacts of Wind-Energy Projects

Committee on Environmental Impacts of Wind-Energy Projects

Board on Environmental Studies and Toxicology

Division on Earth and Life Studies

NATIONAL RESEARCH COUNCIL
OF THE NATIONAL ACADEMIES

THE NATIONAL ACADEMIES PRESS
Washington, D.C.
www.nap.edu

THE NATIONAL ACADEMIES PRESS   500 Fifth Street, NW   Washington, DC 20001

NOTICE: The project that is the subject of this report was approved by the Governing Board of the National Research Council, whose members are drawn from the councils of the National Academy of Sciences, the National Academy of Engineering, and the Institute of Medicine. The members of the committee responsible for the report were chosen for their special competences and with regard for appropriate balance.

This project was supported by Contract No. EC25C001 between the National Academy of Sciences and Executive Office of the President, Council on Environmental Quality. Any opinions, findings, conclusions, or recommendations expressed in this publication are those of the author(s) and do not necessarily reflect the view of the organizations or agencies that provided support for this project.

International Standard Book Number-13:   978-0-309-10834-8 (Book)
International Standard Book Number-10:   0-309-10834-9 (Book)
International Standard Book Number-13:   978-0-309-10835-5 (PDF)
International Standard Book Number-10:   0-309-10835-7 (PDF)

Library of Congress Control Number 2007931763

Cover design by Liza Hamilton, National Research Council. Photograph of 1.5 mW turbines at San Gorgonio, CA, by David Policansky, National Research Council. Graph adapted from Figure 1-1, which was reproduced with permission from the American Wind Energy Association.

Additional copies of this report are available from

The National Academies Press
500 Fifth Street, NW
Box 285
Washington, DC 20055

800-624-6242
202-334-3313 (in the Washington metropolitan area)
http://www.nap.edu

THE NATIONAL ACADEMIES
Advisers to the Nation on Science, Engineering, and Medicine

The **National Academy of Sciences** is a private, nonprofit, self-perpetuating society of distinguished scholars engaged in scientific and engineering research, dedicated to the furtherance of science and technology and to their use for the general welfare. Upon the authority of the charter granted to it by the Congress in 1863, the Academy has a mandate that requires it to advise the federal government on scientific and technical matters. Dr. Ralph J. Cicerone is president of the National Academy of Sciences.

The **National Academy of Engineering** was established in 1964, under the charter of the National Academy of Sciences, as a parallel organization of outstanding engineers. It is autonomous in its administration and in the selection of its members, sharing with the National Academy of Sciences the responsibility for advising the federal government. The National Academy of Engineering also sponsors engineering programs aimed at meeting national needs, encourages education and research, and recognizes the superior achievements of engineers. Dr. Charles M. Vest is president of the National Academy of Engineering.

The **Institute of Medicine** was established in 1970 by the National Academy of Sciences to secure the services of eminent members of appropriate professions in the examination of policy matters pertaining to the health of the public. The Institute acts under the responsibility given to the National Academy of Sciences by its congressional charter to be an adviser to the federal government and, upon its own initiative, to identify issues of medical care, research, and education. Dr. Harvey V. Fineberg is president of the Institute of Medicine.

The **National Research Council** was organized by the National Academy of Sciences in 1916 to associate the broad community of science and technology with the Academy's purposes of furthering knowledge and advising the federal government. Functioning in accordance with general policies determined by the Academy, the Council has become the principal operating agency of both the National Academy of Sciences and the National Academy of Engineering in providing services to the government, the public, and the scientific and engineering communities. The Council is administered jointly by both Academies and the Institute of Medicine. Dr. Ralph J. Cicerone and Dr. Charles M. Vest are chair and vice chair, respectively, of the National Research Council.

www.national-academies.org

COMMITTEE ON ENVIRONMENTAL IMPACTS
OF WIND-ENERGY PROJECTS

Members

PAUL RISSER (*Chair*), University of Oklahoma, Norman
INGRID BURKE, Colorado State University, Ft. Collins
CHRISTOPHER CLARK, Cornell University, Ithaca, NY
MARY ENGLISH, University of Tennessee, Knoxville
SIDNEY GAUTHREAUX, JR., Clemson University, Clemson, SC
SHERRI GOODMAN, The Center for Naval Analysis, Alexandria, VA
JOHN HAYES, University of Florida, Gainesville
ARPAD HORVATH, University of California, Berkeley
THOMAS H. KUNZ, Boston University, Boston, MA
LYNN MAGUIRE, Duke University, Durham, NC
LANCE MANUEL, University of Texas, Austin
ERIK LUNDTANG PETERSEN, Risø National Laboratory, Frederiksborgvej,
    Denmark
DALE STRICKLAND, WEST, Inc., Cheyenne, WY
JEAN VISSERING, Jean Vissering Landscape Architecture, Montpelier, VT
JAMES RODERICK WEBB, University of Virginia, Charlottesville
ROBERT WHITMORE, West Virginia University, Morgantown

Staff

DAVID POLICANSKY, Study Director
RAYMOND WASSEL, Senior Program Officer
JAMES ZUCCHETTO, Director, Board on Energy and Environmental Systems
MIRSADA KARALIC-LONCAREVIC, Manager, Technical Information Center
BRYAN SHIPLEY, Research Associate
JOHN H. BROWN, Program Associate
JORDAN CRAGO, Senior Project Assistant
RADIAH ROSE, Senior Editorial Assistant

Sponsor

EXECUTIVE OFFICE OF THE PRESIDENT, COUNCIL ON ENVIRONMENTAL QUALITY

OTHER REPORTS OF THE BOARD ON ENVIRONMENTAL STUDIES AND TOXICOLOGY

Scientific Review of the Proposed Risk Assessment Bulletin from the Office of Management and Budget (2007)
Assessing the Human Health Risks of Trichloroethylene: Key Scientific Issues (2006)
New Source Review for Stationary Sources of Air Pollution (2006)
Human Biomonitoring for Environmental Chemicals (2006)
Health Risks from Dioxin and Related Compounds: Evaluation of the EPA Reassessment (2006)
Fluoride in Drinking Water: A Scientific Review of EPA's Standards (2006)
State and Federal Standards for Mobile-Source Emissions (2006)
Superfund and Mining Megasites—Lessons from the Coeur d'Alene River Basin (2005)
Health Implications of Perchlorate Ingestion (2005)
Air Quality Management in the United States (2004)
Endangered and Threatened Species of the Platte River (2004)
Atlantic Salmon in Maine (2004)
Endangered and Threatened Fishes in the Klamath River Basin (2004)
Cumulative Environmental Effects of Alaska North Slope Oil and Gas Development (2003)
Estimating the Public Health Benefits of Proposed Air Pollution Regulations (2002)
Biosolids Applied to Land: Advancing Standards and Practices (2002)
The Airliner Cabin Environment and Health of Passengers and Crew (2002)
Arsenic in Drinking Water: 2001 Update (2001)
Evaluating Vehicle Emissions Inspection and Maintenance Programs (2001)
Compensating for Wetland Losses Under the Clean Water Act (2001)
A Risk-Management Strategy for PCB-Contaminated Sediments (2001)
Acute Exposure Guideline Levels for Selected Airborne Chemicals (five volumes, 2000-2007)
Toxicological Effects of Methylmercury (2000)
Strengthening Science at the U.S. Environmental Protection Agency (2000)
Scientific Frontiers in Developmental Toxicology and Risk Assessment (2000)
Ecological Indicators for the Nation (2000)
Waste Incineration and Public Health (2000)
Hormonally Active Agents in the Environment (1999)
Research Priorities for Airborne Particulate Matter (four volumes, 1998-2004)
The National Research Council's Committee on Toxicology: The First 50 Years (1997)
Carcinogens and Anticarcinogens in the Human Diet (1996)
Upstream: Salmon and Society in the Pacific Northwest (1996)
Science and the Endangered Species Act (1995)
Wetlands: Characteristics and Boundaries (1995)
Biologic Markers (five volumes, 1989-1995)
Review of EPA's Environmental Monitoring and Assessment Program (three volumes, 1994-1995)
Science and Judgment in Risk Assessment (1994)
Pesticides in the Diets of Infants and Children (1993)
Dolphins and the Tuna Industry (1992)
Science and the National Parks (1992)
Human Exposure Assessment for Airborne Pollutants (1991)
Rethinking the Ozone Problem in Urban and Regional Air Pollution (1991)
Decline of the Sea Turtles (1990)

Copies of these reports may be ordered from the National Academies Press
(800) 624-6242 or (202) 334-3313
www.nap.edu

Preface

The generation of electricity from wind energy is surprisingly controversial. At first glance, obtaining electricity from a free source of energy—the wind—seems to be an optimum contribution to the nation's goal of energy independence and to solving the problem of climate warming due to greenhouse gas emissions. As with many first glances, however, a deeper inspection results in a more complicated story. How wind turbines are viewed depends to some degree on the environment and people's predilections, but not everyone considers them beautiful. Building wind-energy installations with large numbers of turbines can disrupt landscapes and habitats, and the rotating turbine blades sometimes kill birds and bats. Calculating how much wind energy currently displaces other, presumably less-desirable, energy sources is complicated, and predicting future displacements is surrounded by uncertainties.

Although the use of wind energy has grown rapidly in the past 25 years, frequently subsidized by governments at various levels and in many countries eager to promote cleaner alternative energy sources, regulatory systems and planning processes for these projects are relatively immature in the United States. At the national scale, regulation is minimal, unless the project receives federal funding, and the regulations are generic for construction and management projects or are promulgated as guidelines. Regulation at the state and local level is variable among jurisdictions, some with well-developed policies and others with little or no framework, relying on local zoning ordinances. There are virtually no policy or regulatory frameworks at the multistate regional scale, although of course the impacts and benefits of wind-energy installations are not constrained by political boundaries.

This is the complex scientific and policy environment in which the committee worked to address its responsibility to study the environmental impacts of wind energy, including the adverse and beneficial effects. Among the specified considerations were the impacts on landscapes, viewsheds, wildlife, habitats, water resources, air pollution, greenhouse gases, materials-acquisition costs, and other impacts. The committee drew on information from throughout the United States and abroad, but by its charge, focused on the Mid-Atlantic Highlands (a mountainous region in Pennsylvania, Virginia, Maryland, and West Virginia). Using existing information, the committee was able to develop a framework for evaluating those effects; we hope this framework can inform future siting decisions of wind-energy projects. Often, there is insufficient information to provide certainty for these decisions, and thus in the process of its work the committee identified major research needed to improve the assessment of impacts and inform the siting and operational decisions of wind-energy projects.

The committee membership included diverse areas of expertise needed to address the committee's charge. Committee members originated from across the United States, and one hails from Denmark, adding to the international perspective of the study. Members represented the public and private sectors, and numerous natural and social science disciplines. But most important, the committee worked together as a cohesive group in deciding what issues were important and how important, examining issues from multiple perspectives, recognizing and dealing with biases, framing questions and issues in formats that would convey information effectively to decision makers, and considering, respecting, and reconciling differences of opinion, judgment, and interpretation.

The committee broadly defined "environmental" impacts to include traditional environmental measures such as species, habitats, and air and water quality, but attention was also devoted to aesthetic, cultural, recreational, social, and economic impacts. The committee recognized that the planning, policy, and regulatory considerations were paramount if information about impacts was to be translated into informed decision making. Finally, because decision making about wind-energy projects occurs at a variety of geographic and jurisdictional levels, the committee paid careful attention to scale issues as it addressed impacts and benefits.

The benefits of wind energy depend on the degree to which the adverse effects of other energy sources can be reduced by using wind energy instead of the other sources. Assessing those benefits is complicated. The generation of electricity by wind energy can itself have adverse effects, and projecting the amount of wind-generated electricity available in the future is quite uncertain. In addition, the amount of potential displacement of other energy sources depends on characteristics of the energy market, operation of the transmission grid, capacity factor of the wind-energy generators as well as that of other types of electricity generators, and regulatory policies and practices affecting the production of greenhouse gases. Even if the amount of energy that wind energy displaces is small, it is clear that the nation will depend on multiple energy sources for the foreseeable future and reduction of environmental impacts will thereby require multiple approaches.

The committee began its work expecting that there would be measurable environmental impacts, including biological and socioeconomic impacts, and that there would be inadequate data from which to issue definitive, broadly applicable determinations. Given the complexity of the electric-power industry, the dynamics of energy markets, and the rapidity of technological change, we also expected that predicting the environmental benefits of wind energy would be challenging. On the other hand, the lack of any truly coordinated planning, policy, and regulatory framework at all jurisdictional levels loomed larger than expected throughout our deliberations. Although some predictions about future adverse environmental effects of wind-energy use can be made, the committee recognized gaps in our knowledge and recommended specific monitoring studies that will enable more rigorous siting and operational decisions in the future. Similarly, the report includes descriptions of measures of social impacts of wind-energy development, and recommends studies that would improve our understanding of these impacts.

The complexity of assessing the environmental impacts of wind-energy development can be organized in a three-dimensional action space. These dimensional axes include spatial jurisdictions (local, state/regional, and federal), timing of project stages (pre-project, construction, operational, and post-operational) and environmental and human impacts, each of which include their own time and space considerations. The committee evaluated these issues in offering an evaluation guide for organizing the assessment of environmental impacts. We hope that the results of these deliberations and the evaluations and observations in this report will significantly improve the nation's ability to plan, regulate, and assess the impacts of wind-energy development.

This report has been reviewed in draft form by individuals chosen for their diverse perspectives and technical expertise, in accordance with procedures approved by the National Research Council's Report Review Committee. The purpose of this independent review is to provide candid and critical comments that will assist the institution in making its published report as sound as possible and to ensure that the report meets institutional standards of objectivity, evidence, and responsiveness to the study charge. The review comments and draft manuscript remain confidential to protect the integrity of the deliberative process. We thank the following individuals for their review of this report:

Jan Beyea, Consulting in the Public Interest
Dallas Burtraw, Resources for the Future
Michael Corradini, University of Wisconsin-Madison
Samuel Enfield, PPM Atlantic Renewable
Chris Hendrickson, Carnegie Mellon University
Alan Hicks, New York Department of Environmental Conservation
Mark Jacobson, Stanford University
Kevin Porter, Exeter Associates
Paul Kerlinger, Curry & Kerlinger, LLC
Ronald Larkin, Illinois Natural History Survey
Martin Pasqualetti, Arizona State University
John Sherwell, Maryland Department of Natural Resources
Linda Spiegel, California Energy Commission
James Walker, enXco, Inc.

Although the reviewers listed above have provided many constructive comments and suggestions, they were not asked to endorse the conclusions or recommendations, nor did they see the final draft of the report before its release. The review of this report was overseen by the review coordinator, Gordon H. Orians of the University of Washington (emeritus), and the review monitor, Elsa M. Garmire of Dartmouth College. Appointed by the National Research Council, they were responsible for making certain that an independent examination of this report was carried out in accordance with institutional procedures and that all review comments were carefully considered. Responsibility for the final content of this report rests entirely with the authoring committee and the institution.

The committee gratefully acknowledges the following for making presentations to the committee: Dick Anderson (WEST, Inc.), Edward Arnett (Bat Conservation International), Dinah Bear (Council on Environmental Quality), Gwenda Brewer (Maryland Department of Natural Resources), Daniel Boone (Consultant), Steve Brown (West Virginia Department of Natural Resources), Richard Cowart (The Regulatory Assistance Project), Samuel Enfield (PPM Atlantic Renewable), Ken Hamilton (Whitewater Energy), Alex Hoar (U.S. Fish and Wildlife Service), Judith Holyoke Schoyer Rodd (Friends of the Blackwater), Tom Kerr (U.S. Environmental Protection Agency), Julia Levin (California Audubon), Patricia McClure (Government Accountability Office), The Honorable Alan B. Mollohan (U.S. Representative, WV 1st Congressional District), Kevin Rackstraw (American Wind Energy Association Siting Committee), Dennis Scullion (EnXco, Inc.), John Sherwell (Maryland Department of Natural Resources), Craig Stihler (West Virginia Department of Natural Resources), Robert Thresher (National Renewable Energy Laboratory), James A. Walker (EnXco, Inc.), and Carl Zichella (Sierra Club). In addition, John Reynolds and Joseph Kerecman of PJM Interconnection and officials of Dominion Resources provided helpful information to the committee through personal communications; Laurie Jodziewicz of the American Wind Energy Association, Nancy Rader of the California Wind Energy Association, and Linda White of the Kern Wind Energy Association provided helpful information and contacts. We also thank Wayne Barwikowski and his colleagues at enXco, Inc. for their informative and helpful tour of the San Gorgonio (Palm Springs) wind-energy facility.

The committee's work was enhanced in every way by the extraordinary work of the project director, David Policansky, who provided endless sound advice, insightful expertise, and just good sense. The committee offers David its sincere gratitude for his attentive assistance and for his good fellowship throughout the project, which involved five meetings in five different locations with field trips to several wind-energy installations and public hearings. Ray Wassel and James Zucchetto also provided valuable help in framing questions, analyzing literature, and clarifying our thought processes and writings. Bryan Shipley helped to identify relevant literature and to summarize it for the committee. John Brown helped with meeting planning, including arranging field trips and helping to make sure that the committee arrived where it was supposed to be and returned in good condition. Jordan Crago supported the committee in so many ways that I cannot list them all, but they include literature searching and verification (along with Mirsada Karalic-Loncarevic), organizing drafts and committee comments, and keeping the committee housed and fed. Finally, Board Director James Reisa provided his usual wise counsel at difficult times, and his comments have improved the clarity and relevance of this report. We are grateful to them all.

Finally, I want to offer a personal note of appreciation to the committee and the staff. This was an extraordinary group of people, all with outstanding credentials but many points of view, who came together over the past two years to address an important and challenging topic. During this time they listened to each other, helped each other, and worked incredibly hard. It has been an honor to chair the committee, and my life has been enriched by the time and talents of my committee colleagues.

Paul G. Risser, Chair
Committee on Environmental Impacts of
Wind-Energy Projects

Contents

Environmental Impacts
of Wind-Energy Projects

Wind energy has a long history, having been used for sailing vessels at least since 3100 BC. Traditionally, windmills were used to lift water and grind grain as early as the 10th century AD. However, significant electricity generation from wind in the United States began only in the 1980s, in California; today, electricity is generated from wind in 36 states, including Alaska and Hawaii.

There has been a rapid evolution of wind-turbine design over the past 25 years. Thus, modern turbines are different in many ways from the turbines that were originally installed in California's three large installations at Altamont Pass, Tehachapi, and San Gorgonio (Palm Springs). A typical modern generator consists of a pylon about 60 to 90 meters (m) high with a three-bladed rotor about 70 to 90 m in diameter mounted atop it. Larger blades and taller towers are becoming more common. Other support facilities usually include relatively small individual buildings and a substation.

This study is concerned with utility-scale clusters of generators often referred to as "wind farms," not with small turbines used for individual agricultural farms or houses. Some of the installations contain hundreds of turbines; the wind installation at Altamont Pass in California consists of more than 5,000, and those at Tehachapi and Palm Springs contain at least 3,000 each, ranging from older machines as small as 100 kilowatts (kW) to more modern 1.5 MW turbines. The committee that produced this report focused only on installations onshore. There were no offshore wind-energy installations in the United States as of the beginning of 2007.

THE PRESENT STUDY

Statement of Task

The National Research Council was asked to establish an expert committee to carry out a scientific study of the environmental impacts of wind-energy projects, focusing on the Mid-Atlantic Highlands[1] (MAH) as a case example. The study was to consider adverse and beneficial effects, including impacts on landscapes, viewsheds, wildlife, habitats, water resources, air pollution, greenhouse gases, materials-acquisition costs, and other impacts. Using information from wind-energy projects proposed or in place in the MAH and other regions as appropriate, the committee was charged to develop an analytical framework for evaluating those effects to inform siting decisions for wind-energy projects. The study also was to identify major areas of research and development needed to better understand the

---

[1]The MAH refers to elevated regions of Virginia, West Virginia, Maryland, and Pennsylvania.

Environmental Impacts
of Wind-Energy Projects

Summary

INTRODUCTION

In recent years, the growth of capacity to generate electricity from wind energy has been rapid, growing from almost none in 1980 to 11,603 megawatts (MW) in 2006 in the United States and about 60,000 MW in 2006 globally. Despite this rapid growth, wind energy amounted to less than 1% of U.S. electricity generation in 2006.

Generation of electricity by wind energy has the potential to reduce environmental impacts caused by use of fossil fuels to generate electricity because, unlike fossil fuels, wind energy does not generate atmospheric contaminants or thermal pollution, thus being attractive to many governments, organizations, and individuals. Others have focused on adverse environmental impacts of wind-energy facilities, which include aesthetic and other impacts on humans and effects on ecosystems, including the killing of wildlife, especially birds and bats. Some environmental effects of wind-energy facilities, especially those from transportation (roads to and from the plant site) and transmission (roads or clearings for transmission lines), are common to all electricity-generating plants; other effects, such as their aesthetic impacts, are specific to wind-energy facilities.

This report provides analyses to help to understand and evaluate positive and negative environmental effects of wind-energy facilities. The committee was not asked to consider, and therefore did not address, non-environmental issues associated with generating electricity from wind energy, such as energy independence, foreign-policy considerations, resource utilization, and the balance of international trade.

1

Wind energy has a long history, having been used for sailing vessels at least since 3100 BC. Traditionally, windmills were used to lift water and grind grain as early as the 10th century AD. However, significant electricity generation from wind in the United States began only in the 1980s, in California; today, electricity is generated from wind in 36 states, including Alaska and Hawaii.

There has been a rapid evolution of wind-turbine design over the past 25 years. Thus, modern turbines are different in many ways from the turbines that were originally installed in California's three large installations at Altamont Pass, Tehachapi, and San Gorgonio (Palm Springs). A typical modern generator consists of a pylon about 60 to 90 meters (m) high with a three-bladed rotor about 70 to 90 m in diameter mounted atop it. Larger blades and taller towers are becoming more common. Other support facilities usually include relatively small individual buildings and a substation.

This study is concerned with utility-scale clusters of generators often referred to as "wind farms," not with small turbines used for individual agricultural farms or houses. Some of the installations contain hundreds of turbines; the wind installation at Altamont Pass in California consists of more than 5,000, and those at Tehachapi and Palm Springs contain at least 3,000 each, ranging from older machines as small as 100 kilowatts (kW) to more modern 1.5 MW turbines. The committee that produced this report focused only on installations onshore. There were no offshore wind-energy installations in the United States as of the beginning of 2007.

THE PRESENT STUDY

Statement of Task

The National Research Council was asked to establish an expert committee to carry out a scientific study of the environmental impacts of wind-energy projects, focusing on the Mid-Atlantic Highlands[1] (MAH) as a case example. The study was to consider adverse and beneficial effects, including impacts on landscapes, viewsheds, wildlife, habitats, water resources, air pollution, greenhouse gases, materials-acquisition costs, and other impacts. Using information from wind-energy projects proposed or in place in the MAH and other regions as appropriate, the committee was charged to develop an analytical framework for evaluating those effects to inform siting decisions for wind-energy projects. The study also was to identify major areas of research and development needed to better understand the

---

[1]The MAH refers to elevated regions of Virginia, West Virginia, Maryland, and Pennsylvania.

environmental impacts of wind-energy projects and to reduce or mitigate negative environmental effects.

Current Guidance for Reviewing Wind-Energy Proposals

The United States is in the early stages of learning how to plan for and regulate wind-energy facilities. Federal regulation of wind-energy facilities is minimal if the facility does not have a federal nexus (that is, receive federal funding or require a federal permit), which is the case for most energy development in the United States. The Federal Energy Regulatory Commission regulates the interstate transmission of electricity, oil, and natural gas, but it does not regulate the construction of individual electricity-generation (except for nonfederal hydropower), transmission, or distribution facilities. Apart from Federal Aviation Administration guidelines, federal and state environmental laws protecting birds and bats are the main legal constraints on wind-energy facilities not on federal lands or without a federal nexus.

Wind energy is a recent addition to the energy mix in most areas, and regulation of wind energy is evolving rapidly. In evaluating current regulatory review processes, the committee was struck by the minimal guidance offered to developers, regulators, or the public about (1) the quantity and kinds of information to be provided for review; (2) the degrees of adverse or beneficial effects of proposed wind developments to consider critical for approving or disallowing a proposed project; and (3) the competing costs and benefits of a proposed project to weigh, and how to weigh them, with regard to that single proposal or in comparison with likely alternatives if that project is not built. Such guidance, and technical assistance with gathering and interpreting information needed for decision making, would be enormously useful. This guidance and technical assistance cast at the appropriate jurisdictional level could be developed by state and local governments working with groups composed of wind-energy developers and nongovernmental organizations representing all views of wind energy, in addition to other government agencies. The matrix of government responsibilities and the evaluation guide in Chapter 5 of this report should help the formulation of such guidance.

The committee judges that material in Chapter 5 could be a major step in the direction of an analytic framework for reviewing wind-energy proposals and for evaluating existing installations. If it were followed and adequately documented, it would provide a basis not only for evaluating an individual project but also for comparing two or more proposed projects and for undertaking an assessment of the cumulative effects of other human activities. It also could be used to project the likely cumulative effects of additional wind-energy facilities whose number and placement are identified in various projections. Finally, following this material would allow for a rational documentation of the most important areas for research.

Environmental Benefits of Wind Energy

The environmental benefits of wind energy accrue through its displacement of electricity generation that uses other energy sources, thereby displacing the adverse environmental effects of those generators. Because the use of wind energy has some adverse impacts, the conclusion that a wind-energy installation has net environmental benefits requires the conclusion that all of its adverse effects are less than the adverse effects of the generation that it displaces. However, this committee's charge was to focus on the use of wind energy; it was not able to evaluate fully the effects of other energy sources. The committee also did not fully evaluate so-called life-cycle effects, those effects caused by the development, manufacture, resource extraction, and other activities affiliated with all energy sources. Thus, in assessing environmental benefits of wind-energy generation of electricity, the committee focused on the degree to which it displaces or renders unnecessary the electricity generated by other sources, and hence on the degree to which it displaces or reduces atmospheric emissions, which include greenhouse gases, mainly carbon dioxide (CO_2); oxides of nitrogen (NO_x); sulfur dioxide (SO_2); and particulate matter. This focus on benefits accruing through reduction of atmospheric emissions, especially of greenhouse-gas emissions, was adopted because those emissions are well characterized and the information is readily available. It also was adopted because much of the public discourse about the environmental benefits of wind energy focuses on its reduction of atmospheric emissions, especially greenhouse-gas emissions. The restricted focus on benefits accruing through reduction of atmospheric emissions also was adopted because the relationships between air emissions and the amount of electricity generated by specified types of electricity-generating sources are well known. However, relationships between incremental changes in electricity generation and other environmental impacts, such as those on wildlife, viewsheds, or landscapes, generally are not known and are unlikely to be proportional. In addition, wind-powered generators of electricity share some kinds of adverse environmental impacts with other types of electricity generators (for example, some clearing of vegetation is required to construct either a wind-energy or a coal-fired power plant and its access roads and transmission lines). Therefore, calculating the extent to which wind energy displaces other sources of electricity generation does not provide clear information on how much, or even whether, those other environmental impacts will be reduced. This report does, however, provide a guide to the methods and information needed to conduct a more comprehensive analysis.

Projections for future wind-energy development, and hence projections for future wind-energy contributions to reduction of air-pollutant emissions in the United States, are highly uncertain. Recent model projections by the U.S. Department of Energy (DOE) for U.S. onshore installed wind-energy capacity in the next 15 years range from 19 to 72 gigawatts, or 2 to 7% of projected U.S. onshore installed electricity-generation capacity. In the same period, wind-energy development is projected to account for 3.5 to 19% of the *increase* in total electricity-generation capacity. If the average wind-turbine size is assumed to be 2 MW (larger than most current turbines), 9,500 to 36,000 wind turbines would be needed to achieve that projected capacity.

Because the wind blows intermittently, wind turbines often produce less electricity than their rated maximum output. On average in the mid-Atlantic region, the capacity factor of turbines—the fraction of their rated maximum output that they produce on average—is about 30% for current technology, and is forecast to improve to nearly 37% by the year 2020. Those are the fractions the committee used in estimating how much wind energy would displace other sources. Other factors, such as how wind energy is integrated into the electrical grid and how quickly other energy sources can be turned on and off, also affect the degree to which wind displaces other energy sources and their emissions. Those other factors probably further reduce the 30% (or projected 37%) figure, but the reduction probably is small, at least for the projected amount of onshore wind development in the United States. The net result in the mid-Atlantic region is unclear. Because the amount of atmospheric pollutants emitted varies from one energy source to another, assumptions must be made about which energy source will be displaced by wind. However, even assuming that all the electricity generation displaced by wind in the mid-Atlantic region is from coal-fired power plants, as one analysis has done, the results do not vary dramatically from those based on the assumption that the average mix of electricity sources in the region is displaced.

In addition to CO_2, coal-fired power plants also are important sources of SO_2 and NO_x emissions. Those two pollutants cause acid deposition and contribute to concentrations of airborne particulate matter. NO_x is an important precursor to ozone pollution in the lower atmosphere. However, because current and upcoming regulatory controls on emissions of NO_x and SO_2 from electricity generation in the eastern United States involve total caps on emissions, the committee concludes that development of wind-powered electricity generation using current technology probably will not result in a significant reduction in total emission of these pollutants from the electricity sector in the mid-Atlantic region.

Conclusions

- Using the future projections of installed U.S. energy capacity by the DOE described above, the committee estimates that wind-energy development probably will contribute to offsets of approximately 4.5% in U.S. emissions of CO_2 from electricity generation by other electricity-generation sources by the year 2020. In 2005, electricity generation produced 39% of all CO_2 emissions in the United States.

- Wind energy will contribute proportionally less to electricity generation in the mid-Atlantic region than in the United States as a whole, because a smaller portion of the region has high-quality[2] wind resources than the portion of high-quality wind resources in the United States as a whole.

- Electricity generated in the MAH—including wind energy—is used in a regional grid in the larger mid-Atlantic region. Electricity generated from wind energy in the MAH has the potential to displace pollutant emissions, discharges, wastes, and other adverse environmental effects of other sources of electricity generation in the grid. That potential is estimated to be less than 4.5%, and the degree to which its beneficial effects would be realized in the MAH is uncertain.

- If the future were to bring more aggressive renewable-energy-development policies, potential increased energy conservation, and improved technology of wind-energy generation and transmission of electricity, the contribution of wind energy to total electricity production would be greater. This would affect our analysis, including projections for development and associated effects (for example, energy supply, air pollution, and development footprint). On the other hand, if technological advances serve to reduce the emissions and other negative effects of other sources of electricity generation or if fossil-fuel prices fall, the committee's findings might overestimate wind's contribution to electricity production and air-pollution offsets.

- Electricity generated from different sources is largely fungible. Depending on factors such as price, availability, predictability, regulatory and incentive regimes, and local considerations, one source might be preferentially used over others. The importance of the factors changes over varying time scales. As a result, a more complete understanding of the environmental and economic effects of any one energy source depends on a more complete understanding of how that energy source displaces or is displaced by other energy sources, and it depends on a more complete understanding of the environmental and economic effects of all other available

---

[2]The quality of a wind resource refers to the amount of wind available for wind-powered generation of electricity.

energy sources. Developing such an understanding would have great value in helping the United States make better-informed choices about energy sources, but that was beyond this committee's charge. Nonetheless, the analyses in this report have value until such time as a more comprehensive understanding is developed.

Ecological Impacts

Wind turbines cause fatalities of birds and bats through collision, most likely with the turbine blades. Species differ in their vulnerability to collision, in the likelihood that fatalities will have large-scale cumulative impacts on biotic communities, and in the extent to which their fatalities are discovered. Probabilities of fatality are a function of both abundance and behavioral characteristics of species. Among bird species, nocturnal, migrating passerines[3] are the most common fatalities at wind-energy facilities, probably due to their abundance, although numerous raptor fatalities have been reported, and raptors may be most vulnerable, particularly in the western United States. Among bats, migratory tree-roosting species appear to be the most susceptible. However, the number of fatalities must be considered in relation to the characteristics of the species. For example, fatalities probably have greater detrimental effects on bat and raptor populations than on most bird populations because of the characteristically long life spans and low reproductive rates of bats and raptors and because of the relatively low abundance of raptors.

The type of turbines may influence bird and bat fatalities. Newer, larger turbines appear to cause fewer raptor fatalities than smaller turbines common at the older wind-energy facilities in California, although this observation needs further comparative study to better account for such factors as site-specific differences in raptor abundance and behavior. However, the data are inadequate to assess relative risk to passerines and other small birds. It is possible that as turbines become larger and reach higher, the risk to the more abundant bats and nocturnally migrating passerines at these altitudes will increase. Determining the effect of turbine size on avian risk will require more data from direct comparison of fatalities from a range of turbine types.

The location of turbines within a region or landscape influences fatalities. Turbines placed on ridges, as many are in the MAH, appear to have a higher probability of causing bat fatalities than those at many other sites.

The overall importance of turbine-related deaths for bird populations is unclear. Collisions with wind turbines represent one element of the cumu-

---

[3]Passerines are small to medium mainly perching songbirds; about half of all U.S. birds are passerines.

lative anthropogenic impacts on these populations; other impacts include collisions with other structures and vehicles, and other sources of mortality. As discussed in Chapter 3, those other sources kill many more birds than wind turbines, even though precise data on total bird deaths caused by most of these anthropogenic sources are sparser and less reliable than one would wish. Chapter 3 also makes clear that any assessment of the importance of a source of bird mortality requires information and understanding about the species affected and the likely consequences for local populations of those species.

The construction and maintenance of wind-energy facilities also alter ecosystem structure through vegetation clearing, soil disruption and potential for erosion, and noise. Alteration of vegetation, including forest clearing, represents perhaps the most significant potential change through fragmentation and loss of habitat for some species. Such alteration of vegetation is particularly important for forest-dependent species in the MAH. Changes in forest structure and the creation of openings alter microclimate and increase the amount of forest edge. Plants and animals throughout an ecosystem respond differently to these changes. There might also be important interactions between habitat alteration and the risk of fatalities, such as bat foraging behavior near turbines.

Conclusions

- Although the analysis of cumulative effects of anthropogenic energy sources other than wind was beyond the scope of the committee, a better analysis of the cumulative effects of various anthropogenic energy sources, including wind turbines, on bird and bat fatalities is needed, especially given projections of substantial increases in the numbers of wind turbines in coming decades.
- In the MAH, preliminary information indicates that more bats are killed than was expected based on experience with bats in other regions. Not enough information is available to form a reliable judgment on whether the number of bats being killed will have overall effects on populations, but given a general region-wide decline in the populations of several species of bats in the eastern United States, the possibility of population effects, especially with increased numbers of turbines, is significant.
- At the current level of wind-energy development (approximately 11,600 MW of installed capacity in the United States at the end of 2006, including the older California turbines), the committee sees no evidence that fatalities caused by wind turbines result in measurable demographic changes to bird populations in the United States, with the possible exception of raptor fatalities in the Altamont Pass area, although data are lacking for a substantial portion of the operating facilities.

- There is insufficient information available at present to form a reliable judgment on the likely effect of all the proposed or planned wind-energy installations in the mid-Atlantic region on bird and bat populations. To make such a judgment, information would be needed on the future number, size, and placement of those turbines; more information on bird and bat populations, movements, and susceptibility to collisions with turbines would be needed as well. Lack of replication of studies among facilities and across years makes it impossible to evaluate natural variability.

Recommendation

- Standardized studies should be conducted before siting and construction and after construction of wind-energy facilities to evaluate the potential and realized ecological impacts of wind development. Pre-siting studies should evaluate the potential for impacts to occur and the possible cumulative impacts in the context of other sites being developed or proposed. Likely impacts could be evaluated relative to other potentially developable sites or from an absolute perspective. In addition, the studies should evaluate a selected site to determine whether alternative facility designs would reduce potential environmental impacts. Post-construction studies should focus on evaluating impacts, actual versus predicted risk, causal mechanisms of impact, and potential mitigation measures to reduce risk and reclamation of disturbed sites. Additional research is needed to help assess the immediate and long-term impacts of wind-energy facilities on threatened, endangered, and other species at risk. Details of these recommendations, including the frequency and duration of recommended pre-siting, pre-construction, and post-construction studies and the need for replication, are in Chapter 3.

Impacts on Humans

The human impacts considered by the committee include aesthetic impacts; impacts on cultural resources, such as historic, sacred, archeological, and recreation sites; impacts on human health and well-being, specifically from noise and from shadow flicker; economic and fiscal impacts; and the potential for electromagnetic interference with television and radio broadcasting, cellular phones, and radar. This is not an exhaustive list of all possible human impacts from wind-energy projects. For example, the committee did not address potentially significant social impacts on community cohesion, such as cases where proposed wind-energy facilities might cause rifts between those who favor them and those who oppose them. Psychological impacts—positive as well as negative—that can arise in confronting a controversial project also were not addressed.

There has been relatively little dispassionate analysis of the human impacts of wind-energy projects in the United States. In the absence of extensive data, this report focuses mainly on appropriate methods for analysis and assessment and on recommended practices in the face of uncertainty. Chapter 4 contains detailed conclusions and recommendations concerning human impacts, including guides to best practices and descriptions of information needs. General conclusions and recommendations concerning human impacts follow.

Conclusions

• There are systematic and well-established methods for assessing and evaluating human impacts (described in Chapter 4); they allow better-informed and more-enlightened decision making.

• Although aesthetic concerns often are the most-vocalized concerns about proposed wind-energy projects, few decision processes adequately address them. Although methods for assessing aesthetic impacts need to be adapted to the particular characteristics of wind-energy projects, such as their visibility, the basic principles (described in Chapter 4 and Appendix D) of systematically understanding the relationship of a project to surrounding scenic resources apply and can be used to inform siting and regulatory decisions.

Recommendations

• Because relatively little research has been done on the human impacts of wind-energy projects, when wind-energy projects are undertaken, routine documentation should be made of processes that allow for local interactions concerning the impacts that arise during the lifetime of the project, from proposal through decommissioning, as well as processes for addressing the impacts themselves. Such documentation will facilitate future research and therefore improve future siting decisions.

• Human impacts should be considered within the context of the environmental impacts discussed in Chapter 3 and the broader contextual analysis of wind energy—including its electricity-production benefits and limitations—presented in Chapter 2. Moreover, the conclusions and recommendations concerning human impacts presented by topic in Chapter 4 should not be considered in isolation; instead, they should be treated as part of a process. Questions and issues concerning human impacts should be covered in assessments and regulatory reviews of wind-energy projects.

Analyzing Adverse and Beneficial Impacts in Context

The committee's charge included the development of an analytical framework for evaluating environmental and socioeconomic effects of wind-energy developments. As described in Chapter 1, an ideal framework that addressed all effects of wind energy across a variety of spatial and temporal scales would require more information than the committee could gather, given its time and resources, and probably more information than currently exists. In addition, energy development in general, and wind-energy development in particular, are not evaluated and regulated in a comprehensive and comparative way in the United States, and planning for new energy resources also is not conducted in this manner. Instead, planning, regulation, and review usually are done on a project-by-project basis and on local or regional, but not national, scales. In addition, there are few opportunities for full life-cycle analyses or consideration of cumulative effects.

There also are no agreed-on standards for weighting of positive and negative effects of a proposed energy project and for comparing those effects to those of other possible or existing projects. Indeed, the appropriate standards and methods of conducting such comparisons are not obvious, and it is not obvious what the appropriate space and times scales for the comparisons should be. Therefore, a full comparative analysis has not been attempted here.

The committee approached its task—to carry out a scientific study of the adverse and beneficial environmental effects of wind-energy projects—by analyzing the information available and identifying major knowledge gaps. Some of the committee's work was made difficult by a lack of information and by a lack of consistent (or even any) policy guidance at local, state, regional, or national levels about the importance of various factors that need to be considered. In particular, the committee describes in Chapter 1 and Chapter 5 the reasons that led us to stop short of providing a full analytic framework and instead to offer an evaluation guide to aid coordination of regulatory review across levels of government and across spatial scales and to help to ensure that regulatory reviews are comprehensive in addressing the many facets of the human and nonhuman environment that can be affected by wind-energy development.

Framework for Reviewing Wind-Energy Proposals

Conclusion

• A country as large and as geographically diverse as the United States and as wedded to political plurality and private enterprise is unlikely to plan for wind energy at a national scale in the same way as some European countries are doing. Nevertheless, national-level energy policies (implemented through such mechanisms as incentives, subsidies, research agendas, and federal regulations and guidelines) to enhance the benefits of wind energy while minimizing the negative impacts would help in planning and regulating wind-energy development at smaller scales. Uncertainty about what policy tools will be in force hampers proactive planning for wind-energy development. More-specific conclusions and recommendations follow.

Conclusion

- Because wind energy is new to many state and local governments, the quality of processes for permitting wind-energy developments is uneven in many respects.

Recommendation

- Guidance on planning for wind-energy development, including information requirements and procedures for reviewing wind-energy proposals, as outlined in Chapter 5, should be developed. In addition, technical assistance with gathering and interpreting information needed for decision making should be provided. This guidance and technical assistance, conducted at appropriate jurisdictional levels, could be developed by working groups composed of wind-energy developers; nongovernmental organizations with diverse views of wind-energy development; and local, state, and federal government agencies.

Conclusion

- There is little anticipatory planning for wind-energy projects, and even if it occurred, it is not clear whether mechanisms exist that could incorporate such planning in regulatory decisions.

Recommendation

- Regulatory reviews of individual wind-energy projects should be preceded by coordinated, anticipatory planning whenever possible. Such planning for wind-energy development, coordinated with regulatory review of wind-energy proposals, would benefit developers, regulators, and the public because it would prompt developers to focus proposals on locations and site designs most likely to be successful. This planning could be implemented at scales ranging from state and regional levels to local levels. Anticipatory planning for wind-energy development also would help researchers to target their efforts where they will be most informative for future wind-development decisions.

Conclusion

• Choosing the level of regulatory authority for reviewing wind-energy proposals carries corresponding implications for how the following issues are addressed:

(1) cumulative effects of wind-energy development;
(2) balancing negative and positive environmental and socioeconomic impacts of wind energy; and
(3) incorporating public opinions into the review process.

Recommendation

• In choosing the levels of regulatory review of wind-energy projects, agencies should review the implication of those choices for all three issues listed above. Decisions about the level of regulatory review should include procedures for ameliorating the disadvantages of a particular choice (for example, enhancing opportunities for local participation in state-level reviews).

Conclusion

• Well-specified, formal procedures for regulatory review enhance predictability, consistency, and accountability for all parties to wind-energy development. However, flexibility and informality also have advantages, such as matching the time and effort expended on review to the complexity and controversy associated with a particular proposal; tailoring decision criteria to the ecological and social contexts of a particular proposal; and fostering creative interactions among developers, regulators, and the public to find solutions to wind-energy dilemmas.

Recommendation

• When consideration is given to formalizing review procedures and specifying thresholds for decision criteria, this consideration should include attention to ways of retaining the advantages of more flexible procedures.

Conclusion

- Using an evaluation guide such as the one recommended in Chapter 5 to organize regulatory review processes can help to achieve comprehensive and consistent regulation coordinated across jurisdictional levels and across types of effects.

Recommendation

- Regulatory agencies should adopt and routinely use an evaluation guide in their reviews of wind-energy projects. The guide should be available to developers and the public.

Conclusion

- The environmental benefits of wind-energy development, mainly reductions in atmospheric pollutants, are enjoyed at wide spatial scales, while the environmental costs, mainly aesthetic impacts and ecological impacts, such as increased mortality of birds and bats, occur at much smaller spatial scales. There are similar, if less dramatic, disparities in the scales of realized economic and other societal benefits and costs. The disparities in scale, although not unique to wind-energy development, complicate the evaluation of tradeoffs.

Recommendation

- Representatives of federal, state, and local governments should work with wind-energy developers, nongovernmental organizations, and other interest groups and experts to develop guidelines for addressing tradeoffs between benefits and costs of wind-energy generation of electricity that occur at widely different scales, including life-cycle effects.

1

Introduction

In recent years, the growth of capacity to generate electricity from wind energy has been extremely rapid, increasing from 1,848 megawatts (MW) in 1998 to 11,603 MW in the United States by the end of 2006 (AWEA 2006a) (Figures 1-1, 1-2). Some of that growth was fueled by state and federal tax incentives (Schleede 2003), as well as by state renewable portfolio standards and targets. Despite that rapid growth, wind energy amounted to less than 1% of U.S. electricity generation in 2006. To the degree that wind energy reduces the need for electricity generation using other sources of energy, it can reduce the adverse environmental impacts of those sources, such as production of atmospheric and water pollution, including greenhouse gases; production of nuclear wastes; degradation of landscapes due to mining activity; and damming of rivers. Generation of electricity by wind energy has the potential to reduce environmental impacts, because unlike generators that use fossil fuel, it does not result in the generation of atmospheric contaminants or thermal pollution, and it has been attractive to many governments, organizations, and individuals. But others have focused on adverse environmental impacts of wind-energy facilities, which include visual and other impacts on humans; and effects on ecosystems, including the killing of wildlife, especially birds and bats. Some environmental effects of wind-energy facilities, especially those concerning transportation (roads to and from the plant site) and transmission (roads and clearings for transmission lines), are common to all electricity-generating facilities; others, such as their specific aesthetic impacts, are unique to wind-energy facilities. This report provides analyses to understand and evaluate those environmental effects, both positive and negative.

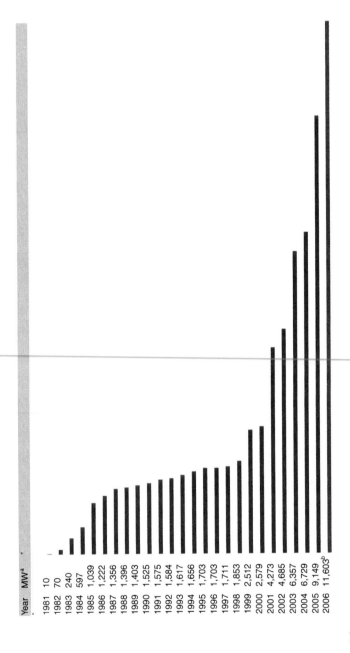

Year	MW[a]
1981	10
1982	70
1983	240
1984	597
1985	1,039
1986	1,222
1987	1,356
1988	1,396
1989	1,403
1990	1,525
1991	1,575
1992	1,584
1993	1,617
1994	1,656
1995	1,703
1996	1,703
1997	1,711
1998	1,853
1999	2,512
2000	2,579
2001	4,273
2002	4,685
2003	6,357
2004	6,729
2005	9,149
2006	11,603[b]

[a]Megawatts
[b]American Wind Energy Association total based on project completion data reported by developers.

FIGURE 1-1 Wind power: U.S. installed capacity (megawatts), 1981-2006.
NOTE: Due to project decommissioning and re-powering, the end-of-year cumulative capacity total does not always match the previous year's year-end total plus additions.
SOURCES: U.S. Department of Energy Wind Energy Program and American Wind Energy Association 2006. Reprinted with permission; copyright 2006, American Wind Energy Association.

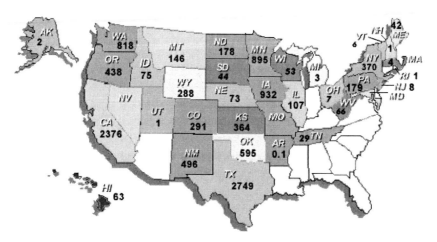

FIGURE 1-2 Total installed U.S. wind-energy capacity: 11,603 MW as of December 31, 2006.
SOURCE: AWEA 2007. Reprinted with permission; copyright 2007, American Wind Energy Association.

Like all sources of energy exploited to date, wind-energy projects have effects that may be regarded as negative. These potential or realized adverse effects have been described not only in the Mid-Atlantic Highlands (MAH) (Schleede 2003) but also in other parts of the country, such as California (CBD 2004) and Massachusetts (almost any issue of the Cape Cod Times, where the proposed and controversial wind-energy installation in Nantucket Sound is discussed).

GENERATING ELECTRICITY FROM WIND ENERGY

Two percent of all the energy the earth receives from the sun is converted into kinetic energy in the atmosphere, 100 times more than the energy converted into biomass by plants. The main source of this kinetic energy is imbalance between net outgoing radiation at high latitudes and net incoming radiation at low latitudes. The global temperature equilibrium is maintained by a transport of heat from the equatorial to the polar regions by atmospheric movement (wind) and ocean currents. The earth's rotation and geographic features prevent the wind from flowing uniformly and consistently.

The kinetic energy of moving air that passes the rotor of a turbine is proportional to the cube of the wind speed. Hence, a doubling of the wind speed results in eight times more wind energy. A modern 1.5 MW wind turbine with a hub height (center of rotor) and tower height of 90 meters (m), operating in a near-optimum wind speed of 10 m/sec (36 km/h) at hub height will create more than 1.4 MW of electricity; in eight hours it will produce the amount of electricity used by the average U.S. household in one year (about 10,600 kilowatt-hour [kWh]).

There is an upper theoretical limit (the Betz limit of 59%) to how much of the available energy in the wind a wind turbine can actually capture or convert to usable electricity. Modern wind turbines potentially can reach an efficiency of 50%. Almost all wind turbines operating today have a three-bladed rotor mounted upwind of the hub containing the turbine. The blades have an aerodynamic profile like the wing of an aircraft. The force created by the lift on the blades result in a torque on the axis; the forces are transmitted through a gearbox, and a generator is used to transform the rotation into electrical energy, which is then distributed through the transmission grid (Figure 1-3).

Human use of wind energy has a long history (the following summary is taken from Pasqualetti et al. 2004). Wind energy has been used for sailing vessels at least since 3100 BC. Windmills were used to lift water and grind grain as early as the 10th century AD. The first practical wind turbine was built by Charles Brush in 1886; it provided enough electricity for 100 incandescent light bulbs, three arc lights, and several electric motors. However, the turbine was too expensive at that time for commercial development.

By the 1920s, some farms in the United States generated electricity by wind turbines, and by the 1940s wind turbines sold by Sears Roebuck and Company were providing electricity for small appliances in rural American homes; in Denmark, 40 wind turbines were generating electricity. The first wind-powered turbine to provide electricity into an American electrical transmission grid was in October 1941 in Vermont. However, significant electricity generation from wind in the United States began only in the 1980s in California. Today (2006), it amounts to less than 1% of U.S. electricity generation.

There has been a rapid evolution of wind-turbine design over the past 25 years. Thus, modern turbines are different in many ways from the turbines that were installed in California's three large installations at Altamont Pass, Tehachapi, and San Gorgonio (Palm Springs) in the early 1980s. A typical turbine structure consists of a pylon (tower or monopole) that can produce electricity at wind speeds as low as 12-14 km/h (3.3-3.9 m/sec). Generators typically reach peak efficiency at wind speeds of approximately 45 km/h (12.5 m/sec) and shift to a safety mode when the wind exceeds a particular speed, often on the order of 80-100 km/h (22-28 m/sec). Smaller generators are used for individual buildings or other uses.

This report is concerned with utility-scale clusters of generators or wind-energy installations (often referred to as "wind farms"), not with small turbines used for individual agricultural farms or houses. Some of

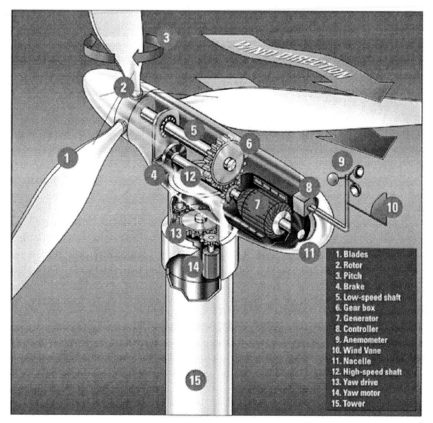

FIGURE 1-3 Structure of a wind turbine.
SOURCE: Alliant Energy 2007. Reprinted with permission; copyright 2007, Alliant Energy.

the utility-scale installations contain hundreds of turbines; for example, the wind-energy facility at Altamont Pass in California consists of more than 5,000 and those at Tehachapi and Palm Springs contain at least 3,000 turbines each, ranging from older machines as small as 100 kW installed more than 20 years ago to modern turbines of 1.5 MW or more (information available at www.awea.org).

Adverse effects of wind turbines have been documented: a recent Final Programmatic Environmental Impact Statement (BLM 2005a) lists the following: use of geologic and water resources; creation or increase of geologic hazards or soil erosion; localized generation of airborne dust; noise generation; alteration or degradation of wildlife habitat or sensitive or unique habitat; interference with resident or migratory fish or wildlife species, including protected species; alteration or degradation of plant communities, including occurrence of invasive vegetation; land-use changes; alteration of visual resources; release of hazardous materials or wastes; increased traffic; increased human-health and safety hazards; and destruction or loss of paleontological or cultural resources. These impacts can occur at the various stages of planning, site development, construction, operation, and decommissioning or abandonment (if applicable), although different phases tend to be associated with different impacts. Any or all of the impacts have the potential to accumulate over time and with the installation of additional generators. Beneficial environmental effects result from the reduction of adverse impacts of other sources of energy generation, to the degree that wind energy allows the reduction of energy generation by other sources. This committee's task includes an evaluation of the importance and frequency of these effects.

The killing of bats and birds has been among the more obvious and objectively quantifiable effects. Birds can be electrocuted along transmission and distribution lines or killed by flying into them (Bevanger 1994; Erickson et al. 2001, 2002; Stemer 2002). Thousands of birds die each year from collisions with wind-energy installations (BLM 2005a). The Altamont facility in California has caused the deaths of many raptors, which were members of protected species (CBD 2004; BLM 2005a). Several species of bats in North America also have been reported killed by collisions with wind-energy installations (Johnson 2005; Kunz et al. 2007). There were no fatalities of federally protected bat species known to this committee at this writing (early 2007).

Another widely cited impact of wind turbines is their visible effect on viewsheds and landscapes. The scale of modern turbines makes them impossible to screen from view, often making aesthetic considerations a major basis of opposition to them (Bisbee 2004). Well-established systematic methods for evaluating aesthetic impacts are available (Smardon et al. 1986; USFS 2003), but they often are misunderstood or poorly implemented, and they will need to be adapted for assessing the unique attributes of wind-energy projects. Methods also are available for identifying the particular values and sensitivities associated with recreational and cultural resources, as discussed in Chapter 4.

The regulatory system for siting and installing wind-energy projects in the United States varies widely, from a fairly thorough process in parts of California to much less rigorous processes in some other states (GAO 2005). In California, as well as in other states, the processes for evaluating and regulating wind-energy installations are evolving. In many areas of the United States, wind-energy installations have been controversial, sometimes strongly so.

THE PRESENT STUDY

Congress asked the National Academies to conduct an assessment of the environmental impacts of wind-energy installations, using the MAH (Pennsylvania, Virginia, Maryland, and West Virginia) as a case study.

Statement of Task

The National Academies was asked to establish an expert committee to carry out a scientific study of the environmental impacts of wind-energy projects, focusing on the MAH as a case example. The study was to consider adverse and beneficial effects, including impacts on landscapes, viewsheds, wildlife, habitats, water resources, air pollution, greenhouse gases, materials-acquisition costs, and other impacts. Using information from wind-energy projects proposed or in place in the MAH and other regions as appropriate, the committee was asked to develop an analytical framework for evaluating those effects that can inform siting decisions for wind-energy projects. The study also was to identify major areas of research and development needed to better understand the environmental impacts of wind-energy projects and reduce or mitigate negative environmental effects.

The committee was not asked to consider, and therefore did not address, nonenvironmental issues associated with generating electricity from wind energy, such as energy independence, foreign-policy considerations, resource utilization, and the balance of international trade.

The Process for This Study

The committee held five meetings: on September 19-20, 2005, in Washington D.C.; on December 15-16 in Charleston, West Virginia; on March 5-7, 2006, in southern California; on May 18-20 in West Virginia; and on July 17-19 in Woods Hole, Massachusetts. The first three meetings included presentations from experts and provided opportunities for public comment; at its third meeting the committee toured the wind-energy installation at San Gorgonio, near Palm Springs, California; and at its fourth meeting it viewed the Mountaineer Wind Energy Center and the proposed Mount Storm projects near Davis, West Virginia, from nearby public highways (access to the Mountaineer site was not permitted). The committee's final meeting was held in closed session and was devoted to finalizing this report. The committee gained familiarity with the relevant body of scientific knowledge through briefings and review of literature, databases, and existing studies of wind farms, both in the MAH and elsewhere, in addition to its own expertise.

Estimating Environmental Benefits of Wind Energy: Focus on Air Emissions

It is not conceptually difficult to estimate the adverse environmental effects of wind-energy projects, although it can be difficult in practice to quantify them. The estimation of the environmental benefits of wind energy is more difficult, because the benefits accrue through its displacement of energy generation using other energy sources, thereby displacing the adverse environmental effects of those generators. To estimate those benefits requires knowledge of what other electricity-generating sources will be displaced by wind energy, so that their adverse effects can be calculated and the offsetting advantages of wind energy can be determined. As described in detail in Chapter 2, the committee has restricted its estimates of the environmental benefits of wind energy to the reduction of air emissions that results from using wind energy for electricity instead of using other sources of electricity generation. The rationale for and limitations of this approach are discussed in detail in Chapter 2, but briefly the approach was adopted because much of the discourse about the advantages of wind energy focuses on reduction of air emissions, including greenhouse gases; because information about air emissions is extensive and readily accessible; and because wind energy has some of the same kinds of adverse impacts other than air emissions that other sources do (for example, some clearing of vegetation is required to construct either a wind-energy or a coal-fired powered plant and their access roads and transmission lines), which complicates the analysis of other adverse impacts. The committee did not conduct a full analysis of life-cycle environmental effects of wind and other sources of electricity generation. This report does, however, provide a guide to the methods and information needed to conduct a more complete analysis.

DEVELOPING AN ANALYTICAL FRAMEWORK

Part of the committee's charge was to develop an analytical framework for reviewing environmental and socioeconomic effects of wind-energy projects. For reasons described in detail in Chapter 5, and summarized below, the committee has stopped short of a complete analytical framework, both in the report itself and in its recommendations. Instead, the committee offers an evaluation guide in Chapter 5 that, if followed, will aid coordination of regulatory review across levels of government and across spatial scales (Figure 5-1) and will help to ensure that regulatory reviews are comprehensive in addressing the many facets of the human and nonhuman environment that can be affected by wind-energy development (Box 5-4).

One reason the committee stopped short is practical: even considering only the environmental effects of wind, some effects are better documented and easier to evaluate than others. Another reason for stopping short of a full analytical treatment is that other types of energy development, and indeed most types of construction, are not currently regulated in a comprehensive and comparative way in the United States. Finally, there is no social consensus at present on how all the effects of wind-energy generation of electricity on various aspects of the human and nonhuman environments should be evaluated as positive or negative, how the advantages and disadvantages should be traded off, or whose value systems should prevail in making such judgments. For all of these reasons, the committee focused its efforts on incrementally improving the way wind-energy decisions are made today. The evaluation guide in Chapter 5 reflects the result of those efforts.

Placing Environmental Effects in Context

Related to the above discussion of an analytical framework is the issue of placing environmental effects of particular electricity-generation units and other human activities in context. For example, although wind-energy projects kill tens of thousands of birds each year in the United States, other human structures and activities, including allowing domestic cats to hunt outside, are responsible for hundreds of millions, if not billions, of bird deaths each year (see Chapter 3 for more discussion of these numbers). Although wind turbines may cause visual impairments, oil-drilling rigs, coal-fired power plants, roads, buildings, and cell-telephone relay towers also may cause visual impairments. To make comparative evaluations of those impacts would imply some sort of weighting of positive and negative effects in an explicit, objective, and systematic way, but that is not done nationally or regionally, and indeed it is not obvious what methods one would use to perform such an analysis. In addition, choosing the proper standard of comparison is difficult: should effects be calculated per turbine or structure, per energy installation, per kWh of electricity generated, or against some other standard?

It is not even obvious that doing such an analysis on a national scale would provide a useful guide to action. Our society does not always weight effects from different causes equally. To understand, evaluate, and compare various environmental impacts of a variety of human structures and activities, such as bird or bat deaths, requires an understanding of the exposures to the dangers, the societal benefits that accrue from the circumstances that lead to exposure, and many other factors, some of which might be unrecognized or unexpressed. Therefore, any systematic comparison of the environmental effects of various methods of generating electricity, especially if it is to include a broader context, would require a depth of analysis and information gathering that would be beyond this committee's charge, although it might have great value in helping the United States make better-informed choices about energy sources. Although a complete, systematic comparison has not been attempted in this report, the analyses that are provided here should have value pending a more comprehensive analysis.

For similar reasons, the committee also has not addressed environmental benefits related to human health. For example, wind-powered electricity generation may lessen the need for electricity generation from coal-fired power plants and thereby reduce the amount of sulfur dioxide (SO_2) and nitrogen oxides (NO_x) emissions produced from coal combustion. SO_2 and NO_x emissions are important contributors to concentrations of airborne particulate matter and are precursors to acid deposition, and NO_x is an important precursor to ozone. Particulate matter and ozone are of considerable concern because of the risk they pose to public health. However, the extent to which emissions from specific electric power plants might be displaced by wind-energy facilities is unknown. Therefore, making health-effects assessments of potential displacement of emissions from electricity-production facilities of unknown location would be highly uncertain (e.g., NRC 2006).

TEMPORAL AND SPATIAL SCALES OF ANALYSIS

Analysis of the environmental impacts of any type of project is complicated enough, but it is exceptionally challenging for wind-energy projects. One obvious problem is how to choose the appropriate temporal and spatial scales for the analysis. A wind facility has local effects at scales of hundreds of meters to one or two kilometers: vegetation is cleared to install the turbines, local drainage patterns can be altered, and animals can be killed by coming into contact with moving turbine blades. At the range of one or two kilometers to a few tens of kilometers, there are visual effects on people; potential but currently unknown population effects on animals that are killed, such as bats and birds; roads are built or modified to allow the carriage of very large and heavy turbine components; and power lines are erected to transmit electricity from the turbine to the grid. At even larger scales, migratory birds and bats, which can travel hundreds to thousands of kilometers or more each way annually, suffer mortality with potential but currently unknown effects on their regional and global populations. Positive effects—the reduction of adverse effects of power generated by burning of fossil fuel, hydroelectric dams, and nuclear reactors—are more difficult to assess, because of regional and national power grids that all are influenced by the availability of wind energy and because some effects of electricity generation are truly global (the emission of greenhouse gases that influence climate change, for example). In addition, the presence or the possible construction of wind-energy installations affects people's decisions and behavior at many levels of organization and at many spatial and temporal scales (see for example the discussion of "opportunity and threat effects" in a National Research Council [NRC] report on the cumulative effects of oil and gas activities on Alaska's North Slope [NRC 2003]). Finally, effects accumulate over space and time, both as a function of the number and locations of wind-energy installations, and as a function of their interactions with other perturbations (NRC 2003).

UNDERSTANDING AND ASSESSING CUMULATIVE ENVIRONMENTAL EFFECTS

When numerous small decisions about related environmental issues are made independently, the combined consequences of those decisions often are not considered (Odum 1982). As a result, the patterns of the environmental perturbations or their effects over large areas and long periods are not adequately analyzed. This is the basic issue of cumulative effects assessment. The general approach to identifying and assessing cumulative effects evolved after passage of the National Environmental Policy Act (NEPA) of 1969, and this committee, like an earlier NRC committee (NRC 2003), has followed that approach. This discussion is adapted from that committee's report.

The NEPA requires environmental review for all federal actions and Environmental Impact Statements (EISs) for federal actions with potentially significant environmental effects. In 1978, the Council on Environmental Quality promulgated regulations implementing the NEPA that are binding on all federal agencies (40 CFR Parts 1500-1508 [1978]). A cumulative effect was defined as "the incremental impact of the action when added to other past, present, and reasonably foreseeable future actions. . . . Cumulative impacts can result from individually minor but collectively significant actions taking place over a period of time." For example, an EIS might conclude that the environmental effects of a single power plant on an estuary might be small and, hence, judged to be acceptable. But the effects of a dozen plants on the estuary are likely to be substantial, and perhaps of a different nature than the effects of a single plant—in other words, the effects are likely to accumulate and may interact. Even a series of EISs might not identify or predict the cumulative effects that result from the interaction of multiple activities.

The accumulation of effects can result from a variety of processes (NRC 1986). The most important ones are:

• Time crowding—frequent and repeated effects on a single environmental medium. An example related to wind-energy development might be repeated effects on multiple individuals within a local population of birds or bats before the population had time to recover.

• Space crowding—high density of effects on a single environmental medium, such as a concentration of turbines or installations in a small region so that the areas affected by individual turbines or installations overlap. Space crowding can result even from actions that occur at great distances from one another. An example related to wind energy might be that impacts from widely separated wind facilities could accumulate on a single migratory population of birds or bats.

• Compounding effects—effects attributable to multiple sources on a single environmental medium, such as the combined effects of turbines, cell-phone towers, transmission lines, and other structures that could kill flying animals.

• Thresholds—effects that become qualitatively different once some threshold of disturbance is reached, such as when eutrophication exhausts the oxygen in a lake, converting it to a different type of lake. The first industrial structure in an otherwise undeveloped environment might cross a visual threshold or a threshold of wilderness values. Another example might be the existence of a threshold in terms of the number of turbines and risk of bird and bat fatalities, or habitat fragmentation.

• Nibbling—progressive loss of habitat resulting from a sequence of activities, each of which has fairly innocuous consequences, but the consequences on the environment accumulate, perhaps causing the extirpation of a species from the area.

These examples illustrate why recognizing and measuring the accumulation of effects depends on the correct choice of domain—temporal and spatial—for the assessment. Although the assessment of cumulative effects has a history of several decades (e.g., NRC 1986), it still is a complex task. The responses of the many components of the environment likely to be affected by an action or series of actions differ in nature and in the areas and periods over which they are manifest. An action or series of actions might have effects that accumulate on some receptors (e.g., target organisms or populations) but not on others, or on a given receptor at one time of the year but not at another. Therefore, a full analysis of where, when, how, and why effects accumulate requires multiple assessments.

To address this problem, an earlier NRC committee (NRC 2003) attempted to identify the essential components of such an assessment:

• Specify the class of actions whose effects are to be analyzed.
• Designate the appropriate temporal and spatial domain in which the relevant actions occur.
• Identify and characterize the set of receptors to be assessed.

- Determine the magnitude of effects on the receptors and whether those effects are accumulating.

These criteria cannot always be applied because of data limitations. Also, the effects of individual actions range from brief or local to widespread, persistent, and sometimes irreversible.

To conduct an analysis of how effects accumulate, one must understand what would occur in the absence of a given activity. The accumulated effects are the difference between that probable history and the actual history. To predict how effects may accumulate for a proposed action, it is essential to have good baseline data and data about the same kinds of receptors in similar areas that were and were not influenced by comparable actions. In some cases, the lack of such information prevented the committee from identifying and assessing possible cumulative effects of some activities or structures related to wind-energy development. Even if accumulating effects are identified, their magnitude and their biological, economic, and social importance must be assessed.

As noted above, it is difficult to assess cumulative effects in the absence of a comprehensive, broad-scale regulatory and assessment framework. The discussion above is presented in the expectation that it, along with the recommendations for development of an evaluation guide presented in Chapter 5, will be useful for future planning and assessment efforts.

ORGANIZATION OF THE REPORT

Chapter 2 sets the context for wind energy in the United States and analyzes the committee's approach to estimating the environmental benefits of wind energy. It describes the considerations involved in understanding under what conditions and to what degree wind energy can displace electricity generation by other sources, and hence reduce the adverse environmental effects of those sources, in particular their air emissions. Chapter 3 provides an evaluation of the literature on the effects of wind turbines on ecosystems and their components, and discusses methods that would be valuable in future evaluations; it also identifies research needs. Chapter 4 deals with effects on humans of wind-energy projects, including aesthetic, noise, cultural, health, economic, and related effects. Chapter 5 compares a variety of extant regulatory and evaluative regimes and extracts their strong points for consideration in other places and at larger (e.g., national) scales, and draws the information together in an evaluation guide that would be most useful for evaluating the effects of existing wind-energy installations and for assessing—and managing—the effects of proposed installations at various scales.

2

Context for Analysis of Effects of Wind-Powered Electricity Generation in the United States and the Mid-Atlantic Highlands

ESTIMATING THE ENVIRONMENTAL BENEFITS OF GENERATING ELECTRICITY FROM WIND ENERGY

This chapter provides an assessment of the environmental benefits of generating electricity from wind energy (current and future development) in the United States and its Mid-Atlantic Highlands (MAH), with specific attention to the potential contribution to the electricity supply and air quality improvement as indicated by emission reductions. For context, a general overview is provided describing issues that should be considered when assessing potential wind-energy development and environmental benefits. This is followed by a more detailed treatment and quantitative analysis of potential development and benefits. We end with a set of conclusions derived from the analysis; that analysis is simplified by including only the most robust assumptions.

Introduction and Overview

The committee's statement of task requires it to consider the beneficial environmental effects of electricity generation by wind-energy facilities. Wind-powered electricity-generating units (EGUs), like EGUs using other sources of energy, have no significant intrinsic environmental benefits; for example, none of their effects directly enhance ecosystem values or services. Indeed, every source of energy used to generate electricity on a large scale has at least some effects that most people would identify as adverse. The environmental and human-health risk reduction benefits of wind-powered electricity generation accrue through its displacement of electricity genera-tion using other energy sources (e.g., fossil fuels), thus displacing the adverse effects of those other generators. Moreover, the only way to fully evaluate the environmental effects of generating electricity from wind energy is to understand all the adverse life-cycle effects of those electricity sources, and to compare them to all the adverse effects of wind energy. Because wind energy has some adverse impacts, the conclusion that a wind-powered EGU has net environmental benefits requires the conclusion that all its adverse effects are less than the adverse effects of the generation that it displaces. This committee's charge was to focus on the generation of electricity from wind energy, however, and so it has not fully evaluated the effects of other electricity sources. In addition, it has not fully evaluated life-cycle effects (see discussion later in this chapter). Thus, in assessing environmental ben-efits, this committee has focused on the degree to which wind-generated electricity displaces or renders unnecessary electricity generated by other sources that produce atmospheric emissions, and hence the degree to which it displaces or reduces atmospheric emissions, which include greenhouse gases, mainly CO_2 (carbon dioxide), NO_x (oxides of nitrogen), SO_2 (sulfur dioxide), and particulate matter. This focus on benefits accruing through reduction of atmospheric emissions, especially of greenhouse gas emissions, was adopted because those emissions are well characterized and the infor-mation is readily available; it also was adopted because much of the public discourse about the environmental benefits of wind energy focuses on its reduction of atmospheric emissions, especially greenhouse gas emissions. Finally, the focus on benefits accruing through reduction of atmospheric emissions was adopted because the relationships between air emissions and the amount of electricity generated by specified types of electricity-generat-ing sources are well known. However, relationships between incremental changes in electricity generation and other environmental impacts, such as those on wildlife, viewsheds, or landscapes, are generally not known and are unlikely to be proportional. In addition, wind-powered generators of electricity share many kinds of adverse environmental impacts with other kinds of electricity generators. Therefore, calculating how much wind en-ergy displaces other sources of electricity generation does not provide clear information on how much, or even whether, those other environmental impacts will be reduced. This report does, however, provide a guide to the methods and information needed to conduct a fuller analysis.

Although most evaluations of the beneficial effects of wind-generated electricity, including the present one, have addressed the degree to which they reduce (through displacement) atmospheric emissions, other important effects are potentially displaced as well. For example, obtaining fossil fuel through mining, drilling, and chemical modification of one form to another (e.g., gasification of coal) has a variety of environmental effects including loss of habitat for terrestrial and aquatic species. Operation of thermal EGUs, which generate heat to drive turbines, produces heated water, either from cooling or in the form of steam to drive the turbines, or both. If the energy from the heated water is not recovered, the water is usually discharged into the environment; in closed cooling systems, its heat is discharged. All forms of generation have associated life-cycle emissions and wastes along with other environmental effects that are affected by the design, materials provision (including mining), manufacture, construction, transportation, assembly, operation, maintenance, retrofits, and decommissioning of the generators and their associated infrastructure. Some of these stages of the life cycle—most notably, mining—have adverse effects on human health as well. For the reasons given above, this committee has not considered all these effects in this study, but a full analysis would include them.

The issue of how much generation of emissions and waste is displaced by production of electricity generation through wind energy also is complex, but it needs to be understood to properly evaluate the environmental effects of wind energy. The primary purpose of this chapter, then, is to analyze the complex array of interacting factors that affect the extent to which wind displaces other energy sources. The analysis will provide a framework for evaluating the environmental effects of wind-energy facilities.

Although the direct and indirect environmental impacts of fossil-fuel generation of electricity are not well understood, the atmospheric emissions of fossil-fuel generators are fairly well characterized. It would seem straightforward to simply subtract the amount of energy generated by wind-energy facilities from the amount generated by fossil-fuel-fired EGUs, multiply by the amount of emissions per unit of energy, and attribute that amount of emission reduction to the wind EGUs. In practice, however, it is extremely difficult to perform the correct calculation. The following sections briefly discuss emissions from fossil-fuel-fired EGUs; the factors involved in calculating the extent to which wind energy reduces those emissions, today and in the future; and the committee's approach to the problem. In all cases, we are discussing generators of electricity.

Atmospheric Emissions from Fossil-Fuel Plants

Currently, most of the electricity used in the United States is generated from fossil fuels. Figure 2-1 shows U.S. electricity generation by fuel type. Wind is part of the "other renewables" category. Fossil-fuel-fired plants emit (among other atmospheric constituents) the so-called criteria pollutants, their precursor gases, and greenhouse gases (GHGs), mainly CO_2. Criteria pollutants are those regulated by the U.S. Environmental Protection Agency (EPA) under the Clean Air Act through the establishment of National Ambient Air Quality Standards (NAAQS). The standards, which

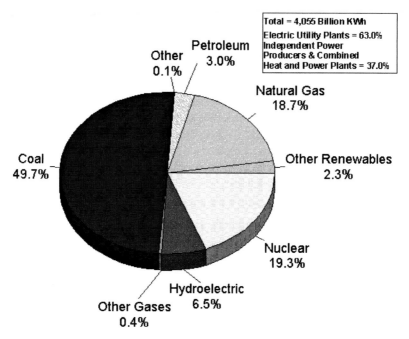

FIGURE 2-1 U.S. electric power industry net generation 2004. Note: Conventional hydroelectric power and hydroelectric pumped storage facility production minus energy used for pumping.
SOURCE: EIA 2005a.

are designed mainly to protect public health, apply to ozone (O_3), particulate matter, carbon monoxide (CO), NO_x, SO_2, and lead. NAAQS are also intended to protect against adverse public-welfare effects, such as damage to agricultural crops from acid deposition. Hazardous air pollutants, such as mercury, also are of environmental concern (see for example NESCAUM 2003). On March 15, 2005, the EPA issued the Clean Air Mercury Rule to permanently cap and reduce mercury emissions from coal-fired power plants for the first time (EPA 2006a).

CO_2 is not currently regulated by any federal authority in the United States, although it is of concern because it is increasing in concentration in the upper atmosphere largely due to emissions from the burning of fossil fuel and has been implicated in climate change (NRC 2001). Various policies and initiatives, mainly from states, seek to reduce atmospheric emissions of CO_2. For example, California established statewide GHG emissions reduction targets to reduce current emissions to 2000 emissions levels by 2010, then to reduce emissions to 1990 levels by 2020, and reduce emissions to 80% below 1990 levels by 2050. In general, coal-fired plants have the largest emissions per unit of energy generated, followed by gas-turbine generators, followed by combined-cycle gas-turbine generators (Denholm et al. 2005; DeCarolis and Keith 2006). Some data on emissions are provided in Appendix B, Table B-1.

The control technologies and regulatory regimes for reducing emissions of criteria pollutants, their precursors, and CO_2 can differ considerably, and therefore the costs of reducing them can be different. The question this section attempts to address, without considering costs, is to what extent will emissions be reduced through replacement of fossil-fuel-fired EGUs with wind-driven EGUs. The committee addresses this question by examining the potential for wind-energy development to achieve reductions in emissions of three major pollutants associated with fossil-fuel-fired EGUs. We focus on NO_x and SO_2, as examples of regulated pollutants. Coal-fired power plants are important sources of SO_2 and NO_x emissions. Those two pollutants cause acid rain and contribute to concentrations of airborne particulate matter. NO_x is an important precursor to ozone pollution in the lower atmosphere. Also, we focus on CO_2, as an example of a generally unregulated pollutant.

Factors that Affect Potential Emissions Reductions by Wind Energy

Emissions can be reduced in two basic ways: current electricity generation by emitting EGUs can be replaced on an immediate basis by generation from nonemitting EGUs (operating displacement), and emitting EGUs can be replaced, or not be built, when capacity is available from nonemitting EGUs (building displacement). The complex array of factors that affect how wind energy displaces other energy sources has been discussed in numerous publications (e.g., Smith et al. 2006). The following discussion is not a comprehensive review, but instead is an attempt to distill the most important issues. Some of these factors are further discussed in the section below, which provides a quantitative evaluation of wind-energy benefits.

There are three major aspects to any EGU. The first is capacity, or the amount of electric power an EGU can produce at its maximum output. This is usually referred to as "nameplate capacity," and it is expressed in some multiple of watts (usually megawatts, MW, one million watts). Electricity customers care about (and are charged for) power consumed during a unit of time, usually expressed as the number of kilowatt-hours (kWh), or one thousand watts for a one-hour period. The average productive output of a power plant is almost always less than its nameplate capacity, and the fraction of nameplate power that the average actual output represents is called the capacity factor. For wind EGUs, because the wind often does not blow at speeds that allow maximum power generation, the capacity factor is much less than nameplate capacity. Cumulative or annual average capacity factors are commonly about 30% and often much lower for shorter time intervals. Also, the capacity factor can be influenced by the accumulation of insects on turbine blades (see Corten and Veldkamp 2001).

The second aspect, dispatchability, is closely related to intermittency, and refers to the degree that a system operator can rely on a power source to be dispatched when it is needed. Electricity customers and electricity system operators also care about intermittency, because customers expect appliances to work when they turn on the switch, and system operators need to balance capacity against expected and realized demand for power. No electric power generator is 100% reliable (i.e., has zero intermittency)—lacking an effective means of electricity storage—but thermal (fossil-fuel and nuclear) and hydroelectric EGUs are generally less intermittent, and hence more dispatchable, than wind-energy facilities. Dispatchability also is related to a power plant's ability (or not) to be ramped up and down quickly. In general, coal-fired EGUs cannot be ramped up and down very easily, and their variable dispatch capacity is limited. Thus, they are more suited to baseload production (i.e., long periods of continuous power production) rather than to providing variable production to balance short-term variation in load and demand. (They also produce more emissions, such as SO_2 and NO_x, when they are not operating at optimum efficiency.) Natural-gas-fired EGUs and wind-driven EGUs (if the wind is blowing) are more capable than coal-fired EGUs of being ramped up and down quickly, as are many hydropower plants.

The third aspect of a power plant is the marginal cost of producing a unit of electric power, or its operating cost. Because the "fuel" for hydro-electric and wind-energy plants is free, they typically have low operating costs.

In addition to the characteristics of EGUs, electricity grids and trans-mission systems also have characteristics that affect the potential of wind energy to replace fossil fuel for generating electricity. Wind-powered EGUs are widely distributed in space, and to make matters more difficult, exclud-ing offshore locations, the highest-quality largest-scale wind resources usu-ally are far from the main centers of demand, i.e., where people live and work (DeCarolis and Keith 2006). Constructing transmission lines is expen-sive, and transporting electrical energy over long distances can be inefficient or costly. In addition, any new power source, including wind, needs to fit into the existing transmission and dispatching infrastructure.

This brings us to the most complex aspect of the entire estimation pro-cedure, and that is modeling the electricity grid. Most existing electricity grids in the United States are large, covering many states in the east and several of the larger states in the west, and are built around existing supply (fossil fuels and hydropower) and demand for electricity. The usefulness of additional generation capacity is affected by the price of that power and by the availability of transmission capacity and interfaces. System operators must deal with transmission constraints as they try to balance load and generation (Keith et al. 2004). As a result, the available generation, the load, and the units available change often, if not constantly, making it difficult to characterize the interactions in a general way.

The reliability of wind forecasts declines rapidly with time, and a variety of techniques are being investigated to improve medium- and long-range forecasts (e.g., Brundage et al. 2001; Gow 2003). As a result, if electricity derived from wind energy is to be incorporated into a dispatch system, a certain amount of backup or reserve power is required. In addition, the marginal cost of electricity generation by different kinds of power plants is more or less dependent on the plant type. Finally, some power plants can be ramped up and down faster and more efficiently than others.

Typically, a new power source is added to the grid by system operators in order of increasing operating costs, or the closely related but not identical "bid prices." Thus any new power source, including wind, displaces generation that costs more than it does, in the dispatch order. More-expensive power sources that are on the margins (for example, at peak demand times) would be displaced by less-expensive sources, depending, of course, on when the new power sources become available. As an example, the wind in the eastern United States averages lower speeds during summer afternoons—the normal times of highest peak demand for electricity there—than it does in winter, when peak demands are typically lower. Thus, to understand the extent to which any power source, including wind, would replace other generation sources, information is needed on demand and availability of the power source throughout the year at fairly small time increments. However, sometimes transmission constraints cause dispatch to be out of economic merit order (Keith et al. 2004). In addition, multiple years of data are examined to account for year-to-year variation. The committee cannot do much more here than to summarize the complexities of the electric-power production, distribution, and dispatching system. To quote DeCarolis and Keith (2006): "Intermittency can affect system operation on three timescales [minute-to-minute, intrahour, and hour- to day-ahead scheduling], but the impact depends on the transmission and generation infrastructure, and the resulting costs are not well understood in cases where wind serves more than a small fraction of demand. While Denmark and parts of Germany have wind serving more than 20% of demand, their experience does little to resolve uncertainties about the costs imposed by intermittent wind resources for at least two reasons. First, both countries are connected to large power pools that serve as capacity reserve for wind. Second, the multiplicity of wind-energy subsidies and absence of efficient markets . . . makes it difficult to disentangle costs." The authors emphasize that the cost of intermittency (in terms of back-up or reserve requirements) will be less if the generation mix is dominated by power plants with fast ramp rates (gas, hydropower) than if it is dominated by coal or nuclear plants, which have high capital costs and slow ramp rates.[1]

Not only wind energy receives government subsidies; all energy sources in the United States do. However, the subsidies vary from time to time, from one type of generator and its fuel to another, and from place to place, which further complicates understanding of how wind will displace other power sources in the mix. The two calculations of importance here are (1) the degree to which wind can contribute to guaranteed capacity (this allows one to predict the degree to which wind can replace existing power plants or obviate the need to construct new ones), and (2) the degree to which wind can be used in the existing grid structure (allowing prediction of the degree to which wind energy can reduce electricity generation, and hence emissions, from existing power plants that use fossil fuels).

A recent report by E.ON Netz, the transmission operator of a large electric grid in Europe (E.ON Netz 2005), concluded that the average capacity factor for its wind supply was about 20%, rising to 85% for brief periods and remaining below 14% for more than half the year. The minimum capacity factor was well under 1% for a short period. E.ON Netz further reported the results of two German studies on the degree to which wind-energy installations contribute to guaranteed capacity: both studies concluded that the contribution on average was approximately 8% of its installed (nameplate) capacity. (The committee refers to guaranteed capacity in this report as capacity value.)

Life-Cycle Costs

The true, full economic and environmental costs of electricity from various sources have not been adequately calculated. Cost estimates (including capital, operating, fuel, and financing costs) for electricity from various sources (coal, nuclear, etc.) are shown in Table 2-1, but these do not reflect the total private and social costs. The numbers in Table 2-1 include subsidies, but it is unclear how much they are. Estimates of the costs attributable to managing the intermittent nature of electricity supplied by wind energy are provided by DeCarolis and Keith (2006) and Strbac (2002).

Environmental externalities associated with operation of a power plant are a substantial, yet largely unquantified component of total costs. Life-cycle cost assessment can help reveal these externalities. Much effort has previously gone into developing methods and estimating externalities for

---

[1]Denmark, for example, has access to substantial hydroelectric capacity, which it relies on to balance the intermittent output from wind-energy installations (IEA 2006).

TABLE 2-1 Summary Cost Estimates for Electricity-Generation Technologies (in 2003 U.S. dollars per kilowatt-hour)

Technology	Cost estimated by:		
	EIA[a]	University of Chicago[b]	MIT[c]
Municipal solid waste landfill gas	0.0352		
Scrubbed coal, new (pulverized)	0.0382	0.0357	0.0447
Fluidized-bed coal		0.0358	
Pulverized coal, supercritical		0.0376	
Integrated coal gasification combined cycle (IGCC)	0.0400	0.0346	
Advanced nuclear	0.0422	0.0433	0.0711
Advanced gas combined cycle	0.0412	0.0354	0.0416
Conventional gas combined cycle	0.0435		
Wind 100 MW	0.0566		
Advanced combustion turbine	0.0532		
IGCC with carbon sequestration	0.0595		
Wind 50 MW	0.0598		
Conventional combustion turbine	0.0582		
Advanced combined cycle with carbon sequestration	0.0641		
Biomass	0.0721		
Distributed generation, base	0.0501		
Distributed generation, peak	0.0452		
Wind 10 MW	0.0991		
Photovoltaic	0.2545		
Solar thermal	0.3028		

[a]For EIA data, see EIA (2005b, Table 38). The 0.6 rule to adjust for scaling effects was applied to the wind 10 MW and 100 MW units using 50 MW as the base reference. Solar thermal costs exclude the 10% investment tax credit.

[b]For University of Chicago data, see University of Chicago (2004).

[c]For MIT data, see MIT (2003).

NOTE: EIA, Energy Information Administration; MIT, Massachusetts Institute of Technology. Estimates are for newly sited facilities and are based on national data. Data exclude regional multipliers for capital, variable operation and maintenance (O&M). Fixed O&M New York costs are higher. Data exclude delivery costs. Data reflect fuel prices that are New York State-specific. Costs reflect units of different sizes; while some technologies have lower costs than others the total capacity of the lower-cost generation technology may be limited—for example, a 500 MW municipal solid waste landfill gas project is unlikely. MIT calculations assumed a 10-year term; consequently, estimated costs are higher.

SOURCE: Mathusa and Hogan 2006.

particular effects of particular energy sources (see EC 1995; Hagler Bailly Consulting, Inc. 1995; Lee et al. 1995).

Life-cycle cost assessment attempts to compare the full costs of various electricity-generation technologies. Such comparisons take into account fuel life cycles (including extracting, refining, and transporting the fuel) and power-plant life cycles (including designing, constructing, operating, maintaining, renovating, decommissioning the power plant) as well as specific environmental issues (e.g., wildlife and human-health impacts of fuel extraction, nuclear waste disposal issues with nuclear power plants, reservoir issues with hydroelectric power plants).

Life-Cycle Assessment

In the past, Life Cycle Assessment (LCA) has become a widely recognized method for comprehensively identifying and quantifying the environmental effects of diverse products, processes, and services (Hendrickson et al. 2006). This is typically a large task: a variety of environmental, human-health, and ecological effects must be identified, quantified, and evaluated for all the life-cycle stages, often scattered geographically and over time. LCA has been embraced by a number of industrial goods manufacturers and service organizations. The use of LCA in public policy making has not been as well publicized, but it can be expected that LCA may be used increasingly to reveal the benefits and costs of new public investments in infrastructure.

At present, LCA methods are commonly used: process-analysis-based LCA; economic input-output analysis-based LCA (EIO-LCA); and hybrid LCA, which combines elements of the former methods.

In process-based LCA, all inputs (e.g., raw materials, energy, and water) and outputs (e.g., air emissions, water discharges, noise) of processes associated with the life-cycle phases of a product or service are assessed. This approach enables very specific analyses, but the data needs may be so large as to make the LCA costly and time-consuming, especially when several process steps are included in the supply chain. Selecting the boundary and scope of analysis is not always straightforward, making comparisons between LCAs difficult.

EIO-LCA helps address the challenges of boundary selection and data intensity by creating a consistent analytical framework for the economy of a country or region based on standard, government-compiled economic input-output tables of commodity production and use data, coupled with material and energy use, and emission and waste generation factors per monetary unit of economic output (Hendrickson et al. 1998, 2006). While EIO-LCA can be used for comprehensive analyses of many products and services, it may not provide the level of detail in a process-based LCA.

To overcome the shortcomings of the above two LCA approaches, but also provide the most comprehensive and relatively cost- and time-effective studies, hybrid LCA has been developed (Suh et al. 2004; Hendrickson et al. 2006).

A hybrid LCA for wind-energy projects might consider:

- Inputs into the life-cycle stages, such as energy (e.g., to manufacture and install the turbines), raw materials (e.g., iron ore), and water.
- Outputs from the life-cycle stages such as emissions to air; and a variety of potential impacts, such as:
  - Bird and bat fatalities,
  - Habitat degradation or destruction,
  - Noise,
  - Visual impacts,
  - Physical impacts (e.g., projectiles resulting from icing of turbine blades), and
  - Other impacts (e.g., shadow, flicker, glare, intrusion into commercial and military airspace).

Of course, the impacts of the above on the environment and on humans (e.g., global warming potential) would need to be analyzed as well.

When conducting an LCA, it is critical to assess uncertainties in the available data and methods used for analysis. Some, but not many, peer-reviewed LCAs of wind-energy technologies have been published. Lenzen and Munksgaard (2002) note that "despite the fact that the structure and technology of most modern wind turbines differ little over a wide range of power ratings, results from existing life-cycle assessments of their energy and CO_2 intensity show considerable variations" due to different LCA approaches, scope, boundary assumptions, geographical distribution, and information used for embedded energy calculations of turbine and tower materials, recycling or overhaul of turbines after the service life, and national energy mixes. They review 72 studies focusing on energy and CO_2 emissions associated with the life cycle of wind turbines and find that the energy intensity (kWh of energy input per kWh of electricity generated) is between 0.02 and 1.016, and the CO_2 intensity (in grams of CO_2 per kWh of electricity generated) is between 8.1 and 123.7.

Pacca and Horvath (2002) introduce the concept of global warming effect (GWE) as a combination of global warming potential and LCA and apply it to the construction and operation phases of several comparable electric power plants: hydroelectric, wind, solar, coal, and natural gas. In detail, their analysis focuses on the GWE of construction, burning of fuels, flooded biomass decay in the reservoir, loss of net ecosystem production, and land use. They find that a wind plant and a hydroelectric power plant in an arid zone (such as the one at Glen Canyon in the Upper Colorado River Basin) have lower GWE than the other power plants that were compared. This is the only region in the United States where the five electricity-generation technologies have been compared in an LCA framework.

Factors that Drive Wind-Energy Development

Forecasts for future wind-energy development presented in this chapter are based on a range of expectations concerning technological, economic, and policy factors that will determine the rate and magnitude of wind-energy development. These factors may be subject to change, as briefly described here.

Technological Changes

Research continues on the development of wind-turbine technology. Modern turbines are more efficient than earlier ones, and that trend is likely to continue. Transmissions (devices for transmitting the rotational kinetic energy of turbine blades to electric generators) also are likely to improve. A major impediment to the incorporation of wind generation into grids is the lack of ability to store electricity for times when the wind is unfavorable. Various approaches to storage are being considered, including storage batteries, hydrogen production and storage, compressed-air energy storage, hydraulic storage (using wind to pump water to use later for generation of hydroelectric power), and perhaps other devices (see, e.g., Fingersh 2004; Denholm et al. 2005; DeCarolis and Keith 2006). No storage system currently is economically viable, although research and development on this topic continue.

In addition to technology applied to the generation and storage of electricity by wind energy, efforts continue in the development of better transmission lines and improved grid management, which would improve the incorporation into the grid of intermittent power sources like wind. Some research focuses on computer and modeling technology. Also, weather forecasting continues to improve, and more reliable wind forecasts could enhance the ability of system operators to include wind into the management of grids. Of course, other sources of electric power, both renewable and nonrenewable, are also subject to continuing technological improvements; those improvements also could change ecological as well as other environmental effects of operating them.

Economic Changes

In early July 2006, the price of crude oil was about $75 per barrel; natural gas was about $5.60 per thousand cubic feet, down from a high of $8.66 in January of 2006. As recently as 2000, crude oil was selling for around $20 per barrel and natural gas for about $3.68 per thousand cubic feet. Coal prices have fluctuated between about $5/ton and $65/ton in recent years, depending on quality, and were climbing toward $100/ton as of

August 2006. Prices of those fossil fuels affect the price of electricity and hence the competitiveness of wind energy. Prices of fossil fuels are notoriously hard to predict, but it is at least plausible that recent trends towards higher prices will continue over the next decade.

Regulatory and Policy Changes

Changing regulatory and policy approaches towards energy production and consumption can have significant impact on wind-energy development. Such approaches might include the production-tax credit (PTC), renewable-portfolio-standard (RPS) legislation, carbon taxes, emissions cap-and-trade programs, emissions regulations, incentives to reduce energy consumption, and others. The approaches used vary from place to place and from time to time, and their effectiveness in reducing emissions (as well as in achieving several other policy goals) are being researched and debated. Their use is increasing, however, and it appears likely that to the degree that air quality and global climate change are considered to warrant governmental action, their use will continue to increase and they are likely to evolve.

The federal PTC and RPS legislation enacted in various states are major drivers of wind-energy development in the United States. The PTC is a federal support that is a direct credit against a company's federal income tax based on the generation of electricity with renewable resources, such as wind. As discussed in the following section, most of the wind resource in the United States could not be profitably developed without incentives such as the PTC (NREL 2006a). RPS legislation has been enacted by 20 states and the District of Columbia, specifying that utilities operating in those states supply a fixed percentage of their power from renewable sources (AWEA 2006b). Because RPS legislation generally allows purchase of renewable energy produced in other states, RPS legislation enacted in states with little wind-energy potential can drive development in other states that have more wind-energy potential. It has been estimated that if state RPS laws remain at current levels, they will be responsible for triggering about 80% of renewable power development in the United States in the next 10 years (Ihle 2005).

Various organizations and even government programs are presently advocating policy changes and initiatives that may dramatically increase the rate of wind-energy development in the United States. A number of organizations, for example, are actively promoting national RPS legislation in the 10-20% range (e.g., Clemmer et al. 2001; AWEA 2006c), and the American Wind Energy Association (AWEA) and the U.S. Department of Energy have jointly committed to pursue a goal of supplying 20% of U.S. electricity needs from wind energy (AWEA 2006d). As discussed later in this chapter, these goals greatly exceed the projections provided by three different organizations within the U.S. Department of Energy (Energy Information Administration, Office of Energy Efficiency and Renewable Energy, and the National Renewable Energy Laboratory).

Analysis of Effects and Benefits in a Context of Change

All of the factors described in this section—technological advances, economic changes, and regulatory and policy changes—will continue to evolve. Some of the evolution, through increased energy efficiency, improved technology for reducing emissions from fossil-fuel plants, and possible improvements in the handling of nuclear waste products, might reduce the economic competitiveness of wind energy. Other changes could increase its competitiveness and penetration into the mix of electric-power generators. The current trend appears to be in the direction of increased penetration and cost-effectiveness of wind energy, and therefore any assessment of the environmental benefits and consequences of wind energy should take at least a decade-long perspective. As described in the following section, the committee has examined a range of 15-year forecasts for wind-energy development based on modeling conducted by several U.S. Department of Energy programs. Although the range of forecast results that we examined was broad, it is still possible that technological, economic, or policy changes as discussed above could result in substantially different outcomes, with cumulative effects that are outside the range of our analysis. Before discussing wind energy in the United States and the MAH, we briefly describe the global status of wind energy to provide context.

WIND ENERGY GLOBALLY

The use of wind energy for electricity generation, which began on a utility scale in about 1980, grew relatively slowly at first with only about 3 gigawatts (GW, one billion watts) installed by 1993. However, by 2003, the world's wind-energy capacity was 39.4 GW, and by 2005 it was more than 59 GW (GWEC 2006). The United States had more wind-energy installed capacity than any other country until 1996, when it was surpassed by Germany; at the end of 2005, with 9.1 GW installed, it was third, behind Germany (18.4 GW) and Spain (10.0 GW). (The United States surpassed 11 GW of installed wind energy capacity in 2006.) India (4.4 GW) and Denmark (3.1 GW) rounded out the top 5; all other countries accounted for 13.9 GW (Florence 2006), and there was wind energy installed in all continents except for South America, but Brazil and Argentina have wind-energy projects in various stages of development (WWEA 2006).

Factors that affect the use of wind energy for electricity generation in other countries are similar to those in the United States in broad outline, but there are local differences among the different countries. For example, the European Wind Energy Association (EWEA) attributes the decision to develop wind energy in Denmark and Germany—among Europe's leaders in the amount of wind-energy capacity—to the nuclear accident at Chernobyl in 1986 and the Brundtland Commission's report on sustainability in 1987. Today, the growing evidence of rapid climate change driven by GHG emissions is an important motivator (EWEA 2006).

As is the case for the United States (see Chapter 1 and this chapter, below), global wind-energy generating capacity is widely expected to continue to grow; for example, the Global Wind Energy Council forecasts it to reach 134.8 GW by 2010, with the strongest growth in the United States, but significant growth elsewhere as well (GWEC 2006).

QUANTIFYING WIND-ENERGY BENEFITS IN THE UNITED STATES AND THE MID-ATLANTIC HIGHLANDS

Generation of electricity on a utility scale in the United States using wind energy has undergone increasingly rapid and geographically widespread development in recent years. The Energy Information Administration (EIA) Annual Energy Outlook 2006 (EIA 2006a) indicates that 9.646 GW of wind-energy capacity was installed by the end of 2005 and forecasts that total installed capacity (onshore) will exceed 11.5 GW in 2006; AWEA reports that the 11 GW mark for the United States was reached in 2006 (AWEA 2007). Based on data provided for the EIA Annual Electric Generator Report (EIA 2004a), installed capacity in 2004 included about 17,000 wind turbines associated with more than 200 separate projects distributed in 26 states. Based on a comparison of installed capacity for wind-powered electricity generation in 1999 and 2005 (Figure 2-2), more than two-thirds of the installed wind-energy capacity in the United States was developed in the first five years of this decade.

High rates of growth in the wind-powered electricity-generating industry are projected to continue well into the future. The following sections of this chapter examine projected wind-energy development for the contiguous United States, and in particular for its MAH. The potential contributions to electricity supply and reduction of air-pollution emissions are estimated based on projections through 2020.

Wind-Energy Potential

Estimates of U.S. wind-energy potential for electricity generation have changed as models have improved, more and better data have been collected and analyzed, and land-use exclusions have been considered. In particular, there has been an increase in the geographic resolution of wind-

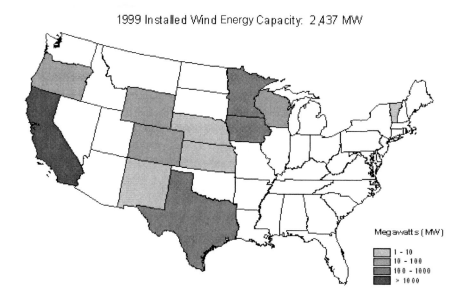

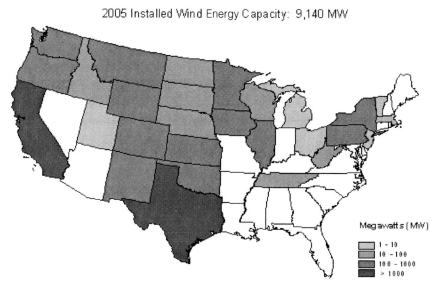

FIGURE 2-2 Installed wind-energy capacity in the contiguous United States in 1999 and 2005. SOURCE: Modified from Flowers 2006.

energy maps. The grid cell resolution of the Wind Energy Resource Atlas of the United States (Elliott et al. 1986) was about 25 km². Current maps of U.S. wind-energy potential have grid cell resolutions ranging from 200 m² to 1 km² for individual states. The National Renewable Energy Laboratory (NREL) has assembled these more current maps and accounted for land use and other exclusions (technical, legal, and environmental) as a basis for estimating both total and practical wind-power capacity and for projecting future wind-capacity development for electricity generation in the United States.

Wind class represents the potential for an area to generate electricity, based on mean wind-power density (in units of W/m²) or equivalent mean wind speed at specified height(s) (Table 2-2). Class 1 is the lowest wind-power class; Class 7 is the highest wind-power class. Commercial wind-turbine applications are generally limited to areas with Class 3 or better winds (Figure 2-3). Profitable development in areas with less than Class 5 wind, which represent more than 90% of total estimated potential wind-energy capacity, depends on incentives such as the federal PTC (NREL 2006a). Although wind-energy development tends to focus on areas with higher-class winds, some areas with lower-class winds will likely be developed sooner due to proximity to demand and availability of transmission lines. Class 4 wind sites, for example, are on average 5 times closer to load centers and represent 20 times more wind resource than sites with Class 5 and higher winds (NREL 2006a). Box 2-1 illustrates the distribution of winds rated Class 3 and higher in the MAH based on wind-power density at 50 m above the ground.

TABLE 2-2 Estimates of Total Potential U.S. Wind-Energy Capacity: GW[a]

	Class 3	Class 4	Class 5	Class 6	Class 7	Sum of Classes 3-7
Without Exclusions[b]	5,984	2,648	465	129	61	9,286
% of Classes 3-7	64.4%	28.5%	5.0%	1.4%	0.7%	
With Exclusions[c]	5,137	2,348	392	79	23	7,979
% of Classes 3-7	64.4%	29.4%	4.9%	1.0%	0.3%	

[a]Based on data provided on March 15, 2006, by NREL, Golden, CO; assumes 5 MW/km².
[b]No exclusions except slope >20%.
[c]Standard exclusions applied by NREL for defining available windy land, including environmental criteria, land-use criteria, and other criteria. See Appendix B, Tables B-2 and B-3, for description of the wind resource database.

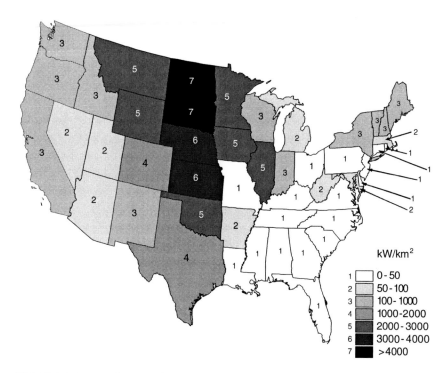

FIGURE 2-3 Distribution of potential onshore wind-energy capacity by state based on wind-resource coverages assembled by NREL. Land-use exclusions have been applied; see Appendix B, Table B-2. Wind-energy capacity is depicted as density (kW/km^2) assuming that each km^2 of area with Class 3 winds and better has a wind-energy capacity of 5 MW. Note: 93.2% of potential wind-energy capacity occurs west of the Mississippi River.

Development Projections

A number of approaches have been used to forecast future wind-capacity development for electricity generation in the United States (Table 2-3). The National Energy Modeling System (NEMS) was developed by the EIA for forecasts of energy supply, demand, and prices. NEMS is a modular system that takes a market-based approach to balancing supply and demand among energy production and end-use sectors. Wind-capacity forecasts are generated for 13 energy-market regions through application of a Wind Energy Submodule (Table 2-3).

The NEMS-GPRA07 model is a modified version of NEMS used to develop benefits projections for the Office of Energy Efficiency and Renewable Energy (EERE). The results are used to evaluate the performance of

BOX 2-1
Mid-Atlantic Highlands: Wind-Energy Potential

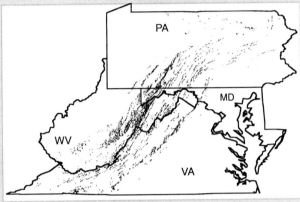

Distribution of winds rated Class 3 and higher in the MAH region based on wind-energy density at 50 m (NREL 2003). Class 3 and higher winds in the MAH are predominantly associated with mountain ridge crests.

Estimates of Potential MAH Wind-Energy Capacity: GW[a]

	Class 3	Class 4	Class 5	Class 6	Class 7	Sum
Maryland	0.55	0.12	0.04	0.01	0.00	0.72
Pennsylvania	2.00	0.51	0.18	0.06	0.00	2.76
Virginia	0.61	0.29	0.14	0.14	0.05	1.23
West Virginia	2.13	0.65	0.27	0.21	0.06	3.31
Total	5.30	1.57	0.62	0.41	0.11	8.01
% of Classes 3-7	66.2%	19.6%	7.8%	5.2%	1.3%	

[a]Based on data provided on March 15, 2006, by NREL, Golden, CO.; assumes 5 MW/km². Standard exclusions applied by NREL for defining available windy sources, including environmental criteria, land-use criteria, and other criteria. See Appendix B for description of the wind resource database.

the Wind Technologies Program, including efforts to solve institutional problems and research to improve the cost and performance of wind generation of electricity.

The Wind Energy Deployment System (WinDS) model was developed by NREL (NREL 2006b) to provide a detailed approach to forecasting wind-energy development in the United States. WinDS uses a Geographic Information System database involving 356 different electricity supply and demand regions to address market issues related to wind-energy development, including access to and cost of transmission, and the intermittency

TABLE 2-3 Projected U.S. Electricity-Generation Capacity with Three Forecasts for Wind-Capacity Development (GW)

	2005[a]	2010	2015	2020
Total U.S. Capacity[b]	955.6	988.4	964.7	1027.4
	Model Projections of Installed Wind Capacity			
EIA-AEO 2006[b]	9.6	16.3	17.7	18.8
% of Total U.S. Capacity	1.0%	1.6%	1.8%	1.8%
EERE-GPRA07[c]	—	8.9	18.9	59.0
% of Total U.S. Capacity		0.9%	2.0%	5.7%
NREL-WinDS[d]	11.9	25.6	43.7	72.2
% of Total U.S. Capacity	1.2%	2.6%	4.5%	7.0%

[a]Values for 2005 are model results based on historic data available at the time of the analysis.

[b]Based on application of the NEMS. Reported in the Annual Energy Outlook for 2006 and in Supplemental Tables 73 and 89, EIA, Office of Integrated Analysis and Forecasting, U.S. Department of Energy (EIA 2006a).

[c]Based on application of the NEMS-GPRA07 model, a modified version of the NEMS. Reported in Projected Benefits of Federal Energy Efficiency and Renewable Energy Programs FY 2007 Budget Request (NREL 2006a).

[d]Based on application of the WinDS model developed by the NREL. Modeled national capacity totals provided to the committee on March 16, 2006, by NREL Energy Analysis Office, Golden, CO. For model information, see NREL 2006b.

of wind. Table 2-3 provides reference case forecasts for WinDS model output provided to the committee by NREL. Figure 2-4 indicates the distribution of installed U.S. wind-capacity forecast for 2020 based on this model output. Box 2-2 illustrates the distribution of future installed wind-power capacity in the MAH based on the WinDS model. Box 2-3 shows MAH estimates related to onshore wind-capacity development. As shown in Table 2-3, estimates of onshore installed U.S. wind-energy capacity in the next 15 years range from 19 to 72 GW, or 2-7% of projected onshore U.S. installed electricity-generation capacity. If the average turbine size is 2 MW—larger than most current turbines—between 9,500 and 36,000 wind turbines would be needed to achieve that projected capacity.

The three modeling approaches represented in Table 2-3 differ in degree of geographic aggregation, in the methods for accounting for transmission and intermittency constraints, and in assumptions about future technology and development costs. Much of the difference in forecast results appears to be related to different expectations for future wind-project performance and capital costs. For example, there are large differences in expectations for decreasing capital costs with increasing market penetration. Whereas the NEMS-GPRA07 and WinDS forecasts are based on an 8% decrease in

Projected 2020 Installed Wind-Energy Capacity: 72,146 MW

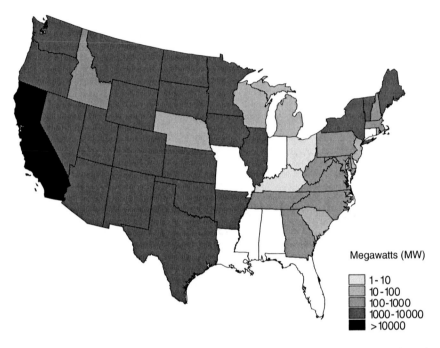

FIGURE 2-4 Projected distribution of installed wind-energy capacity in 2020 based on the WinDS model reference case. Results are shown as the state-level aggregation of 356 supply and demand regions included in the model.

capital costs for every doubling of installed wind-energy capacity worldwide, the EIA-AEO forecasts are based on a 1% decrease in capital costs for every doubling of installed capacity nationwide. Fully understanding the differences in forecasts among the models, however, will be difficult without a carefully designed model comparison study. In the absence of such a study, the committee simply concludes that any forecast of future wind-energy development involves substantial uncertainty.

Although the development projections (Table 2-3 and Figure 2-4) are based on current policies and expectations for technical advancement, other scenarios could be considered that involve technical breakthroughs or major policy changes (incentives and mandates) that would result in forecasts for substantially more development (Short et al. 2006). However, major changes in technology (e.g., much larger or more efficient turbines) or major changes in policy (e.g., discounting environmental concerns and land-use constraints) may create conditions outside the range of our analy-

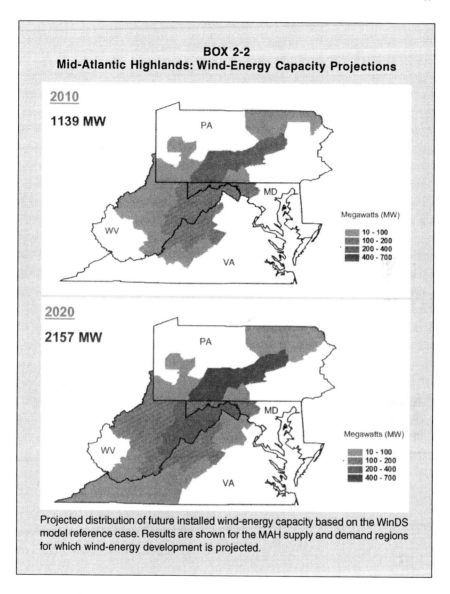

BOX 2-2
Mid-Atlantic Highlands: Wind-Energy Capacity Projections

Projected distribution of future installed wind-energy capacity based on the WinDS model reference case. Results are shown for the MAH supply and demand regions for which wind-energy development is projected.

sis of effects (see Chapter 3). The range of forecast results (Table 2-3) is broad. There is more than a three-fold difference between the high and low projections of installed capacity in 2020. The highest projection in the table estimates about a seven-fold increase in installed capacity in 15 years. Given that only limited data are available for evaluation of both beneficial and adverse effects of existing development, especially in the MAH region, the

---

BOX 2-3
Mid-Atlantic Highlands: Development

Estimates Related to Onshore Wind-Capacity Development

Basis for Estimate	Capacity (MW)	1.5 MW Turbines
NREL estimate of total technical capacity[a]	8015	5344
NREL WinDS model reference case projection for 2020[b]	2158	1439
Operating projects[c]	219	146
Approved but not operating[c]	925	617
PJM (electricity grid operator) interconnection queue[c]	3856	2571

[a]Wind-capacity potential for MD, PA, VA, and WV provided on March 16, 2006, by NREL, Golden, CO. Estimate limited to Class 3 and better wind areas above 1,000 feet elevation. Standard exclusions applied by NREL for defining available wind resource, including environmental, land-use, and other criteria. See Appendix B for description of the wind resource database and exclusion criteria.

[b]Modeled onshore capacity totals for MD, PA, VA, and WV provided on March 16, 2006, by NREL, Golden, CO. Based on application of the WinDS model. (For model information, see NREL 2006b.)

[c]Based on assembled information for projects that are in service, that have state or local-level approval, or that are listed in the PJM interconnection queue (Boone 2006).

This comparison suggests that the WinDS forecast may be low for the MAH. The projects that are already in operation or permitted (with state and local-level approvals) represent more than half of the capacity forecast for 2020 by the WinDS model. The sum of the operating or permitted capacity and the capacity of projects in the connection queue is more than twice the capacity forecast for 2020 by the WinDS model. Although some percentage of the projects that have applied for grid connection may not go forward, it is apparent that the WinDS forecast for the MAH may be exceeded before 2020. Other analyses suggest that recently enacted renewable portfolio standard legislation by mid-Atlantic states will result in substantially more MAH wind development. Ihle (2005), for example, projects that 7,600 MW of wind capacity will be installed in the mid-Atlantic states by 2016. Most of this development would occur on MAH ridges.

---

committee has not conducted analysis of effects associated with scenarios that estimate even greater increases.

Contribution of Wind-Powered Generation to Meeting Projected Electricity Demand

Between 2005 and 2020, based on the WinDS model application (Table 2-3 and Figure 2-4), installed wind-power capacity for generating electricity is projected to increase from 1 to 7% of the total installed U.S.

capacity of all electricity generator types. Projections of installed capacity alone, however, do not provide a sufficient basis for evaluating the potential contribution of wind energy to the electricity supply. As discussed earlier in this chapter, due to the intermittency of wind, installed wind-power capacity is not continuously available for electricity production. Unlike other sources of electricity, wind-generated electricity is not very dispatchable (see discussion earlier in this chapter).

Factors that Limit Wind Energy

The relatively low capacity factor of wind-powered EGUs and other intermittency-related issues affect the extent that wind energy can contribute to the electricity supply. The capacity factor, for any electric-power source, represents the amount of electricity produced in a specified period of time relative to the hypothetical maximum production for the installed capacity. For 2,624 wind turbines installed in the United States since 2000, the cumulative annual capacity factor in 2004 was 30.0% (EIA 2004a, 2004b). In contrast, the annual capacity factors for thermal power plants serving base load are typically much higher. Capacity factors for coal-fueled EGUs designed to run continuously, for example, are typically in the 70-90% range. Power plants serving peak loads, commonly fueled by natural gas, have lower capacity factors because they are dispatched on a variable basis to match variation in demand.

Because wind-powered generators have an inherently low capacity factor, the percentage of total electricity generation from wind energy is substantially less than the percentage of total installed capacity. Based on records assembled for the EIA Annual Energy Outlook 2006 (EIA 2006a), the percentage of total U.S. installed capacity provided by wind energy in 2005 was 1.0% (see Table 2-3). In contrast, the percentage of total electricity generation provided by wind energy was 0.6%. Consideration of future wind-energy contributions to electricity generation thus requires assumptions about the potential for change in capacity factor as well as projections of installed capacity.

The extent to which wind energy can contribute as a source of electricity generation also is affected by limitations related to integration with the electricity-distribution system or grid. The significance of these limitations may both increase in time as more wind-generated electricity is introduced to electricity grids and decrease as improvements to the grids are achieved. Reserve requirements, in particular, can reduce the effective load-carrying capacity of installed facilities to produce wind-generated electricity. Reserve requirements are determined by the need for dispatchable generation to respond to both variations in demand and to generation and transmission outages. To the degree that wind generation is not dispatchable, it does not directly contribute to reserve requirements, and because fluctuations in wind-powered generation introduce additional load variance into the grid, it can increase the reserve requirement. The effective amount of electricity generation from installed wind-powered EGUs may thus be less than indicated by a simple capacity-factor adjustment.

Reserve requirements are generally met through control of conventional generators that have some amount of variable dispatch capacity and by maintenance of stand-by generators with quick-start capacity. At low wind-penetration levels, the load variance introduced by wind-generated electricity is generally small in relation to both normal operating variance and variable dispatch or quick-start capacity in the grid. This means that the need for additional reserves is generally low with initial wind-energy development, and the effective load-carrying capacity of wind-generated electricity is not necessarily reduced by the need for additional reserves. But this may change as more wind capacity is installed and a larger percentage of grid capacity is represented by wind. Estimates provided by Biewald (2005) and UWIG (2006) suggest that additional reserves are not required until the percentage of total generation provided by wind-generated electricity-generating facilities reaches 10-20%, a range that greatly exceeds the 0.6% of U.S. generation currently provided by wind energy (see Table 2-4). Experience in other areas with more wind development indicates that loss

TABLE 2-4 Projected U.S. Electricity Generation Based on Three Forecasts of Wind-Capacity Development: Billions of kWh (thousands of GWh)

	2005[a]	2010	2015	2020
Total U.S. Generation[b]	4065.7	4387.7	4727.1	5107.5
	Projections of Wind Generated Electricity[b]			
EIA-AEO 2006	23.2	50.9	56.0	59.8
% of Total U.S. Generation	0.6%	1.2%	1.2%	1.2%
EERE-GPRA07[c]		27.8	59.8	187.6
% of Total U.S. Generation		0.6%	1.3%	3.7%
NREL-WinDS[b]	28.7	80.0	138.1	229.4
% of Total U.S. Generation	0.7%	1.8%	2.9%	4.5%

[a]Values for 2005 are model results based on historical data available at the time of the analysis.

[b]Total generation from all sources in the contiguous United States, based on application of the NEMS. Reported in the Annual Energy Outlook 2006, EIA, Office of Integrated Analysis and Forecasting, U.S. Department of Energy (EIA 2006a).

[c]Based on forecasts of installed wind-generation capacity provided in Table 2-3. Capacity factors for calculation of electricity generation are based on installed capacity and generation data for wind energy provided in the Annual Energy Outlook 2006 (EIA 2006a).

of effective load-carrying capacity and the need for additional reserves may become important as wind development expands, as discussed below.

The following examples illustrate the difficulty of translating installed capacity of wind-powered electricity generation, even modified by capacity factor, into a displacement of other energy sources. Germany, for example, has more installed wind-powered generation capacity than any other country in the world (E.ON Netz 2006). Installed wind capacity was equal to about 14% of Germany's total installed generating capacity in 2004.[2] The contribution of wind energy to the "guaranteed capacity" of the German electric generation system in 2004 was only 8% of installed wind-energy capacity (less than half of the annual capacity factor) and it is projected to decrease to 4% of installed wind capacity in 2020, given an expected three-fold increase in installed wind-energy capacity (E.ON Netz 2005).

Seasonal and diurnal variation in wind energy also affect the contribution of wind-powered electricity generation relative to other power sources, and annual capacity factors do not account for this temporal variation in the contribution of wind energy. In many areas of the United States, the availability of wind energy is lowest in the afternoon hours of summer months when both the demand and the rate of growth in demand for electricity are the highest. As indicated above, for 2,624 wind turbines installed in the United States since 2000, the cumulative average annual capacity factor in 2004 was 30.0%. For the same turbines, the cumulative average August capacity factor was 22.7%, or about 25% less than the annual capacity factor (EIA 2004a, 2004b). Box 2-4 presents monthly variability in electricity demand and wind capacity factor in the MAH states.

Estimating the Effective Electricity Generation from Installed Wind-Energy Capacity

In the absence of information concerning the need for increased reserve capacity or other effects of temporal variation in wind energy, annual average capacity factors provide a reasonable basis for approximating the effective amount of electricity generated from installed wind-energy capacity. However, this approximation may prove unreliable for specific projects or regions, and we acknowledge uncertainty concerning the effect of rapidly expanding wind development. Perhaps of more importance, although current capacity factors for wind development can be calculated based on available capacity and generation data (e.g., EIA data reports), the estimation of future capacity factors involves assumptions and unspecified uncertainty.

---

[2]Based on E.ON Netz (2005) estimates of wind capacity (16,400 MW) and EIA (2006b) estimates of total capacity (118,850 MW).

BOX 2-4
Mid-Atlantic Highlands: Wind-Capacity Factor versus Electricity Demand

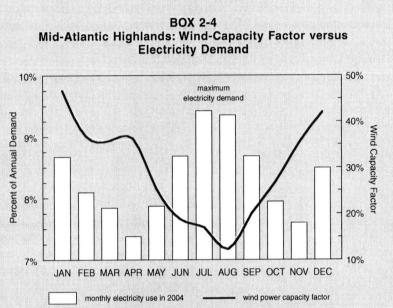

Electricity demand and wind-power profiles in the MAH states. The electricity demand profile is based on 2004 monthly sales data (EIA 2004c). The wind-power profile is based on average monthly capacity factors determined using the available 2002-2004 wind-generation data for four operating wind projects in the MAH (EIA 2004b).

The correlation between monthly wind-capacity factors and monthly electricity demand in the four MAH states is generally negative. The electricity-grid system that includes the MAH is managed by PJM Interconnection, a summer-season peaking system, with a greater rate of growth in demand in summer than in winter (PJM 2005a). PJM has developed rules for determination of the capacity value for wind-powered EGUs and other "intermittent capacity resources" (PJM 2005b). When wind-powered EGUs are first connected to the PJM grid they are assigned an initial "capacity credit," which represents the percentage of a project's installed capacity that can be traded in the PJM electricity market. The initial capacity credit for new wind-energy projects is 20%, which approximates the average summer capacity factor for wind-energy projects in the region. As data for a wind-energy project become available, the capacity credit is adjusted by calculating a three-year running average capacity factor based on afternoon hours in the summer months. The expected amount of electricity provided by a wind-powered EGU in the PJM system is thus specifically determined for the time when the availability of wind energy is the least and the demand for electricity is the greatest. The relationship between wind-capacity factor and electricity demand may differ for other regions of the country.

Projections for both installed wind-energy capacity and wind-powered electricity generation included in the EIA Annual Energy Outlook (reference-case forecasts; EIA 2006a) indicate that the annual average capacity factor for all installed wind capacity (not just new projects) will increase from 27.5% in 2005 to 36.2% in 2020. In contrast, the EERE has assumed that future capacity factors will be substantially higher, given projected results of the EERE Wind Technologies Program (NREL 2006a).[3] The committee has used the EIA estimates of capacity factor to assess the effective amount of electricity generated from wind-powered EGUs in both the United States and the MAH subregion. The EIA capacity-factor estimates allow for moderate improvement in technology, they account for the fact that future installed capacity will be a mix of both older and newer turbines, and they are intermediate between currently observed capacity factors and the most optimistic forecasts of future capacity factors. It also seems reasonable to expect additional constraints on wind-powered EGU performance as accessible areas with higher-class winds are exploited and development expands into areas with lower-class winds.

Table 2-4 provides forecasts for wind-generated electricity through 2020 in relation to forecasts of total electricity generation through 2020. These forecasts are based on the model projections of installed onshore wind-energy capacity in the contiguous United States provided in Table 2-3 and on projections of total U.S. generation capacity provided in the EIA Annual Energy Outlook for 2006 (EIA 2006a). The forecasts for both wind generation and total generation are for onshore wind-energy development in the contiguous United States. As discussed above, the forecasts for wind-energy generation are adjusted for capacity-factor limitations, but not for other potential effects of the temporal variation in wind.

As with the range of forecasts for installed wind-powered EGU capacity in Table 2-3, the range of forecasts for their effective electricity generation in Table 2-4 suggests a high degree of uncertainty. The forecasts, however, provide a context for evaluating both the electricity supply and air-quality benefits of future wind-energy development in the United States. The highest forecast for 2020 indicates that wind-energy development will provide 7.0% of total installed electricity-generation capacity, and 4.5% of electricity generation, which is consistent with the fact that wind turbines generally have a lower capacity factor than other electricity-generation sources. It is also significant that the forecast growth in wind-energy development will

---

[3]For input to the NEMS-GPRA07 model, EERE estimated different capacity factors depending on program support for research and development (NREL 2006a). The estimated capacity factors for new onshore wind projects with Class 4 winds in 2020 were 46.9 and 37.2%, with and without projected program results. EERE did not report estimates for Class 3 winds, although Class 3 winds are now being developed, and areas with Class 3 winds are far more extensive than areas with higher-class winds (see Table 2-2).

occur in a context of rapidly increasing electricity demand. Although wind-energy development has been identified as one of the fastest-growing energy sources in the United States, this growth has typically been represented in terms of a percentage change in installed wind-energy capacity (e.g., GAO 2004; EERE 2006). In order to evaluate the potential contribution of wind-energy development to the electricity supply, we have examined projected growth in wind-powered electricity generation in relation to projected growth in total electricity generation. Based on the EIA forecasts in Table 2-4, total electricity generation from all sources is projected to increase by more than 1,000,000 gigawatt-hours (GWh) between 2005 and 2020. As shown in Figure 2-5, the projected increase in wind generation is expected to account for 3.5 to 19.3% of this increase in total generation. Thus, based on projections examined by the committee, 80.7 to 96.5% of the growth in U.S. electricity generation by 2020 is expected to be obtained from generation sources other than wind.

Contribution of Wind Energy to Air-Quality Improvement

Our approach to assessing the benefits of wind-energy development for air-quality improvement focuses on displacement of several of the pollutant emissions from fossil-fueled EGUs (in this case, CO_2, NO_x, and SO_2).

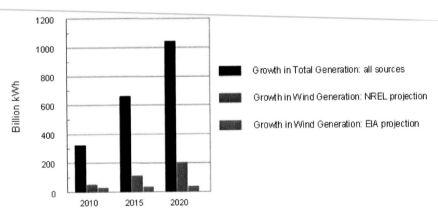

FIGURE 2-5 Cumulative growth in total annual electricity generation between 2005 and 2020, compared with projected growth in onshore wind generation. Total electricity generation in the United States in 2020 is projected to exceed total generation in 2005 by 1041.8 billion kWh. Electricity generation with wind power in 2020 is projected to exceed wind generation in 2005 by 36.6 to 200.7 billion kWh. The projections for growth in total generation and wind generation are based on the data provided in Table 2-4.

A more informative assessment would account for atmospheric residence times, transport patterns, atmospheric chemistry, and the response properties of environmental receptors, all of which are beyond the practical scope of our task.

Estimating Emissions Displacement

The generator types associated with the U.S. electricity supply differ greatly in terms of their contributions to total generation and pollutant emissions (Figure 2-6). Despite the inevitable uncertainties (discussed previously in this chapter), emissions-displacement analysis is needed for policy and regulatory decisions (Appendix B, Table B-1). The wide range of emissions-displacement rates results from different quantitative approaches, as well as differences related to the geographic distribution of generator types and the achievement of emission reductions through air-quality regulation.

A simple approach to evaluation of emissions displacement on a large regional scale is illustrated by a recent Programmatic Environmental Impact Statement (PEIS) prepared for assessment of wind-energy development on western U.S. lands administered by the Bureau of Land Management (BLM 2005a). The BLM-PEIS, which relied in part on emissions data from the early 1990s, compared two extremes, 100% coal displacement and 100%

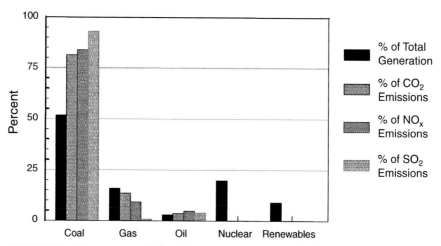

FIGURE 2-6 Percentage of electricity generation provided by generator types in relation to the percentage of CO_2, NO_x, and SO_2 emitted from all electricity generation in the United States. The renewables include hydroelectric, biomass, wood, solar, and wind.
SOURCE: Based on data for 2000 (EPA 2006b).

natural-gas displacement. Although the emissions reductions associated with displacement of coal generation dramatically exceeded the emissions reductions associated with displacement of natural-gas generation (see Appendix B, Table B-1), the BLM-PEIS provided no analysis or other basis for favoring either extreme.

The BLM-PEIS treatment of the emissions-displacement issue may actually be appropriate given the problems and uncertainties associated with more detailed analyses. However, simply providing bounds for the potential emissions-displacement benefits of wind-energy development (or other renewable-energy and energy-efficiency initiatives) is not sufficient for many regulatory and policy purposes. A number of methods for determining specific emissions-displacement rates have thus been developed and applied. These methods can generally be assigned to two categories:

- Methods based on emissions rates associated with affected fossil-fuel-fired EGUs.
- Methods based on system-average emissions rates.

The methods in the first category are potentially the most reliable, although the data requirements are much greater, the analysis is far more complex, and the issue of transparency is more difficult. Identification of affected EGUs generally requires application of a system-dispatch model. This involves accounting for the temporal distribution of wind energy or actual wind generation, the identity and operational properties of EGUs operating on the margin, and transmission limits or other dispatch constraints. Analysis of long-term displacement must also consider the introduction of new EGUs to meet increasing baseload (continuous demand over a long period) and peaking demand, as well as the retirement of old EGUs.

System-dispatch models can be used either to determine emissions displacement from specific fossil-fuel-fired EGUs or to determine emission-displacement rates associated with fossil-fuel-fired EGUs on the operating margin. The focus on the operating margin is based on economic-dispatch order, which means that the most expensive EGUs are the first to be displaced when less expensive generation is available. The most expensive units are generally those that provide peaking power or respond to short-term variation in demand (e.g., such as by natural-gas-fueled generators), rather than those that provide baseload power (e.g., such as by nuclear and coal-fueled generators). Strict adherence to economic-dispatch order, however, may be compromised by transmission limitations and requirements to maintain an acceptably low risk of loss of supply.

Although system-dispatch modeling often is identified as the preferred method for estimating emissions displacement, its use and acceptance are limited by the problem of access to necessary, but proprietary, technical and information resources. System-dispatch models are generally owned and used by utility companies, grid operators, or private consultants. The input data required by such models, including information on grid structure and performance, costs and dispatch properties of EGUs, and detailed wind-energy information, are generally not available to either the public or resource-management agencies. Box 2-5 discusses the issues of transparency in developing emission-reduction estimates for MAH. Given this lack of transparency, it can be difficult or impossible for independent parties to objectively review and verify emissions-displacement estimates based on system-dispatch modeling.

System-average emission rates are commonly used for analysis of emissions displacement when the data and resources needed for system-dispatch modeling are unavailable (NESCAUM 2002; Keith et al. 2003; UNFCCC

BOX 2-5
Mid-Atlantic Highlands: Transparency and Emissions Reduction Estimates

Transparency has been identified as an accounting principle that must be applied for credible quantification and public acceptance of emissions reductions claims (WBCSD and WRI 2005). Although system-dispatch modeling potentially offers the most reliable approach for estimating the emissions-displacement benefits of wind-energy projects, actual model applications have varied with respect to transparency.

For example, default emissions-displacement rates were determined for use in an emissions-benefit workbook developed by the Ozone Transport Commission for evaluating renewable-energy and energy-efficiency projects in three eastern U.S. grid regions, including the PJM grid region that includes the MAH (OTC 2002). In this case, the estimated emissions-displacement rates were attributed to EGUs operating on the margin. Although the associated documentation identified the data sources, as well as important assumptions, minimal information was provided about the proprietary simulation model that was applied to identify displaced units and estimate emissions-displacement rates (Keith et al. 2002).

In other examples, emissions-displacement rates were developed as a basis for crediting municipal wind-energy purchases with emission reductions (Hathaway et al. 2005) or for assessing the air-quality benefits of specific wind-project proposals (High and Hathaway 2006). Model results in these cases indicated that displaced emissions would either exclusively or predominantly be associated with coal-fueled generating units. Again, however, only minimal details concerning the model applications were provided. Moreover, although the associated documentation identified some of the data sources, critical data, including proprietary or confidential information related to both wind-energy performance and identification of the displaced generating units, were not provided.

2006). System-average emission rates are calculated by dividing total system emissions by total system generation, providing a single emission rate expressed as mass of pollutant per unit of electricity generation (e.g., lbs/ MWh). Given an estimate of potential electricity generation from wind-energy development or any other source, this rate can be applied to estimate the mass of pollutant that would be emitted to obtain the same generation from the existing mix of EGUs in the system. The advantage of the system-average emission rate is that it can be applied relatively easily, using emissions and generation data that are publicly available. A disadvantage is that it tends to be weighted toward emissions from fossil-fuel-fired EGUs that supply baseload rather than fossil-fuel-fired EGUs operating on the margin. This means that use of the system-average rate will overestimate emissions displacement in grid regions where baseload is dominated by high-emission generation (e.g., coal-fueled EGUs) and underestimate emissions where baseload is dominated by low-emission generation (e.g., nuclear and hydro-electric EGUs). A potentially useful modification is to use the marginal-average emission rates instead of the system-average emission rate. This may work well in grid regions where the fossil-fuel-fired EGUs operating on the margin are relatively uniform with respect to emissions. However, in other areas, such as the PJM grid region, which has both coal-fueled and natural-gas-fueled EGUs operating on the margin (PJM 2006a), marginal-average emission rates would be weighted toward the higher emission rates associated with the coal-fueled EGUs, regardless of which type of EGU would actually be displaced by wind-energy generation.

Emissions Displacement in Context

In this section, the committee examines the potential for obtaining reductions in emissions of NO_x, SO_2, and CO_2 through the increased use of wind energy to generate electricity. The comparative lack of air-pollutant emissions has been identified as the most important environmental benefit of wind energy (AWEA 2006e). Evaluation of these benefits, however, is complicated by a number of contextual factors in addition to the problem of identifying emissions-displacement rates. These factors include the presence of emissions from sources other than fossil-fuel-fired EGUs, continuing growth in demand for electricity, and changing emission rates for fossil-fuel-fired EGUs. Other differences in environmental impacts of various sources of energy are potentially important (e.g., species and habitat impacts), although they are not addressed here for the reasons given earlier (see Chapter 3 for a discussion of ecological impacts of wind energy).

Wind development can only displace emissions from electricity-generation sources. It is expected that emissions associated with most industry, home heating, and transportation will not be affected by changes in sources of electricity generation. In 2001 about 68% of anthropogenic SO_2 emissions, but only about 23% of anthropogenic NO_x emissions, in the United States were associated with the burning of fossil fuels for electricity generation (EPA 2005).

The largest source of SO_2 emissions is coal combustion; the largest source of NO_x emissions is transportation (EPA 2006c). About 39% of anthropogenic CO_2 emissions in the United States in 2001 resulted from electricity generation, while the balance was derived from other sources (EIA 2006d).

The task of evaluating air-quality benefits of wind-powered electricity generation is complicated by increasing electricity use and changing emission rates for fossil-fuel-fired EGUs. Reference case projections provided in the EIA Annual Energy Outlook for 2006 (EIA 2006a) indicate that generation of electricity in the United States will increase at an average rate of 1.6% per year between 2004 and 2030. Despite this growth, emissions from fossil-fuel-fired EGUs of NO_x and SO_2, which are subject to regulatory controls, are projected to decrease by an average of 4.0 and 2.1% per year. Emissions from fossil-fuel-fired EGUs of CO_2, which are largely uncontrolled, are projected to increase by an average of 1.4% per year. The opposing changes in emissions influence projections of future trends in system-average emission rates (in units of lbs/MWh) between 2000 and 2020 (Table 2-5). The table shows that emissions of all three pollutants are expected to decrease on a per unit of energy basis. However, whereas system-average emission rates for NO_x and SO_2 are projected to decline by 72 and 74%, system-average emission rates for CO_2 are projected to decline by only 12%. As indicated in Figure 2-7, the projected increase in electricity generation, the concurrent decrease in emissions of NO_x and SO_2 from fossil-fuel-fired EGUs, and the concurrent increase in emissions of CO_2 from those EGUs, all represent continuations of pre-existing trends.

TABLE 2-5 Observed and Projected System-Average Emission Rates for U.S. Electricity Generation (lbs/MWh)

	2000[a]	2005[b]	2010[b]	2015[b]	2020[b]
CO_2	1392	1287	1272	1241	1223
NO_x	2.96	1.92	1.07	0.89	0.83
SO_2	6.04	5.28	2.69	1.96	1.58

[a]Based on total electrical generation and associated emissions in 2000, reported in the eGRID database (EPA 2006b).

[b]Based on forecasts of total electrical generation and associated emissions provided in the Annual Energy Outlook 2006, EIA, Office of Integrated Analysis and Forecasting, U.S. Department of Energy (EIA 2006a). The committee has not assessed the uncertainty associated with these estimates.

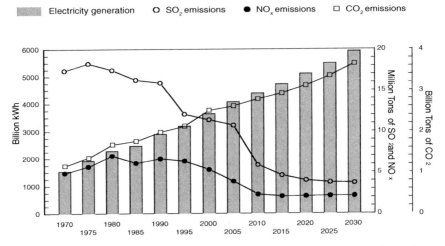

FIGURE 2-7 Past and projected changes in emissions of CO_2, NO_x, and SO_2 from EGUs in relation to the past and projected increase in electricity generation. Data through 2000 are observed; data for 2005-2030 are projected. Generation data were obtained from EIA (2006c, 2006a); emissions data were obtained from EPA (2005) and EIA (2004c, 2006a).

U.S. emissions data for 1970-2003 indicate that emissions of SO_2 from fossil-fuel-fired EGUs declined 37%, while emissions of NO_x from those EGUs declined 9% (EPA 2005). These past declines in emissions of NO_x and SO_2, as well as the projected future declines, can be attributed to implementation of the Clean Air Act and related regulatory programs. Future declines are also expected to result from the upcoming implementation of the Clean Air Interstate Rule (CAIR). Because both pollutants are subject to emissions caps and allowance trading, there is only limited opportunity to achieve additional emissions reductions with wind-energy development.[4] In the context of a "cap-and-trade" program, a reduction in emissions requires a reduction in the emissions cap. One means for wind-energy projects to achieve this is through allowance "set-asides," whereby a percentage of the allowed emissions under the cap are available

---

[4]A national cap on SO_2 emissions from EGUs was initially established under Title IV of the Clean Air Act. Additional controls in 28 eastern states are required by CAIR, including reductions in NO_x emissions that are expected to be achieved primarily through a cap-and-trade program for emissions from EGUs. At this time, the extent to which emission reductions in addition to those expected from CAIR would be sought by some eastern states is unknown (see, for example, *Clean Air Report*, February 22, 2007, Inside Washington Publishers, Arlington, VA).

for retirement commensurate with emission reductions credited to renewable-energy or energy-efficiency projects (see Keith et al. 2003; EPA 2004;
Bluestein et al. 2006). At present there is no set-aside program for SO_2
allowances, although set-asides for NO_x allowances can be established by
states affected by the CAIR. Emissions-displacement rates for NO_x set-
asides of 1.5 lbs/MWh through 2015 and 1.25 lbs/MWh after 2015 have
been proposed by the EPA (Bluestein et al. 2006). However, the potential
for emissions-cap reductions due to wind development remains uncertain.
In the six states that have established NO_x set-asides, only 1 to 5% of total
NO_x allowances are reserved for set-asides, and this amount can be allo-
cated to either renewable-energy or energy-efficiency projects. The National
Research Council (NRC 2006) pointed out that cap-and-trade programs
have potential pitfalls and that such programs can result in emission trades
from one location to another and from one period to another with potentially detrimental consequences. However, analytical tools are not sufficient
to assess the potential effect of cap-and-trade programs on local air quality
or the extent to which wind-powered EGUs might alter those effects. In
contrast to NO_x and SO_2, emissions of CO_2 from fossil-fuel-fired EGUs
or other sources are not subject to national regulatory controls or emissions caps, although subregional control efforts have been initiated (RGGI
2006) and various national controls have been proposed (see Johnston
et al. 2006). Thus, to the extent that CO_2-emitting sources of electricity
generation are displaced, wind-energy development can achieve displacement of CO_2 emissions. As indicated above, however, CO_2 emissions from
fossil-fuel-fired EGUs are projected to increase an average of 1.4% per year
between 2004 and 2030 (reference-case forecast; EIA 2006a). Moreover,
as also indicated above, fossil-fuel-fired EGUs accounted for about 39% of
anthropogenic CO_2 emissions in the United States in 2005 (EIA 2006d).
Compared with just the projections for CO_2 emissions from fossil-fuel-fired
EGUs, the potential for offsetting emissions with wind-energy development
is illustrated by Figure 2-8, which compares projected annual emissions of
CO_2 from fossil-fuel-fired EGUs in the United States with offsets that might
be achieved through wind-energy development. The estimated offsets are
based on the maximum forecasts for wind-powered generation of electricity
provided in Table 2-4 and on the system-average emission rates for CO_2
listed in Table 2.5. Based on this comparison, the effect of wind development by 2020 is expected to offset CO_2 emissions from fossil-fuel-fired
EGUs in the United States by 4.5%. If fossil-fuel-fired EGUs continue to
account for less than half of anthropogenic CO_2 emissions in the United
States, then the effect of projected wind-energy development in 2020 would
be to offset total anthropogenic CO_2 emissions by less than 2.25%. However, potential technological improvements in emission controls, and other
factors that will affect total CO_2 emissions, are as hard to predict for the

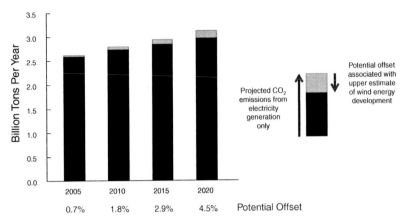

FIGURE 2-8 Projected increase in U.S. CO_2 emissions from EGUs and potential offsets associated with wind-energy development. CO_2 emissions are based on forecasts reported in the Annual Energy Outlook 2006 (EIA 2006a). CO_2 offset estimates are based on the maximum forecasts for U.S. wind-powered generation of electricity provided in Table 2-4 and on the system-average emission rates provided in Table 2-5.

transportation and industrial sectors as for the electricity-generation sector, and so the total reduction of U.S. anthropogenic CO_2 emissions by wind-energy development in 2020 could be more or less than 2.25%.

A range of potential offsets can be estimated based on emissions rates for coal-fueled and natural-gas-fueled EGUs. In 2000, the average CO_2 emissions rate for coal-fueled EGUs in the United States was about 157% of the system-average emissions rate; the average CO_2 emissions rate for natural-gas-fueled EGUs was about 85% of the system-average emissions rate (see Appendix B, Table B-1). Based on these values, the range of potential CO_2 emissions offsets given the maximum forecast for U.S. wind-powered generation of electricity in 2020 is 3.8 to 7.1% of projected emissions from EGUs.

CONCLUSIONS

Electricity generated from different sources is fungible. Depending on factors such as price, availability, predictability, regulatory and incentive regimes, and local considerations, one source might be preferentially used over others. The importance of the factors changes over varying time and space scales. As a result, a more complete understanding of the environmental and economic effects of any one energy source depends on a more complete understanding of how that energy source displaces or is displaced by other energy sources, and on a more complete understanding of the environmental and economic effects of all other available energy sources. LCA can be used to help fulfill that need.

- Projections for future development of wind-powered electricity generation, and hence projections for future wind-energy contributions to reduction of air-pollutant emissions in the United States, are highly uncertain. However, some insight can be gained from recent model projections by the U.S. Department of Energy. Estimates for onshore installed U.S. wind-energy capacity in the next 15-year range from 19 to 72 GW, or 2 to 7% of projected U.S. installed electricity-generation capacity.[5] In part because the wind blows intermittently, wind turbines often produce less electricity than their rated maximum output. On average in the MAH, the capacity factor of wind turbines is about 30% for current technology, forecast to improve to near 37% by the year 2020. The projections the committee has used in this chapter suggest that onshore wind-energy development will contribute about 60 to 230 billion kWh, or 1.2 to 4.5% of projected U.S. electricity generation in 2020. In the same period, wind-energy development is projected to account for 3.5 to 19% of the *increase* in total electricity-generation capacity. If the average turbine size is 2 MW—larger than most current turbines—between 9,500 and 36,000 wind turbines would be needed to achieve that projected capacity.

- Projections for future wind-energy contributions to air-pollution emissions reductions in the United States also are uncertain. However, given that current and future regulatory controls on emissions of NO_x and SO_2 from electricity generation in the eastern United States involve total caps on emissions, the committee concludes that development of wind-powered electricity generation using current technology probably will not result in a significant reduction in total emission of these pollutants from EGUs in the mid-Atlantic region. Using the future projections of installed U.S. energy capacity by the U.S. Department of Energy, we further conclude that development of wind-powered electricity generation probably will contribute to offsets of about 4.5% in emissions of CO_2 from electricity generation sources in the United States by the year 2020. In 2005, emissions of CO_2 from electricity generation were estimated to be 39% of all CO_2 emissions in the United States.

- Although the wind resource in the MAH is closer to electricity markets and transmission lines than much of the wind resource in the United

---

[5]There was no installed offshore wind-energy generating capacity in the United States as of mid-2006, although several projects have been proposed and at least two projects are currently in the permitting stage of development. Department of Energy projections for total installed offshore wind capacity in the next 15 years range up to 12 GW.

States, a smaller portion of the mid-Atlantic region has high-quality wind resources than does the United States as a whole. As a result, wind energy will likely contribute proportionally less to electricity generation in the mid-Atlantic region than in the United States as a whole.

- Electricity generated in the MAH—including wind energy—is used in a regional grid in the larger mid-Atlantic region. Electricity generated from wind energy in the MAH has the potential to displace pollutant emissions, discharges, wastes, and other adverse environmental effects over the life cycle of other sources of electricity generation in the grid, but that potential is less than 4.5%, and the degree to which its beneficial effects would be felt in the MAH is uncertain.

- In the presence of more aggressive renewable-energy-development policies, potential increased energy conservation, and improving technology of wind-energy electricity generation and transmission, the above findings may underestimate wind energy's contribution to total electricity production. This would affect the committee's analysis, including projections for development and associated effects (e.g., energy supply, air pollution, development footprint). On the other hand, if technological advances serve to reduce the emissions and other negative effects of other sources of electricity generation, or if fossil-fuel prices fall,[6] our findings may overestimate wind energy's contribution to electricity production and air-pollution offsets.

---

[6]Although it may appear unlikely that fossil-fuel prices will fall very far for a long period, so many geopolitical, technological, and economic factors affect fuel prices that it remains difficult to predict the future trajectory of those prices with confidence.

3

Ecological Effects of Wind-Energy Development

CHAPTER OVERVIEW

At regional to global scales, the effects of wind energy on the environment often are considered to be positive, through the production of renewable energy and the potential displacement of mining activities, air pollution, and greenhouse gas emissions associated with nonrenewable energy sources (see Chapter 2). However, wind-energy facilities have been demonstrated to kill birds and bats and there is evidence that wind-energy development also can result in the loss of habitat for some species. To the extent that we understand how, when, and where wind-energy development most adversely affects organisms and their habitat, it will be possible to mitigate future impacts through careful siting decisions. In this chapter, we review the effects of wind-energy development on ecosystem structure and functioning, through direct effects of turbines on organisms, and on landscapes through alteration and displacement. We recommend a research and monitoring framework for reducing these impacts. Although the focus of our analysis is the Mid-Atlantic Highlands, we use all available information to assess general impacts. Although other sources of development on sites that are suitable for wind-energy development affect wildlife and their habitats (e.g., mineral extraction, cutting of timber), and there are other sources of anthropogenic mortality to animals, as stated previously, this committee was charged to focus on wind energy, and therefore did not conduct a comprehensive comparative analysis of impacts from other sources of development.

Wind turbines cause fatalities of birds and bats through collision, most likely with the turbine blades. Species differ in their vulnerability to collision, in the likelihood that fatalities will have large-scale cumulative impacts on biotic communities, and in the extent to which their fatalities are discovered and publicized. This chapter reviews information on the probabilities of fatalities, which are affected by both abundance and behavioral characteristics of each species.

Factors such as the type, location, and operational schedules of turbines that may influence bird and bat fatalities are reviewed in this chapter. The overall importance of turbine-related deaths for bird populations is unclear. Collisions with wind turbines represent one element of the cumulative anthropogenic impacts on bird populations; other impacts include collisions with tall buildings, communications towers, other structures, and vehicles, as well as other sources of mortality such as predation by house cats (Erickson et al. 2001, 2005). While estimation of avian fatalities caused by wind-power generation is possible, the data on total bird deaths caused by most anthropogenic sources, including wind turbines, are sparse and less reliable than one would wish, and therefore it is not possible to provide an accurate estimate of the incremental contribution of wind-powered generation to cumulative bird deaths in time and space at current levels of development.

Data on bat fatalities are even sparser. While there have been a few reports of bat kills from other anthropogenic sources (e.g., through collisions with buildings and communications towers), the recent bat fatalities from wind turbines appear to be unprecedentedly high. More data on direct comparisons of turbine types are needed to establish whether and why migratory bats appear to be at the greatest risk of being killed. Clearly, a better understanding of the biology of the populations at risk and analysis of the cumulative effects of wind turbines and other anthropogenic sources on bird and bat mortality are needed.

The construction and maintenance of wind-energy facilities alter ecosystem structure, through vegetation clearing, soil disruption, and potential for erosion, and this is particularly problematic in areas that are difficult to reclaim, such as desert, shrub-steppe, and forested areas. In the Mid-Atlantic Highlands forest clearing represents perhaps the most significant potential change through fragmentation and loss of habitat for forest-dependent species. Changes in forest structure and the creation of openings alter microclimate and increase the amount of forest edge. There may also be important interactions between habitat alteration and the risk of fatalities, such as bat foraging behavior near turbines.

The recommendations in this chapter address the types of studies that need to be conducted prior to siting and prior to and following construction of wind-energy facilities to evaluate the potential and realized ecological impacts of wind-energy development. The recommendations also address assessing the degree to which a particular site is acceptable for wind-energy development and the types of research and monitoring needed to help inform decision makers.

INTRODUCTION

There are two major ways that wind-energy development may influence ecosystem structure and functioning—through direct impacts on individual organisms and through impacts on habitat structure and functioning. Environmental influences of wind-energy facilities can propagate across a wide range of spatial scales, from the location of a single turbine to landscapes, regions, and the planet, and a range of temporal scales from short-term noise to long-term influences on habitat structure and influences on presence of species. In this chapter, we review the documented and potential influences of wind-energy development on ecosystem structure and functioning, focusing on scales of relevance to siting decisions and on influences on birds, bats, and other vertebrates.

Construction and operation of wind-energy facilities directly influence ecosystem structure. Site preparation activities, large machinery, transportation of turbine elements, and "feeder lines," transmission lines that lead from the wind-energy facility to the electricity grid, all can lead to removal of vegetation, disturbance, and compaction of soil, soil erosion, and changes in hydrologic features. Although many of these activities are relatively local and short-term in practice (e.g., construction), there may be substantial effects on habitat quality for a variety of organisms. These changes will likely be detrimental to some species and beneficial to others. Wind-energy development that is focused on specific topographic features (e.g., ridgelines) that represent key habitat features for some species may have disproportionately detrimental impacts on those species that depend on or are closely associated with these habitats.

Recent reviews of available literature have clearly documented direct impacts of wind turbines on birds and bats (GAO 2005; Barclay and Kurta 2007; Kunz et al. 2007), including death from colliding with turbine blades. As discussed below, little is known about the circumstances contributing to fatalities, but issues such as turbine height and design, rotor velocity, number and dispersion of turbines, location of the turbine on the landscape, and the abundance, migration, and behavioral characteristics of each species present are likely to influence fatality rates. In addition, non-flying organisms may be affected by turbine construction and operation, because of alteration of habitat and behavioral avoidance, possibly due to noise, vibration, motion of turbines, or their mere presence in the landscape.

We can make three general predictions about the large-scale and long-term impacts of individual fatalities. First, life-history theory predicts that characteristics of populations of affected species determine the consequences of increased mortality: organisms whose populations are characterized by low birth rate, long life span, naturally low mortality rates, a high trophic level, and small geographic ranges are likely to be most susceptible to cumulative, long-term impacts on population size, genetic diversity, and ultimately, population viability (e.g., McKinney 1997; Purvis et al. 2000). Bats are unusual among mammals with respect to their life-histories, because they typically have small body sizes but long life spans (Barclay and Harder 2003), and the probability of extinction in bats has been linked to several of these characteristics (Jones et al. 2003). Second, the effects of a decline in one species on entire biotic communities is determined by the role of the species in the larger context: losses of keystone species, organisms that have a disproportionately high impact on ecosystem functioning (Power et al. 1996), and those that provide important ecosystem services (Daily et al. 1997) are of most concern. Species that are important predators and perform critical top-down control over communities, and species that are important prey sources can be keystone species in both natural and human-altered ecosystems (Cleveland et al. 2006). Notably, many raptors and insectivorous bats fill these roles. Finally, we do not know how the migration patterns of affected species will influence regional-scale mortality; we also do not understand the consequences of deaths of individuals of these migrating species to the local populations they originate from. Unfortunately this type of information is nearly impossible to obtain.

The ecological influences of wind-energy facilities are complex, and can vary with spatial and temporal scale, location, season, weather, ecosystem type, species, and other factors. Moreover, many of the influences are likely cumulative, and ecological influences can interact in complex ways at wind-energy facilities and at other sites associated with changed land-use practices and other anthropogenic disturbances. Because of this complexity, evaluating ecological influences of wind-energy development is challenging and relies on understanding factors that are inadequately studied. Despite this, several patterns are beginning to emerge from the information currently available. Increased research using rigorous scientific methods will be critical to filling existing information gaps and improving reliability of predictions.

In this chapter, we review the literature on the ecological effects of wind-energy development, focusing on wildlife and their habitats. We then provide an assessment of projected impacts of future development in the Mid-Atlantic Highland region based on the limited information currently available. Finally, we provide an overview of current methods and metrics for monitoring ecological impacts of wind-energy facilities, and propose research and monitoring priorities.

BIRD DEATHS IN CONTEXT

A primary question that arises from considerations of current and projected cumulative bird deaths from wind turbines is whether and to what degree they are ecologically significant. A related (but nonetheless different) question is how the number of turbine-caused bird deaths compares with the number of all anthropogenically caused bird deaths in the United States. The committee approaches the answer to the latter question with great hesitation, for four reasons. First, the accuracy and precision of data available to answer the question are poor. Although it is clear that more birds are killed by other human activities than by wind turbines, both natural mortality rates for many species and fatalities resulting from many types of human activities are poorly documented. In addition, different sources of human-caused fatalities do not affect all bird species to the same degree. Second, the demographic consequences of various mortality rates are poorly understood for most bird species, as are factors such as the timing of fatalities and sex or age bias in fatalities resulting from different anthropogenic causes, which could have a variety of demographic impacts. Moreover, the demographic and ecological importance of any given mortality rate being considered is relative to population size, which is poorly known for most species. Third, grouping all species together in any estimate provides information that is not ecologically relevant. For example, the ecological consequences and conservation implications of the deaths of 10,000 starlings (*Sturnus vulgaris*) are far different from those of the deaths of 10,000 bald eagles (*Haliaeetus leucocephalus*). Finally, consideration of aggregate bird fatalities across the United States from any cause—including those caused by wind-energy installations—is not the appropriate spatial scale to address the question of interest. Region-specific information about the demographic effects of any cause of mortality on species of interest would be much more informative. Thus, for example, it is more important to know how many raptors of a particular species are killed by turbines and other human mortality sources in a particular region than it is to know how many raptors are killed nationwide.

Having said the above, we provide here estimates summarized by Erickson et al. (2005) and estimates reported by the U.S. Fish and Wildlife Service (USFWS 2002a). Those sources emphasize the uncertainty in the estimates, but the numbers are so large that they are not obscured even by the uncertainty. Collisions with buildings kill 97 to 976 million birds annually; collisions with high-tension lines kill at least 130 million birds, perhaps more than 1 billion; collisions with communications towers kill between 4 and 5 million based on "conservative estimates," but could be as high as 50 million; cars may kill 80 million birds per year; and collisions with wind turbines killed an estimated 20,000 to 37,000 birds per year in

2003, with all but 9,200 of those deaths occurring in California. Toxic chemicals, including pesticides, kill more than 72 million birds each year, while domestic cats are estimated to kill hundreds of millions of songbirds and other species each year. Erickson et al. (2005) estimate that total cumulative bird mortality in the United States "may easily approach 1 billion birds per year."

Clearly, bird deaths caused by wind turbines are a minute fraction of the total anthropogenic bird deaths—less than 0.003% in 2003 based on the estimates of Erickson et al. (2005). However, the committee re-emphasizes the importance of local and temporal factors in evaluating the effects of wind turbines on bird populations, including a consideration of local geography, seasonal bird abundances, and the species at risk. In addition, it is necessary to consider the possible cumulative bird deaths that can be expected if the use of wind energy increases according to recent projections (see Chapter 2).

TURBINES CAUSE FATALITIES TO BIRDS AND BATS

Information on fatalities of birds and bats associated with wind-energy facilities in the Mid-Atlantic Highlands is limited, largely because of the relatively small amount of wind-energy development in the region to date, the modest investments in monitoring and data collection, and in some cases, restricted access to wind-energy facilities for research and monitoring. This lack of information requires the use of information from other parts of the United States (and elsewhere). The following discussion summarizes what is known regarding bird and bat fatalities caused by wind-energy facilities throughout the United States. National and regional results are related to the potential for fatalities in the Mid-Atlantic Highlands where appropriate.

Early industrial wind-energy facilities, most of which were developed in California in the early 1980s, were planned, permitted, constructed, and operated with little consideration for the potential impacts to birds or bats (Anderson et al. 1999). Discoveries of raptor fatalities at the Altamont Pass Wind Resource Area (APWRA) (Anderson and Estep 1988; Estep 1989; Orloff and Flannery 1992) triggered concern about possible impacts to birds from wind-energy development on the part of regulatory agencies, environmental groups, wildlife resource agencies, and wind- and electric-utility industries throughout the country.

Initial discoveries of bird fatalities resulted from chance encounters by industry maintenance personnel with raptor carcasses at wind-energy facilities. Although fatalities of many bird species have since been documented at wind-energy facilities, raptors have received the most attention (Anderson and Estep 1988; Estep 1989; Howell and Noone 1992; Orloff and Flannery

1992, 1996; Howell 1995; Martí 1995; Anderson et al. 1996a,b, 1997, 1999, 2000; Johnson et al. 2000a,b; Thelander and Rugge 2000; Hunt 2002; Smallwood and Thelander 2004, 2005; Hoover and Morrison 2005). This attention is likely because raptors are lower in abundance than many other bird species, have symbolic and emotional value to many Americans, and are protected by federal and state laws. Raptor carcasses also remain much longer than carcasses of small birds, making fatalities of raptors more conspicuous to observers. Raptors occur in most areas with potential for wind-facility development, although raptor species appear to differ from one another in their susceptibility to collisions.

Early studies of wind-energy facility impacts on birds were based on the carcasses discovered during planned searches. However, fatality estimates did not account for potential survey biases, most importantly biases in searcher efficiency and carcass "life expectancy" or persistence. Most current estimates of fatalities include estimates for all species and are based on extrapolation of the number of observed fatalities at surveyed turbines to the entire wind-energy facility, although not all studies adequately correct for observer-detection bias and carcass persistence, the latter usually referred to as scavenger-removal bias (e.g., Erickson et al. 2004).

Until relatively recently, little attention has been given to bat fatalities at wind-energy installations. This is largely because few bat fatalities have been reported at most wind-energy facilities (Johnson 2005). While some bat fatalities were reported beginning in the early 1990s, few of the earliest studies of fatalities at wind-energy facilities were designed to look for or evaluate bat fatalities, and thus did not use systematic search protocols or account for observer bias or scavenging. The scarcity of reported fatalities also may be due in part to the rarity of post-construction studies designed specifically to detect bat fatalities at wind-energy facilities. Recent surveys indicate that some wind-energy facilities have killed large numbers of bats in the United States (Arnett 2005; Johnson 2005), Europe (Dürr and Bach 2004; Hötker et al. 2004; UNEP/EUROBATS 2006), and Canada (R.M.R. Barclay, University of Calgary, personal communication 2006).

BIRD AND BAT FATALITIES

In the following discussion, fatality rate is presented as fatalities/turbine/year or fatalities/MW/year. Because turbine size, and presumably risk, varies from facility to facility, we have chosen to make comparisons of fatalities among turbines using the metric fatalities/MW/year. The MW used in this metric represents the nameplate capacity for the turbines and does not represent the actual amount of MW produced by a turbine or wind-energy plant. The reader is referred to Chapter 2 for a more general discussion of nameplate capacity. A more accurate measure of MW production for individual turbines would provide a much better metric for comparison purposes. For example, two turbines with the same nameplate capacity may operate a much greater percentage of time at a Class 5 wind site than in a Class 4 wind site.

Bird Species Prone to Collisions with Wind Turbines

Songbirds (order Passeriformes) are by far the most abundant bird group in most terrestrial ecosystems, and also the most often reported as fatalities at wind-energy facilities. The number of fatalities reported by individual studies in the eastern United States ranges from 0 during a five-month study at the Searsburg, Vermont facility (Kerlinger 1997) to 11.7 birds per MW during a one-year study at Buffalo Mountain, Tennessee (Nicholson 2003). In a review of bird collisions reported in 31 studies at wind-energy facilities, Erickson et al. (2001) reported that 78% of the carcasses found at facilities outside of California were protected passerines (i.e., songbirds protected by the Migratory Bird Treaty Reform Act of 2005). The remainder of the fatalities included waterfowl (5.3%), waterbirds (3.3%), shorebirds (0.7%), diurnal raptors (2.7%), owls (0.5%), fowl-like (galliform) birds (4.0%), other (2.7%), and non-protected birds (e.g., starling, house sparrow, and rock dove or feral pigeon; 3.3%). Despite the relatively high proportion of passerines recorded, actual fatalities of passerines probably are underrepresented in most studies, because small birds are more difficult to detect and scavenging of small birds can be expected to be higher (e.g., Johnson et al. 2000b). Moreover, given the episodic nature of bird migration, it is possible that many previous studies with relatively long search intervals failed to detect some fatalities of small birds during the migration season, and thus existing estimates of fatalities could be underestimates.

Data allowing accurate estimates of bird fatalities at wind-energy facilities in the United States are limited, particularly in the Mid-Atlantic Highlands region. Of the studies reviewed for this report, 14 were conducted using a survey protocol for all seasons of occupancy for a one-year period (Table 3-1) and incorporated scavenging and searcher-efficiency biases into estimates (Erickson et al. 2000, 2004; Young et al. 2001, 2003a,b; Howe et al. 2002; Johnson et al. 2002, 2003b; Nicholson 2003; Kerns and Kerlinger 2004; Koford et al. 2004). Protocols used in these 14 studies varied considerably, but all generally followed the guidance in Anderson et al. (1999). The wind-energy facilities included in these studies contain turbines that range in size from 600 kW to 1.8 MW. Passerines make up 75% of the fatalities at these facilities and 76% of the fatalities at the two forested facilities in the eastern United States (Table 3-2, Figure 3-1). The greatest difference between fatalities at wind-energy facilities in the eastern United States and those in other regions is the relative abundance of doves, pigeons, and "other" species (e.g., swifts and hummingbirds, cuckoos, woodpeckers) in the east.

Total annual bird fatalities per turbine and per MW are similar for all regions examined in these studies, although data from the two sites evaluated in the eastern United States suggest that more birds may be killed at wind-energy facilities on forested ridge tops than in other regions. It is not known whether this is due to higher risk of collisions at these sites, or higher abundance of birds in the region. Most studies report that passerine fatalities occur throughout facilities, with no identified relationship to site characteristics (e.g., vegetation, topography, turbine density). The relatively high proportion of passerines probably reflects the fact that this group is by far the most abundant of all birds at the facilities where these fatalities occurred. Relative exposure is difficult to measure and there are no data suggesting that fatalities expressed as percentages are proportional to abundance. As discussed below, behavior appears to be important in determining the risk of collision.

The combined average raptor fatality rate for the 14 studies (Table 3-2) is 0.03 birds per turbine/year and 0.04 per MW/year. The regional raptor fatalities per MW/year are similar, ranging from 0.07 in the Pacific Northwest region to 0.02 in the eastern United States. With the exception of the two eastern facilities, Mountaineer and Buffalo Mountain, which are in forest (68 MW combined), the land use/land cover is similar in all regions (Table 3-1). Most of the wind-energy facilities occur in agricultural areas (333 MW combined) and agriculture/grassland/Conservation Reserve Program lands (438 MW combined), and the remainder occur in short-grass prairie (68 MW combined). Landscapes vary from mountains, plateaus, and ridges, to areas of low relief. Aside from the size of the rotor-swept area, each of these facilities used similar technologies. Bird abundance may be an important factor in fatalities (discussed in more detail below), although standard estimates of bird use are not available for all 14 studies.

Interpreting fatalities of breeding and migrating passerines is challenging because of inadequate estimation of exposure of different species to risk. The most common fatalities reported at wind-energy facilities in the western and middle United States are relatively common species, such as horned lark (*Eremophila alpestris*), vesper sparrow (*Pooecetes gramineus*), and bobolink (*Dolichonyx oryzivorus*). These species perform aerial courtship displays that frequently take them high enough to enter the rotor-swept area of a turbine (Kerlinger and Dowdell 2003). The western meadowlark (*Sturnella neglecta*), on the other hand, is quite common and is frequently reported in fatality records, yet is not often seen flying as high as the rotor-swept area of wind turbines. By contrast, crows, ravens, and vultures are among the most common species seen flying within the rotor-swept area of turbines (e.g., Orloff and Flannery 1992; Erickson et al. 2004; Smallwood and Thelander 2004, 2005), yet they seldom are found during carcass

TABLE 3-1 Description of Wind-Energy Facilities Based on Data
Collected During the Period of Bird Occupancy over a Minimum Period
of One Year and Where Standardized Bird Mortality Studies Conducted,
Including Scavenging and Searcher Efficiency Biases. Vegetation
Categories Include Agriculture (AG), Grass Land (Grass), Conservation
Reserve Program (CRP) Grasslands, Short-Grass Steppe, and Forest.
Seasons Include Spring (S), Summer (Su), Fall (F), and Winter (W)

Wind Facility	Vegetation	Dates of Study
Vansycle, OR	Ag/Grass/CRP	1/99-12/99
Nine Canyon, WA	Ag/Grass/CRP	9/02-8/03
Stateline, OR/WA	Ag/Grass/CRP	1/02-12/03
Combine Hills, OR	Ag/Grass/CRP	2/04-2/05
Klondike, OR	Ag/Grass/CRP	2/02-2/03
Foote Creek Rim, WY Phase I	Short-grass Steppe	11/98-12/00
Foote Creek Rim, WY Phase II	Short-grass Steppe	7/99-12/00
Wisconsin	Agriculture	Spring 98-12/00
Buffalo Ridge, MN Phase I	Agriculture	4/94-12/95 3/96-11/99
Buffalo Ridge, MN Phase II	Agriculture	3/98-11/99
Buffalo Ridge, MN Phase III	Agriculture	3/99-11/99
Top of Iowa, IW	Agriculture	4/03-12/03
Buffalo Mountain, TN	Forest	10/01-9/02
Mountaineer, WV	Forest	4/03-11/03

surveys. Clearly, abundance and behavior interact to influence exposure of breeding passerines and other birds to the risk of collisions.

While estimated bird fatalities for these 14 wind-energy facilities are relatively low when compared to other sources of bird fatalities (Erickson et al. 2001), the lack of multiyear estimates of density and other population characteristics at most wind-energy facilities makes it difficult to draw general conclusions about their effects on populations of bird fatalities. In addition, lack of replication of studies among facilities and years makes it impossible to evaluate natural variability and the likelihood of unusual episodic events in relation to bird fatalities.

Influences of Turbine Design on Bird Fatalities

The structure and design of existing wind turbines vary considerably, and it is likely that additional modifications will occur over time. Changes in turbine design result from technological improvements, differences in

Search Interval	Number of Turbines in Facility	Number of Turbines Searched	Reference
28 days	38	38	Erickson et al. (2000)
14 days S, Su, F 28 days W	37	37	Erickson et al. (2003b)
14 days	454	124-153	Erickson et al. (2004)
28	41	41	Young et al. (2005)
28 days	16	16	Johnson et al. (2003b)
28 days	69	69	Young et al. (2001)
28 days	36	36	Young et al. (2003b)
Daily-weekly	31	31	Howe et al. (2002)
7 days	73	50	Johnson et al. (2002)
14 days	73	21	
14 days	143	40	Johnson et al. (2002)
14 days	138	30	Johnson et al. (2002)
2-3 days	89	26	Koford et al. (2004)
2/week-weekly	3	3	Nicholson (2003)
S-11 days Su-28 days F-7 days	44	44	Kerns and Kerlinger (2004)

generation capacity, and in some cases, modifications to meet site-specific needs (such as modification of height because of Federal Aviation Administration [FAA] constraints). Differences in design of turbines could affect fatality rates of birds. For example, as turbine heights increase, nocturnally migrating passerines could be increasingly affected because they tend to migrate at levels above 400 feet (see Appendix C for further discussion).

Much of the early work on fatalities at wind-energy facilities occurred in California, because most wind energy was produced at three wind-resource areas: APWRA, San Gorgonio, and Tehachapi. Not coincidentally, some of the existing concern regarding the impact of wind-energy facilities on birds is rooted in the fatalities that have occurred at the APWRA, and thus although many of the characteristics of APWRA differ from those of the Mid-Atlantic Highlands region, the history of APWRA provides important background and context.

The APWRA currently has between 5,000 and 5,400 turbines of various types and sizes, with an installed capacity of approximately 550 MW

TABLE 3-2 Regional and Overall Bird and Raptor Mortality[a] at Wind-Energy Facilities Based on Data Collected During the Period of Bird Occupancy over a Minimum Period of One Year and Where Standardized Bird Mortality Studies Were Conducted, Including Scavenging and Searcher Efficiency Biases Were Incorporated into the Estimates (additional metadata for these facilities contained in Table 3-1)

Wind Project	Project Size		Turbine Characteristics		
	Number of Turbines	Number of MW	Rotor Diameter (m)	Rotor-Swept Area (m²)	MW
Pacific Northwest					
Stateline, OR/WA[b]	454	300	47	1735	0.66
Vansycle, OR[b]	38	25	47	1735	0.66
Combine Hills, OR[b]	41	41	61	2961	1.00
Klondike, OR[b]	16	24	65	3318	1.50
Nine Canyon, WA[b]	37	48	62	3019	1.30
Totals or simple averages	586	438	56	2554	1.02
Weighted averages	586	438	49	1945	0.808
Rocky Mountain					
Foote Creek Rim, WY Phase I[c]	72	43	42	1385	0.60
Foote Creek Rim, WY Phase II[c]	33	25	44	1521	0.75
Totals or simple averages	105	68	43	1453	0.675
Totals or weighted averages	105	68	43	1428	0.655
Upper Midwest					
Wisconsin	31	20	47	1735	0.66
Buffalo Ridge Phase I[d]	73	22	33	855	0.30
Buffalo Ridge Phase I[d]	143	107	48	1810	0.75
Buffalo Ridge, MN Phase II[d]	139	104	48	1810	0.75
Top of Iowa[d]	89	80	52	2124	0.90
Totals or simple averages	475	333.96	46	1667	0.67
Totals or weighted averages	475	333.96	46	1717	0.53
East					
Buffalo Mountain, TN[e]	3	2	47	1735	0.66
Mountaineer, WV[e]	44	66	72	4072	1.50
Totals or simple averages	47	68	60	2903	1.08
Overall (weighted average)[f]	47	68	70	3922	1.45

[a]Mortality rates are on a per year basis.
[b]Agriculture/grassland/Conservation Reserve Program (CRP) lands.
[c]Shortgrass prairie.
[d]Agricultural.
[e]Forest.
[f]Weighted averages are by megawatt and turbine number.

Raptor Mortality		All Bird Mortality		
Number per Turbine per Year	Number per MW per Year	Number per Turbine per Year	Number per MW per Year	Source
0.06	0.09	1.93	2.92	Erickson et al. (2004)
0.00	0.00	0.63	0.95	Erickson et al. (2000)
0.00	0.00	2.56	2.56	Young et al. (2005)
0.00	0.00	1.42	0.95	Johnson et al. (2003b)
0.07	0.05	3.59	2.76	Erickson et al. (2003b)
0.03	0.03	2.03	2.03	
0.05	0.07	1.98	2.65	
0.03	0.05	1.50	2.50	Young et al. (2001)
0.04	0.06	1.49	1.99	Young et al. (2003b)
0.04	0.05	1.50	2.24	
0.03	0.05	1.50	2.31	
0.00	0.00	1.30	1.97	Howe et al. (2002)
0.01	0.04	0.98	3.27	Johnson et al. (2002)
0.00	0.00	2.27	3.03	Johnson et al. (2002)
0.00	0.00	4.45	5.93	Johnson et al. (2002)
0.01	0.01	1.29	1.44	Koford et al. (2004)
0.00	0.01	2.06	3.13	
0.00	0.00	2.22	3.50	
0.00	0.00	7.70	11.67	Nicholson (2003)
0.03	0.02	4.04	2.69	Kerns and Kerlinger (2004)
0.02	0.01	5.87	7.18	
0.03	0.02	4.27	2.96	

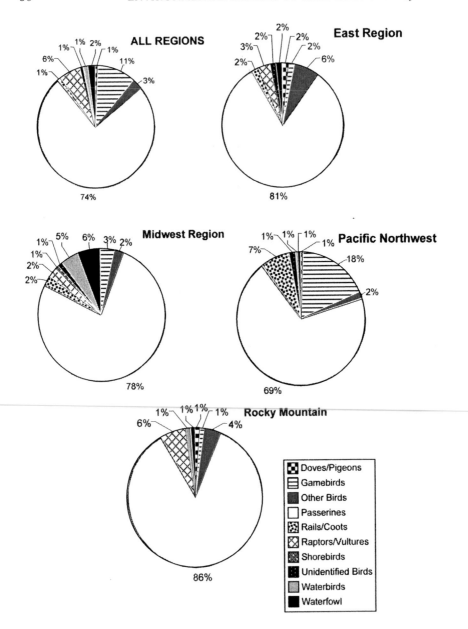

FIGURE 3-1 Composition of bird fatalities at 14 wind-energy facilities in the United States.
SOURCES: Compiled by committee from Erickson et al. 2000, 2003b, 2004; Young et al. 2001, 2003b, 2005; Howe et al. 2002; Johnson et al. 2002, 2003b; Nicholson 2003; Kerns and Kerlinger 2004; Koford et al. 2004.

(~102 kW/turbine); San Gorgonio consists of approximately 3,000 turbines of various types and sizes with an installed capacity of approximately 615 MW (~205 kW/turbine); and Tehachapi Pass has approximately 3,700 turbines with an installed capacity of approximately 600 MW (~162 kW/turbine). The following discussion generally refers to these facilities as "older generation" wind-energy facilities.

While replacement of some smaller turbines with modern turbines has occurred (through repowering), these three wind-resource areas primarily consist of relatively small turbines ranging from 40 to 200-300 kW, with the most common turbine rated at approximately 100 kW. Most of the higher-resource wind sites within each area have a high density of turbines, and the support structures for older turbines are both lattice and tubular, all with abundant perching locations for birds on the tower and nacelle (Figures 3-2a and b). (Figure 3-3 shows a more modern installation, Mountaineer, West Virginia, for comparison.) Additionally, all three areas have above-ground transmission lines. Perching sites for raptors are ubiquitous within all three areas, but particularly at the APWRA. There are different vegetation communities at all three sites, with San Gorgonio being the most arid, and Tehachapi the most montane and with some forest.

McCrary et al. (1986) conducted one of the earliest studies of the impact of wind-energy facilities on birds at San Gorgonio. However, the widely publicized report of bird fatalities at APWRA by Orloff and Flannery

FIGURE 3-2a Turbines at San Gorgonio showing lattice and monopole towers. SOURCE: Photograph by David Policansky.

FIGURE 3-2b Turbines at San Gorgonio showing high density and diversity of types.
SOURCE: Photograph by David Policansky.

(1992) promoted the most scrutiny of the problem. In spite of subsequent industry attempts to reduce raptor fatalities, they remain relatively high at the APWRA and reduction of fatalities was the focus of a recent decision by the Alameda County Board of Supervisors to issue conditional permits for the continued operation of the facility.

Smallwood and Thelander (2004, 2005) investigated the impacts of approximately 1,500 turbines for 4 years and 2,500 turbines for 6 months; the turbines ranged from 40 to 330 kW. While the Smallwood and Thelander (2004, 2005) studies are the most comprehensive to date, due to small sample sizes for turbines greater than 150 kW, extrapolation of fatality rates to all turbines in the AWPRA may not be appropriate. Hunt (2002) completed a four-year radiotelemetry study of golden eagles at the APWRA and concluded that while the population is self-sustaining, fatalities resulting from wind-energy production were of concern because the population apparently depends on immigration of eagles from other subpopulations to fill vacant territories. A follow-up survey was conducted in 2005 (Hunt and Hunt 2006) to determine the proportion of occupied breeding golden-eagle

FIGURE 3-3 Mountaineer Wind Energy Center, West Virginia. The five turbines in this photograph are at the southwest end of the array of 1.5 MW turbines; they are at the lower left of the aerial view in Figure 3-7.
SOURCE: Photograph by David Policansky.

territories in the APWRA. Within a sample of 58 territories all territories occupied by eagle pairs in 2000 were also occupied in 2005.

Contemporary utility-scale wind-energy facilities use different turbines from those at the older wind-energy facilities discussed above. The turbines are larger, with lower rotational rates (~15-27 rpm), although they retain a relatively high tip speed (~80 m/sec); tubular towers; primarily underground electrical service; lighting following FAA guidelines; few perching opportunities; and the rotor-swept area is higher above ground level (agl). In addition, many of the developments have occurred in areas with a different land use than the earlier California wind-energy facilities. Nonetheless, the potential cumulative impacts of these turbines should not be overlooked, especially for resident species.

The estimated fatality rates for raptors at the older California turbines (e.g., Orloff and Flannery 1992; Anderson et al. 2004, 2005; Smallwood and Thelander 2004, 2005) are generally greater than for newer turbines (Figure 3-4), although most of the sites for the newer turbines have much

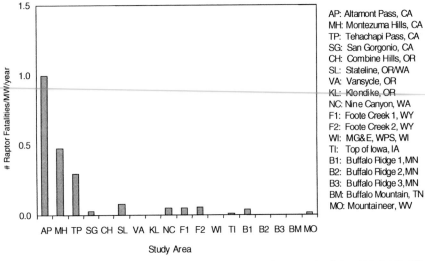

FIGURE 3-4 Fatality rates for raptors at 4 older wind-energy facilities (AP, MH, TP, SG) unadjusted for searcher efficiency, carcass-removal bias, and raptor abundances at the sites, and raptor fatality rates at 14 newer facilities (CH, SL, VA, KL, NC, F1, F2, WI, TI, B1, B2, B3, BM, MO) adjusted for searcher efficiency and carcass-removal bias.

SOURCES: Howell 1997; Erickson et al. 2000, 2003a,b, 2004; Howe et al. 2002; Johnson et al. 2002, 2003b; Nicholson 2003; Young et al. 2003a,c, 2005; Anderson et al. 2004, 2005; Kerns and Kerlinger 2004; Koford et al. 2004; Smallwood and Thelander 2004, 2005.

lower raptor abundance, there are relatively few studies of new wind-energy facilities, and there are major geographic gaps in the available data. Even though the raptor fatalities appear higher at wind-resource areas with the older technology, there is a marked difference among the older facilities. For example, raptor fatalities at the APWRA were higher than at Montezuma Hills, somewhat lower at Tehachapi (Anderson et al. 2004), and very low at the San Gorgonio facility (Anderson et al. 2005). Because the four facilities use similar technology, this difference may be influenced by other factors, most likely raptor abundance and prey availability.

The relationship of raptor abundance and technology will be better addressed when it is possible to study old and new turbines together in areas of varying raptor density. The three wind-energy facilities in northern California—High Winds and Diablo Winds in Solano County and the APWRA in Alameda County—may present such an opportunity when estimates of fatalities are published. The Solano County sites have newer turbines, and with the exception of golden eagles, higher raptor use than the APWRA (Orloff and Flannery 1992; Smallwood and Thelander 2004, 2005). Preliminary data from High Winds (Kerlinger et al. 2006) and Diablo Winds (WEST 2006) indicate they have higher raptor use, and higher raptor mortality than do projects in the Pacific Northwest (e.g., Erickson et al. 2004) and midwest (e.g., Johnson et al. 2000a,b). Alameda County, California, has permitted repowering of a small portion of the APWRA, replacing the MW production of smaller turbines with a smaller number of large newer turbines; fatality data from the APWRA collected before and after repowering can be used in a before/after control/impact (BACI) study, the preferred study design for observational studies (Anderson et al. 1999). Results from this and other repowering efforts in California will help evaluate the relative role of technology in bird fatalities, as would studies of fatalities at wind-energy facilities with large turbines in other areas of the country with relatively high raptor densities (e.g., eastern mountain ridges, coastal areas).

Most bird fatalities at wind-energy facilities are assumed to be caused by collisions with wind turbine blades. Even though there is no evidence indicating that passerines collide with turbine-support structures, numerous studies have documented passerine collisions with other solid structures (Erickson et al. 2001). Several studies have reported fatalities from buildings, and similar structures such as smokestacks and communications towers (Erickson et al. 2001). Bird fatalities associated with communications towers generally increase with height of the tower and lighting, with larger fatality events occurring at towers greater than 152 m (500 feet) in height. (Kerlinger 2000; Longcore et al. 2005). Nevertheless, shorter guyed towers[1] (< 152 m)

---

[1]Most tall towers are guyed (that is, they have cables called guys attached to the ground at some distance from their base to stabilize them); more shorter towers are not guyed.

may also present a risk for birds (Longcore et al. 2005). In a study of bird fatalities associated with 69 turbines and 5 guyed meteorological towers at a wind-energy facility in Carbon County, Wyoming, Johnson et al. (2001) reported that fatalities associated with the 40-m meteorological towers were three times greater than those associated with the 61-m wind turbines.

Although the steady red lights commonly recommended by the FAA have been shown to attract night-migrating birds and have been associated with an increase in bird fatalities at communications towers and other tall structures (Erickson et al. 2001; Manville 2001; Longcore et al. 2005; Gauthreaux and Belser 2006), there is no evidence to suggest a lighting effect on passerine fatalities at wind-energy facilities, with the exception of the Mountaineer Wind Energy Center in West Virginia. Kerns and Kerlinger (2004) reported the largest bird fatality event ever recorded at a wind-energy facility, with 33 documented passerine fatalities discovered on May 23, 2002. These fatalities apparently occurred during heavy fog conditions. All of the carcasses were located at a substation and at three adjacent turbines. The substation was brightly lit with sodium vapor lights. Following the discovery of the fatalities, the bright lights were turned off and no further large events were reported at the site. The second-largest fatality event documented involved 14 warblers, vireos, and flycatchers found during a May 17 carcass search of two adjacent turbines at the Buffalo Ridge, Minnesota wind-energy facility (Johnson et al. 2002). Like the West Virginia example, the event appeared to follow inclement weather, although only one of the turbines was lighted and lighting was not considered important (Johnson et al. 2002).

Influences of Site Characteristics on Bird Fatalities

Site characteristics may influence risk of fatality for birds, including location relative to key habitat resources (such as nesting sites, prey, water, and other resources) or concentration areas during migration, vegetative community in which the turbines are constructed, topographic position, and other factors. Relatively little is known about many of these relationships, but evidence for the importance of some of these variables is becoming clearer. Better understanding of these relationships will likely be helpful in siting decisions for future wind-facility development.

The effect of topography on fatality rates of birds is unclear. Of the 14 studies referred to in Table 3-1, most occurred in agricultural or grassland communities and in a variety of landscapes. Without more data from different plant communities and landscapes it is not possible to evaluate their influence of bird fatalities.

It is generally assumed that nocturnal migrating passerines move in broad fronts, as opposed to following specific and well-defined migration pathways, and rarely respond to topography (Lowery and Newman 1966; Richardson 1972; Williams et al. 1977), but this topic needs further study. A continent-wide study of nocturnal bird migration based on birds crossing the disc of the moon during four nights in October in 1952 (Lowery and Newman 1966) found little or no evidence that migrating birds were influenced by major rivers or mountain ranges in the eastern United States. However, the rugged mountains in the western United States did appear to affect the patterns of migration. Flight responses of migrants to the Great Lakes and the Gulf of Mexico were mixed. Some species flew parallel to the shoreline and appeared to be avoiding a crossing while others were observed departing across the large bodies of water. Bingman et al. (1982) found that on most nights during autumn migration in eastern New York State passerines showed a preferred migration track toward the southwest and in strong winds from the west and northwest the migrants drifted. On reaching the Hudson River some of the migrants changed their headings and followed a track direction that closely paralleled the river, and in doing so partially compensated for the effects of wind drift.

Schüz et al. (1971) and Berthold (2001, pp. 57-60) concluded that most migratory species in Europe show broad-front migration for at least a portion of their journey and suggested that species that have broad breeding ranges (E-W) tend to have broad-front migration pathways that cross all geomorphological features (such as mountains, river valleys, lakes). Hüppop et al. (2006) noted that the migration of birds over the waters of the German Bight also is broad-front. Recent radar studies of migration in the continental United States also support the conclusion that many species of migratory birds show broad-front migration (Gauthreaux et al. 2003). Gauthreaux et al. (2003) used a network of NEXRAD weather radars to quantify nocturnal bird migration over the United States, and the migration maps produced from the study clearly show that large geographical-scale migratory movements occur in response to weather favorable to migration. No evidence of specific flyways can be seen in the migration maps at the scale of surveillance of the radars (240 km range), and the results are in keeping with the findings of Lowery and Newman (1966).

Weather surveillance NEXRAD radar has rather coarse resolution (1 km × 1.0°) and consequently may not detect deviations in migration patterns at smaller spatial scales. Moreover, migrants flying at low altitudes may be missed by Doppler weather surveillance radars. Low-flying migrants could respond to topographic features more readily than migrants flying at higher altitudes. This would explain some of the conflicting findings regarding flight paths reported for migratory birds. Williams et al. (2001) cite work in Europe suggesting migrating birds respond to coastlines, river systems, and the Alps (e.g., Eastwood 1967; Bruderer 1978, 1999; Bruderer and Jenni 1988). While responses of birds to coastlines and major rivers has been noted in North America (e.g., Richardson 1978, Bingman et al. 1982), evidence is limited on the response to major changes in topography (Mc-Crary et al. 1983). Williams et al. (2001) used radar, ceilometers, and daily censuses in a study of passerine migration in the area of Franconia Notch, New Hampshire, a major pass in the northern Appalachian Mountains. They report that what they assumed to be migrating passerines surveyed by marine X-band radar appeared to react to topography in the Franconia Notch area. However, the study design and X-band radar equipment used in the study focused on localized and relatively low-altitude target movements and did not allow assessment of a broader area for movement patterns, and some of the detected targets may have been bats. However, Mabee et al. (2006) reported that for 952 flight paths of targets approaching a high mountain ridge along the Allegheny Front in West Virginia, the vast majority (90.5%) did not alter their flight direction while crossing the ridge. The remaining targets either shifted their flight direction by at least 10 degrees (8.9%) while crossing the ridge or turned and did not cross the ridge (0.6%)—both considered reactions to the ridgeline.

There is considerable agreement that migration patterns of most birds are species-specific. Species with limited breeding and wintering ranges generally have restricted migration pathways, but species with widely dispersed breeding ranges tend to show broad-front migration. A recent discussion of the flyway versus broad-front migration patterns in the United States is in Lincoln et al. (1998, pp. 53-72).

Many of the mountain ridgelines, and in particular those along the eastern edge of the Appalachian Mountains, appear to provide migratory pathways for diurnal fall migrants such as raptors (Bednarz et al. 1990). Raptors concentrate along ridges during migration and during daily hunting flights, presumably to take advantage of rising thermals and favorable winds used for soaring. This relationship was quantified at the Foote Creek Rim (FCR) wind-energy facility in Wyoming (Johnson et al. 2000a). Approximately 85% of the golden eagles, ferruginous hawks, and Swainson's hawks observed flying at the height of the rotor-swept area for the proposed turbines were within 50 m of the edge of the north to south trending mesa. Thus, raptors are likely more at risk when turbines are placed in areas where favorable winds exist for soaring.

Although high raptor fatalities have been documented at the APWRA, studies conducted at San Gorgonio and Tehachapi Pass (Anderson et al. 2004) documented relatively low raptor mortality (McCrary et al. 1983, 1984, 1986; Anderson et al. 2005) in comparison to the APWRA. The unadjusted per-turbine and per-MW raptor fatality rates reported for these sites are 0.006 and 0.03 for San Gorgonio, 0.04 and 0.20 for Tehachapi, and 0.1 and 1-1.23 for the APWRA. The primary difference among the three sites appears to be the abundance of raptors (Erickson et al. 2002).

The APWRA has the most raptors, presumably because of the abundance of prey, particularly small mammals (Smallwood and Thelander 2004, 2005). San Gorgonio has the fewest raptors, while raptor densities at Tehachapi Pass are intermediate (Anderson et al. 2004, 2005). The West Ridge within the Tehachapi Pass study area had the highest raptor use observed during the study, approximately half the estimated use of the APWRA (Anderson et al. 2004). The West Ridge also had the highest reported raptor fatalities among the three geographic subdivisions of Tehachapi Pass studied. These data suggest that differences in site quality, resulting in differences in abundance and exposure to turbines, may play an important role in determining mortality of some species. Smallwood and Thelander (2004, 2005) and Orloff and Flannery (1992) reported more raptor fatalities at wind turbines constructed in canyons at APWRA than at other locations within the area.

It also is usually assumed that nocturnally migrating passerines migrate relatively high agl. In a review of radar studies in the eastern United States, Kerlinger (1995) concluded that three-quarters of passerines (assumed passerines because bats were not considered) migrate at altitudes between 91 and 610 m. Recent marine radar studies conducted with modern X-band equipment capable of estimating target altitude from ~10 m to 1.5 km agl suggest that most nocturnal migrants fly above 125 m agl, the upper reach of most modern wind turbines. For example, using X-band marine radar in a vertical configuration, Mabee and Cooper (2002) for two study areas in the Pacific Northwest reported 3 and 9% of targets were below 125 m agl, while Mabee et al. (2004), also using vertical X-band marine radar, estimated that 13% of targets (birds and bats were not distinguished) detected on a mountain ridge in West Virginia were below 125 m agl. Nevertheless, X-band marine radar studies suggest there is a large amount of nighttime variation in flight altitudes (e.g., Cooper et al. 1995a,b), with targets averaging different altitudes on different nights and at different times during each night. Some of the intra-night variation is due to birds landing at dawn and taking flight at dusk, or bats emerging at dusk or returning at dawn. Kerlinger and Moore (1989) and Bruderer et al. (1995) concluded that atmospheric structure is the primary factor affecting flight direction and height of targets assumed to be migrating passerines. For example, Gauthreaux (1991) found that birds (and possibly bats) crossing the Gulf of Mexico appear to fly at altitudes where favorable winds exist.

In summary, it appears likely that nocturnally migrating passerines fly in broad fronts given the limit of resolution of current methods of detection, and that during migration the vast majority fly at altitudes well above the rotor-swept area of wind turbines. However, when weather conditions (e.g., low ceiling, light precipitation) compress bird migration closer to the surface, migrants may deviate their flights in response to topographical changes and could be at risk of collisions with wind turbines along ridgelines. Under favorable weather conditions migrant birds landing at night or beginning flight at dusk are potentially at risk of collision. This is particularly so if turbines are located adjacent to migratory stopover areas where migrants may be concentrated. Raptors often concentrate along topographic features when updrafts exist that facilitate soaring and may be at greater risk of collision when wind turbines are constructed in these locations. Nevertheless, prey abundance may also strongly influence raptor abundance and thus risk of collisions.

Temporal Pattern of Bird Fatalities at Wind-Energy Facilities

Although additional research is needed for more complete understanding of temporal patterns of fatalities at wind-energy facilities, a number of patterns emerge and it is clear that risk of fatality differs with location, meteorological condition, time of night, and time of year for both birds and bats.

Based on the available data, fatalities of passerines occurred in all months surveyed (Table 3-2). Bird fatalities along the Appalachian ridge have been most common from April through October (Nicholson 2003; Kerns and Kerlinger 2004), although the seasonal timing of fatalities varies somewhat among sites. For example, peak passerine fatalities occurred during spring migration at Buffalo Ridge, Minnesota (Johnson et al. 2002), and during fall migration at Stateline in Washington and Oregon (Erickson et al. 2004). This seasonal pattern suggests that both migrating and breeding resident bird species are being killed at wind-energy facilities (Howe et al. 2002; Johnson et al. 2002, 2003b; Young et al. 2003a, 2005; Koford et al. 2004).

Estimating the importance of fatalities to local populations requires that fatalities be assigned to a source population. However, allocation of fatalities to migrating and non-migrating passerines is problematic. It seems clear that some fatalities occur during migration. For example, a dead bird generally is considered a migrant if the species is not detected during bird surveys conducted during the breeding season and the habitat is unsuitable for nesting or brood rearing for the species. In many cases, however, the species may be present during the breeding season, but may be discovered as a fatality only, or more often during the migration season. Previous studies have not been able to distinguish resident breeders from migrants, although Erickson et al. (2001) provisionally reported a range of 34.4 to 59.9% of the fatalities as nocturnal migrants. Based on the available data, it appears that approximately half the reported fatalities at new wind-energy facilities are nocturnal migrating birds, primarily passerines, and the other half are resident birds. There is some evidence that young birds disperse during the nighttime (Mukhin 2004), and this may account for some "breeding season" mortality.

For example, in a four-year study of summer movements of juvenile reed warblers (*Acrocephalus scirpaceus*) marked as nestlings in Europe, captures by song playback suggest the existence of nocturnal post-fledging movements in this species. The uncertainty as to the geographic source of birds (and bats) killed at wind turbines could possibly be reduced if feather or other tissue samples were taken from carcasses and examined for stable hydrogen isotopes (see Appendix C).

Inclement weather has been identified as an important factor contributing to bird collisions with other obstacles, including power lines, buildings, and communications towers (Estep 1989; Howe et al. 1995), although the effect of weather on fatalities at communications towers is confounded by the height of the tower, type of lighting, and whether the tower is guyed or unguyed. Johnson et al. (2002) estimated that as many as 51 of the 55 bird fatalities discovered at the Buffalo Ridge wind-energy facility in southwestern Minnesota may have occurred in association with thunderstorms, fog, and gusty winds. Estimating the effect of weather is problematic because it is difficult to observe migration in poor visibility and precipitation. Nonetheless, the association of fatalities with episodic weather events recorded at wind-energy facilities (e.g., Johnson et al. 2002) and communications towers (Erickson et al. 2001) suggests that weather could be a factor contributing to bird fatalities at these sites.

Bat Species Are Prone to Collision with Wind Turbines

Data allowing reliable assessments of bat fatalities at wind-energy facilities in the United States are limited. Only six of the studies that we reviewed were conducted using a systematic survey protocol for all seasons of occupancy for a one-year period (Table 3-3) and had scavenging and searcher-efficiency biases incorporated into estimates (Figure 3-4; Arnett 2005; Johnson 2005; Arnett et al. in press). In contrast, protocols for assessing bat fatalities varied considerably and thus make actual fatality rates difficult to compare (Arnett 2005). The wind-energy facilities included in these studies contain turbines that range in size from 600 kW to 1.8 MW. Bat fatalities at wind-energy facilities in the eastern United States are much higher than those in western states.

Of the 45 bat species known from North America (north of Mexico), 11 have been recovered in ground searches at wind-energy facilities (Johnson 2005; Kunz et al. 2007; Arnett et al. in press). Among these, nearly 75% have been foliage-roosting eastern red bats (*Lasiurus borealis*), hoary bats (*Lasiurus cinereus*), and tree-cavity-dwelling silver-haired bats (*Lasionycteris noctivagans*), each of which migrate long distances

TABLE 3-3 Regional Comparison of Characteristics of Monitoring Studies and Factors Influencing the Estimates of Bat Fatalities at 11 Wind-Energy Facilities in the United States

Region	Facility	Landscape[a]	Estimated Fatalities/ MW/Year[b]
Pacific Northwest	Klondike, OR	CROP, GR	0.8
	Stateline, OR/WA	SH, CROP	1.7
	Vansycle, OR	CROP, GR	1.1
	Nine Canyon, WA	GR, SH, CROP	2.5
	High Winds, CA	GR, CROP	2.0
Rocky Mountains	Foote Creek Rim, WY	SGP	2.0
South Central	Oklahoma Wind Energy Center, OK	CROP, SH, GR	0.8
Upper Midwest	Buffalo Ridge, MN-I	CROP, CRP, GR	0.8
	Buffalo Ridge, MN-II (1996-1999)	CROP, CRP, GR	2.5
	Buffalo Ridge, MN-II (2001-2002)	CROP, CRP, GR	2.9
	Lincoln, WI	CROP	6.5
	Top of Iowa, IA	CROP	8.6
East	Meyersdale, PA[i]	DFR	15.3
	Mountaineer, WV (2003)	DFR	32.0
	Mountaineer, WV (2004)[i]	DFR	25.3
	Buffalo Mountain, TN-I	DFR	31.5
	Buffalo Mountain, TN-II	DFR	41.1[i]

[a]CROP = agricultural cropland, CRP = Conservation Reserve Program grassland, DFR = deciduous forested ridge, GR = grazed pasture or grassland, SGP = short grass prairie, SH = shrubland.

[b]Estimated number of fatalities, corrected for searcher efficiency and carcass removal, per turbine divided by the number of megawatts (MW) of installed capacity.

[c]Overall estimated percent searcher efficiency using bat or bird (*) carcasses during bias correction trials to correct fatality estimates.

[d]Number of bats or birds (*) used during bias correction trials and mean number of days that carcasses lasted during trials, the metric used to correct fatality estimates.

[e]Proportion of eight trial bats not scavenged after 7 days were used to adjust fatality estimates.

(Table 3-4). Other bat species killed by wind turbines in the United States include the western red bat (*Lasiurus blossivilli*), Seminole bat (*L. seminolus*), eastern pipistrelle (*Pipistrellus subflavus*), little brown myotis (*Myotis lucifugus*), northern long-eared bat (*M. septentrionalis*), long-eared myotis (*M. evotis*), big brown bat (*Eptesicus fuscus*), and Brazilian free-tailed bat (*Tadarida brasiliensis*).

To date, no fatalities of federally listed bat species have been reported (Johnson 2005), although it is possible that some of the bats that were

Search Interval	Percent Search Efficiency[c]	Carcass Removal Bats/Day[d]	Reference
28 days	75*	32*/14.2	Johnson et al. (2003a)
14 days	42*	171* + 7 / 16.5	Erickson et al. (2003a)
28 days	50*	40*/23.3	Erickson et al. (2000)
14 days	44*	32*/11	Erickson et al. (2003b)
14 days	50*	8[e]	Kerlinger et al. (2006)
14 days	63	10 / 20	Young et al. (2003c), Gruver (2002)
8 surveys[f]	67[g]		Piorkowski (2006)
14 days	29*	40/10.4	Johnson et al. (2003b, 2004)
14 days	29*	40/10.4	Johnson et al. (2003b)
14 days	53.4	48/10.4	Johnson et al. (2004)
1-4 days	70*	50*/~10	Howe et al. (2002)
2 days	72*	156*[h]	Jain (2005)
Daily	25	153/18	Kerns et al. (2005)
7-27 days	28*	30*/6.7	Kerns and Kerlinger (2004)
Daily	42	228/2.8	Kerns et al. (2005)
3 days	37	42/6.3	Fiedler (2004)
7 days	41	48/5.3	Fiedler et al. (2007)

[f]Two searches (one each in late May and late June) conducted at each turbine in 2004, and four searches every 14 days conducted at each turbine between May 15 and July 15 in 2005.

[g]Author used a hypothetical range of carcass-removal rates derived from other studies (0-79%) to adjust fatality estimates.

[h]Number of birds used during six trials. The mean number of days that carcasses lasted was not available; on average 88% of bird carcasses remained 2 days after placement.

[i]Six-week study period from August 1 to September 13, 2004.

[j]Weighted mean number of bat fatalities/MW with weights equal to the proportion of 0.66 MW (n=3 of 18) and 1.8 MW (n=15 of 18) turbines.

SOURCE: Modified from Arnett et al. in press.

overlooked by observers during surveys or taken by scavengers included endangered and threatened species, or in other years not sampled where conditions were conducive to use by listed species. Some wind-energy facilities may be constructed where it would be highly unlikely for endangered species to occur at the site. Search efficiency at these sites ranged from 25 to 75%, suggesting that many of the bats that were killed were never found (Arnett 2005; Johnson 2005; Arnett et al. in press) and that many of the bats that were killed were taken by scavengers. Nonetheless, the dominance

TABLE 3-4 Species Composition of Annual Bat Fatalities Reported for Wind-Energy Facilities in the United States

Species[a]	Pacific Northwest[b]	Rocky Mountains	South Central	Upper Midwest	East	Total
Hoary bat	153 (49.8%)	155 (89.1%)	10 (9.0%)	309 (59.1%)	396 (28.9%)	1,023 (41.1%)
Eastern red bat	—	—	3 (2.7%)	106 (20.3%)	471 (34.4%)	580 (23.3%)
Western red bat	4 (1.3%)	—	—	—	—	4 (0.2%)
Seminole bat	—	—	—	—	1 (0.1%)	1 (0.1%)
Silver-haired bat	94 (30.6%)	7 (4.1%)	1 (0.9%)	35 (6.7%)	72 (5.2%)	209 (8.4%)
Eastern pipistrelle	—	—	1 (0.9%)	7 (1.3%)	253 (18.5%)	261 (10.5%)
Little brown myotis	2 (0.7%)	6 (3.5%)	—	17 (3.3%)	120 (8.7%)	145 (5.8%)
Northern long-eared myotis	—	—	—	—	8 (0.6%)	8 (0.4%)
Big brown bat	2 (0.7%)	2 (1.1%)	1 (0.9%)	19 (3.6%)	35 (2.5%)	59 (2.4%)
Brazilian free-tailed bat	48 (15.6%)	—	95 (85.5%)	—	—	143 (5.7%)
Unknown	4 (1.3%)	4 (2.2%)	—	30 (5.7)	15 (1.1%)	53 (2.1%)
Total	307	174	111	523	1,371	2,486

[a]One confirmed anecdotal observation of a western long-eared myotis (*Myotis evotis*) has been reported in California, but is not included in this table.

[b]Pacific Northwest data from one wind-energy facility in California, three in eastern Oregon, and one in Washington; Rocky Mountain data from one facility each in Wyoming and Colorado; Upper Midwest data from one facility each in Minnesota, Wisconsin, and Iowa; South-Central data from one facility in Oklahoma; East data from one facility each in Pennsylvania, West Virginia, and Tennessee.

SOURCES: Kunz et al. 2007; modified from Johnson 2005. Reprinted with permission; copyright 2007, Ecological Society of America.

of the hoary bat in the reported fatalities appears to be a consistent theme in most studies in the United States to date, whereas fatalities of eastern red bats are highest in the east, and fatalities of silver-haired bats appear to be highest in the Pacific Northwest (Table 3-4).

Migratory tree bats are the commonest reported bat fatalities at wind-energy facilities in the United States. The numbers of bats killed in the eastern United States at wind-energy facilities installed along forested ridge tops have ranged from 15.3 to 41.1 bats/MW/year of installed capacity (Table 3-4). Bat fatalities reported from other regions of the western and midwestern United States have been lower, ranging from 0.8 to 8.6 bats/MW/year. Nonetheless, a recent study designed to assess bat fatalities in southwestern Alberta (Canada) found that fatalities were comparable to those found at wind-energy facilities located in forested ridges of the eastern United States (R.M.R. Barclay and E. Baerwald, University of Calgary, personal communication 2006).

There are, however, geographic differences in fatalities/MW of installed capacity among bat species. Bat fatalities at wind-energy facilities appear to be highest along forested ridge tops in the eastern United States and lowest in relatively open landscapes in the midwestern and in western states (Fiedler 2004; Johnson 2005; Fiedler et al. 2007; Arnett et al. in press), although relatively large numbers of fatalities have been reported in agricultural regions from northern Iowa (Jain 2005) and southwestern Alberta, Canada (R.M.R. Barclay and E. Baerwald, University of Calgary, personal communication 2006). Additionally, in a recent study conducted in mixed-grass prairie with wooded ravines in Woodward County, north-central Oklahoma, Piorkowski (2006) found 111 dead bats beneath wind turbines, 86% of which were pregnant or lactating Brazilian free-tailed bats. Western red bats, hoary bats, silver-haired bats, and Brazilian free-tailed bats also have been reported at wind-energy facilities in northern California (Kerlinger et al. 2006). To date, no assessments of bat fatalities have been reported at wind-energy facilities in the southwestern United States, a region where large numbers of migratory Brazilian free-tailed bats are resident during the warm months (McCracken 2003; Russell and McCracken 2006), and where this species provides important ecosystem services to agriculture (Cleveland et al. 2006). High fatality rates also can be expected for other species in the southwestern United States, where bat fatalities have not been monitored, and at wind-energy facilities in western states where rigorous monitoring for bat fatalities has been limited (Kunz et al. in press a). Despite the relatively high proportion of fatalities of migratory tree-roosting bats in each of the five regions summarized in Table 3-4, the eastern pipistrelle, a non-migratory species, accounted for 18.8% of the fatalities in the eastern United States.

Evaluations of the four sites in the Mid-Atlantic Highlands and elsewhere, where search efficiencies have been assessed, represent the best available data, but even those evaluations are limited because of the highly variable search efforts and carcass-removal studies. Studies where search efficiency and carcass removals are assessed daily provide the best data set for interpreting fatality rates (Mountaineer Wind Energy Center in 2004, Table 3-3). It is not known whether the high fatalities in the Mid-Atlantic Highlands wind-energy facilities and other areas in the eastern United States actually differ from those reported in other regions, or whether instead they reflect higher risk, higher abundance of migratory bats in the region, or limited search efforts in other regions. Most studies report that fatalities occur throughout the facilities, with no identified relationship to site characteristics (e.g., vegetation, topography, or turbine density) (Arnett 2005; Arnett et al. in press). The relatively high proportion of migratory bats may be influenced by the fact that these bats often forage along topographically uniform linear landscapes (i.e., ridgelines, forest edges). Given that there are no reliable abundance data for migratory tree species or, in fact, most other species of bats, it is impossible to determine at this time whether regional differences in fatalities are proportional to abundance. Given the apparent episodic nature of bat migration (Arnett et al. in press), it is possible that many previous studies with relatively long search intervals failed to detect some fatality events involving bats during migration, and thus existing estimates of fatalities may be too low. As discussed further below, the foraging and roosting behavior of migratory tree-roosting bats may provide important insight for estimating risk of collision.

The lack of multiyear studies and previous, possibly biased estimates of fatalities at most existing wind-energy facilities makes it difficult to draw general conclusions about the long-term effects of bat deaths on bat populations. This is partly due to the lack of efforts to look for bats in early studies, since bat fatalities were not recognized as a problem.

In particular, lack of replication of studies to assess bat activity and fatalities among different wind-energy facilities and years makes it impossible to evaluate natural variation, in particular episodic migration events, changing weather conditions, and other stochastic events as they relate to fatalities.

Influences of Turbine Design on Bat Fatalities

Relatively little is known about the influence of wind-turbine design on bat fatalities. To date, most large numbers of turbine-related bat fatalities have been reported from large, onshore utility-scale wind-energy facilities, in which 1 to 1.5 MW turbines are mounted on cylindrical monopoles. Few if any fatalities were reported from older, lattice-tower turbines that were the source of high raptor fatalities at the facilities in California, although search protocols were designed primarily for the detection of raptors (e.g., ≥ 30-day search intervals), and thus bat fatalities were most likely underestimated. Most modern wind turbines are tall and white, extending well above the forest canopies in the eastern United States, and quite likely are visually (if not acoustically) detectable to bats on cloudless nights. These large turbines stand in sharp contrast to the surrounding vegetation, and one hypothesis is that they may function as a visual beacon to bats and their insect prey (many insects are attracted to large white objects [Kunz et al. 2007]), especially during nights with sufficient moonlight.

All wind turbines produce sound that can be detected by most humans, and presumably by bats as well. Some turbines also produce broadband ultrasound (a range of frequencies above 20 kHz, approximately the upper limit of human hearing) as well as infrasound (defined as frequencies below 20 Hz, approximately the lower limit of human hearing). The ears of echolocating insectivorous bats are primarily tuned to a range of ultrasonic frequencies, which they use while navigating and capturing insect prey, although many species also produce and respond to frequencies below 20 kHz. Thus, sounds produced by modern wind turbines, which include audible and ultrasonic frequencies (some sounds are generated by the gear box in the nacelle, whereas others are produced by the rotation of the blades through air—often producing a "swishing" sound), may either attract bats—given their curiosity about novel objects in the environment—or confuse them upon detection. Additional research is needed to quantify the responses of bats to these sounds.

Although FAA lighting is not mandatory, the FAA does make recommendations to developers, which usually are followed. Recent observations summarized by Horn et al. (in press) suggest that bats are not attracted to FAA lights installed on wind turbines, although these blinking lights produce broadband pulsed ultrasonic frequencies (T.H. Kunz and S. Gauthreaux, personal observation 2006) that could function as an attractant to bats if they are used on wind turbines. Nonetheless, because ultrasonic frequencies are highly attenuated, especially in moist air (Griffin 1971; Lawrence and Simmons 1982), it is not likely that these sounds would function as a long-distance beacon that would either attract or repel bats. The functional range of echolocation for insectivorous bats that emit frequencies between 25 and 125 kHz can be as short as 5 m (Stilz and Schnitzler 2005).

Lighting on associated maintenance buildings or power stations at wind-energy facilities appears to attract insects. However, given that some insects are attracted to different types of lighting and light-colored objects, wind-turbine monopoles and blades may attract insects that bats feed on. Moreover, the large numbers of insects struck by moving turbine blades suggest that nocturnally flying insects are common at the height of the rotor-swept area (Corten and Veldkamp 2001). Accumulations of dead or moribund insects on the blades of wind turbines can reduce the efficiency of turbines by up to 50%, at least in some regions. Flying insects may also be attracted to the heat produced by nacelles of wind turbines (Ahlén 2002, 2003; Hensen 2004), and if bats respond to high densities of flying insects near wind turbines, their chances of being struck by turbine blades probably are increased (Kunz et al. 2007).

Wind turbines also produce obvious blade-tip vortices (Figure 3-5), and if bats get temporarily trapped in these moving air masses it may be difficult for them to escape. Rapid pressure changes associated with these conditions may lead to internal injuries, disorientation, and death of bats (Dürr and Bach 2004; Hensen 2004; Kunz et al. 2007).

FIGURE 3-5 Blade-tip vortices created by moving rotor blades in a wind tunnel illustrate the swirling wake that trails downwind from an operating wind turbine. SOURCE: Robert W. Thresher, National Renewable Energy Laboratory.

The causal factors and patterns of bat fatalities at wind turbines remain uncertain. Observations using thermal infrared imaging suggest that sometimes bats are killed by direct impact with turbine blades (Horn et al. in press). However, there are many unanswered questions. Are bats unable to detect rotating wind-turbine blades during migration and when they forage? When blade tips of large wind turbines rotate at speeds up to 80 m/sec (180 mph), a bat flying at speeds ranging from 2 to 27 m/sec (Neuweiler 2000) would not be able to react fast enough to avoid collision in the rotor-swept area. Are bats attracted to moving turbine blades? The turbine and blades produce audible sounds, ultrasound, and infrasonic vibrations, and because some bat species are known to orient to distant sounds (Buchler and Childs 1981), it is possible that bats are attracted to sounds produced by turbines or become disoriented and when they are migrating or feeding in the vicinity of wind turbines (Kunz et al. 2007).

Alternatively, it is conceivable that bats are visually attracted to wind turbines (Kunz et al. 2007). Migratory hoary bats reportedly seek the nearest available trees when daylight approaches (Dalquest 1943; Cryan and Brown in press), thus bats may mistake the large, conspicuous monopoles of wind turbines for roost trees (Kunz and Lumsden 2003). Because bats are curious animals, they may be killed as they explore novel objects in their environment. Observations of bat activity at wind turbines in Iowa (Jain 2005) and in Sweden (Ahlén 2002) suggest that bats were not attracted to turbines. However, if bats were simply colliding with random objects, bat fatalities also would be expected at meteorological towers. To date, no bat carcasses have been found near meteorological towers, even though these towers have been searched in several monitoring projects (Johnson 2005; Arnett et al. in press).

Will major developments of wind-energy facilities pose increased risks to bats in areas where they migrate or commute nightly to and from roosts? Can migratory species sustain high fatality rates, insofar as eastern red bats already appear to be in decline in New York (Mearns 1898) and in three Midwestern states (Whitaker et al. 2002; Carter et al. 2003; Winhold et al. 2005)? Bats are relatively long-lived (Wilkinson and South 2002; Brunet-Rossini and Austad 2004) and have low reproductive rates compared to many other mammals (Barclay and Harder 2003). For example, on average, the maximum recorded life span of a bat is 3.5 times greater than a non-flying placental mammal of similar size. Records now exist for individuals of at least five bat species in the wild surviving more than 30 years (Wilkinson and South 2002). Moreover, bats of the family Vespertilionidae (the family of most bats killed by wind turbines in North America) have average litter sizes of between 1.11 and 1.38 litters per year (Barclay and Harder 2003). These traits may seriously limit their ability to recover from persistent or repeated fatality events.

Given our current knowledge and the projected development of wind-energy facilities in the United States and elsewhere, the potential for biologically significant, cumulative impacts is a major concern (Kunz et al. 2007).

Independent of wind turbines and other anthropogenic structures, the migration period probably is a time of high mortality in bats, mostly during adverse weather and other stochastic events (Griffin 1970; Tuttle and Stevenson 1977; Fenton and Thomas 1985; Fleming and Eby 2003). There are enormous gaps in knowledge about migration in bats and the underlying evolutionary forces that have led to this behavior. If migratory tree bats experience naturally high mortality during migration from such factors as inclement weather, predation, and reduced food supplies, it is possible that with their low reproductive rates they will not be able to adjust to the expected cumulative affects resulting from the development of wind-energy facilities proposed in the United States and elsewhere (Kunz et al. 2007).

Influence of Site Characteristics on Bat Fatalities

Recent studies suggest a geographic pattern to bat fatalities at wind-energy facilities (Table 3-3). The unexpectedly high fatalities of migratory tree bats (*Lasionyceris* and *Lasiurus*) might reflect a risk to their populations, given that large numbers of these bats have been reported from these regions of North America (Cryan 2003; Kunz et al. 2007). While most evidence suggests that bats may be most vulnerable during the migration period, the observations of fatalities of Brazilian free-tailed bats in Oklahoma suggests that some species, in particular those that form large colonies and disperse and feed nightly at high altitudes (Williams et al. 1973; Cleveland et al. 2006), also may be at considerable risk. With relatively recent development of large wind-energy facilities in west Texas in the expected migratory route of Brazilian free-tailed bats from Carlsbad Caverns National Park, and more wind-energy facilities being proposed for west Texas and along the border with Mexico, migrating Brazilian free-tailed bats may be at risk. Regions of the United States where large numbers of bats are believed to concentrate in roosts and disperse and forage nightly at altitudes within the rotor-swept zone of modern wind turbines should be high priorities for investigation.

Temporal Patterns of Bat Fatalities at Wind-Energy Facilities

Much of the uncertainty about spatial and temporal factors responsible for high fatalities, especially those experienced by migratory tree-roosting species, reflects the scarcity of intensive and long-term studies conducted on these species, especially at wind-energy facilities during the maternity periods from May through July, and during migratory periods and when resident bats feed in the vicinity of wind-energy facilities (Kunz et al. 2007). Available data suggest that most bat fatalities at wind-energy facilities occur during fall migration (Table 3-3). However, these observations may be biased because of reduced effort in collection during the spring and summer migration periods, with reduced effort during the intervening periods. For example, spring migration of eastern red bats, hoary bats, and silver-haired bats in North America generally occurs from early April through mid-June, and autumn migration from mid-July through November (Cryan 2003). Moreover, other species killed by wind turbines in the eastern United States—the eastern pipistrelle, big brown bat, little brown myotis, and northern long-eared bats—are resident throughout much of their geographic range from mid-April to mid-October (Barbour and Davis 1969). Tracking with aircraft indicates that migrating Indiana bats (*Myotis sodalis*) usually are traveling directly towards their summer destination shortly after they leave their hibernacula (A. Hicks, New York Department of Environmental Conservation, personal communication 2006) (Figure 3-6).

While most bats in North America migrate from winter to summer roosts (e.g., *Myotis* species), the distances traveled are not comparable to the long-distance movements made by migratory tree-roosting species (Griffin 1970; Fleming and Eby 2003). Wind-energy facilities on mountain ridges in the Mid-Atlantic Highlands and elsewhere in the eastern United States have resulted in the highest reported bat fatalities for tree-roosting species (Nicholson 2003; Fiedler 2004; Arnett 2005; Arnett et al. in press). Thus, seasonal migrations, social behavior, orientation cues, and roosting habits differ markedly among hibernating and long-distance migrating species. However, higher bat fatalities are not confined to forested mountain ridges such as the mid-Atlantic region and elsewhere in the eastern United States. If this is the case, migratory bats could be vulnerable to high mortality from expanded wind-energy development in other regions of North America.

Preliminary observations suggest a strong association of bat fatalities with thermal inversions following frontal passage (Arnett 2005). Thermal inversions create cool, foggy conditions in the valleys with warmer air rising to the ridge tops that remain clear. These conditions could provide strong inducement for both insects and bats, whether migrating or not, to concentrate their activities along ridge tops (Kunz et al. 2007).

Although almost nothing is known about weather conditions that stimulate bat migration, one reasonable assumption is that conditions that are favorable for bird migration would also be favorable for bat migration. According to a review of studies on the timing of bird migration in relation to weather (Richardson 1990), the greatest density of migration occurs with following winds relative to the preferred direction of migration,

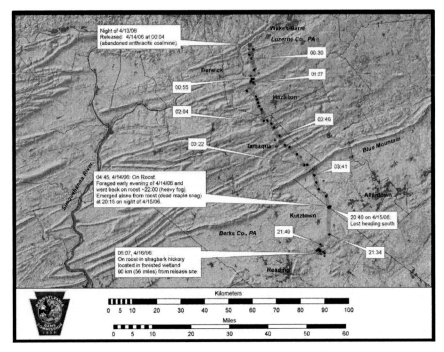

FIGURE 3-6 Migration route of an Indiana bat over forested ridge tops in eastern Pennsylvania (immediately south of Wilkes Barre, Luzerne County). This bat was captured and released at an abandoned coal mine at 00:04 h on April 14, 2006. It was tracked by aircraft traveling in a southeasterly direction, settling in a dead maple snag at 04:45 h. In the early evening of April 14, it foraged briefly and returned to its roost at 20:00 h (due to heavy fog). It emerged from its roost tree at 20:15 on night of April 15, but at 20:40 it was temporarily lost heading south (near Kutztown, Berks County). On April 16, it was located roosting in a shagbark hickory tree in forested wetland 90 km (56 miles) from its release site.
SOURCE: C.M. Butchkoski and G. Turner, Pennsylvania Game Commission, personal communication 2006. Reprinted with permission; copyright 2006, C.M. Butchkoski and G. Turner.

but some migration in headwinds has been recorded for some species and when migrants are flying over extensive bodies of water and cannot land. Because of co-variation among weather variables there is also correlation of peak numbers of migrants with other weather variables (e.g., falling temperatures and rising barometric pressure after a cold front passage in fall), but it is difficult to tell whether the relationships are coincidental or causative. Clearly birds do not typically initiate migration when weather conditions are poor (poor visibility, rain, very low cloud ceiling), but on rare occasions migrants aloft may move into locations with such conditions and either land or continue to fly at low altitudes.

WIND-ENERGY PROJECTS ALTER ECOSYSTEM STRUCTURE

The effects of wind-energy projects on ecosystem structure, and in particular habitats for various species, depend upon the vegetation and other landscape components for resident and migratory species that exist prior to construction. For example, influences of a project on a previously logged and subsequently surface-mined site typically differ from influences at a previously undisturbed forest site. An aerial photograph (Figure 3-7) provides an example of this variation on the Mountaineer Wind Energy Center in Tucker County, West Virginia. The turbines on the northeast end of the turbine string appear to have been constructed in a relatively undisturbed portion of the ridge, while the turbines near the center of the turbine string are constructed in an area of coal- and gravel-mining activity.

FIGURE 3-7 Aerial view of Mountaineer Wind-Energy Facility, which includes 44 1.5 MW turbines. SOURCE: Photograph by David Policansky.

Disturbance is likely dependent on individual site differences with respect to topography, type of vegetation, amount of existing roads, historic land use, and size and dispersion of turbines.

Estimates of direct surface disturbance per turbine vary by source and geographic location. The Bureau of Land Management (BLM 2005a) estimates the potential surface disturbance per turbine to be approximately 3 acres on land administered by the Bureau of Land Management, whereas Nicholson (2003) estimated surface disturbance at 1 acre per turbine for the 16-turbine Buffalo Mountain, Tennessee wind-energy facility. From aerial photography Boone et al. (2005) estimated that disturbance resulting from the construction of eight of the turbines at the Mountaineer Wind Energy Center ranged from 3.9 to 7.1 acres per turbine, not including forest removal for road construction and associated maintenance facilities. However, the sample of turbines was arbitrary and could not be extrapolated to the entire wind-energy facility.

Creating open areas in contiguous forest changes microclimate, by increasing light and wind in newly opened areas (Marsh et al. 2005). This results in increased temperature and reduced relative humidity and soil moisture of affected area (Kapos et al. 1997; Turton and Freiburger 1997), and can lead to elevated rates of wind throw resulting in modified forest structure (Laurance 1997). The intensity of effect varies with topographic features such as slope and elevation, but the fact that wind turbines are often placed on ridge tops, locations of high sustained winds, likely exacerbates the potential for structural damage to vegetation at some sites.

The use of suitable habitat by some forest-dwelling species (e.g., cerulean warbler [*Dendroica cerulean*] and redback salamander [*Plethodon cinereus*]) is influenced by the distance to the forest edge (i.e., the interface of forest and open areas). This "depth of edge influence" is sometimes referred to as the functional edge (Wood et al. 2006). Such an impact may radiate outside of the area actually disturbed by turbine development for some species to a distance of 100 m in all directions from the forest edge of the "footprint" (Reed et al. 1996; Haskell 2000). For certain taxa, however, the edge influence may continue to greater depths (e.g., over 200 m for invertebrates; Didham 1997) or greater than 340 m for cerulean warblers (Wood et al. 2006), resulting in much larger estimates of habitat loss for some species. Thus, the total short-term (i.e., during construction activities) loss of habitat for forest-dependent species is likely greater than that of the actual cleared area (Reed et al. 1996; Boone et al. 2005). The long-term impacts of a created opening will likely vary depending on the sensitivity of a species to depth-of-edge influence and the amount of activity in the open area.

The mechanism causing the loss of habitat due to the depth-of-edge influence may also differ among taxa. For example, some species appear to avoid the edge because the habitat has been modified (e.g., for invertebrates) while other species may avoid the area due to disturbance (i.e., displacement) even though the habitat is not substantially modified. In the case of displacement the impact may be shorter-term if the disturbance is removed (e.g., construction) or the animals become habituated to the disturbance. However, if the effect is due to modification of the habitat so that it becomes less suitable, the impact is expected to be of longer duration.

Forested landscapes in the eastern United States are fragmented over broad geographic regions and species associated with edges generally have not experienced declines (e.g., Bell and Whitmore 1997). Habitat for some species actually has increased with increasing amount of edge, leading to increases in the populations of species in eastern forests such as white-tailed deer (*Odocoileus virginianus*), brown thrasher (*Toxostoma rufum*), northern cardinal (*Cardenalis cardenalis*), northern mockingbird (*Mimus polyglottos*), ruffed grouse (*Bonasa umbellus*), and wild turkey (*Meleagris gallopavo*). Creation of additional habitat for edge-associated species may place some of these species (including some bat species) at higher risk than if the turbines were not present at these sites. Some wildlife-management agencies (e.g., West Virginia Department of Natural Resources) have concluded that a goal of "creating edge" to benefit populations of harvested species may have unintended negative consequences. For example, the overabundance of edge-tolerant species such as white-tailed deer can have detrimental effects on forest productivity and wildlife species richness (Rossel et al. 2005).

Habitat fragmentation can be defined as the breaking up of large contiguous tracts of suitable habitat for a species into increasingly smaller patches that are isolated from each other by barriers consisting of unsuitable or less suitable habitat. There is a substantial literature that examines the effects of fragmentation on the ecology of forest ecosystems (e.g., Laurance and Cochrane 2001; Fahrig 2003), although much of this literature focuses on a larger spatial scale than that represented by the extent of most wind-energy projects. Wind-energy projects in the central Appalachian Mountains can fragment previously contiguous tracks of forest at some scale by road construction, turbine installation, and the presence of ancillary structures.

Habitats for forest species are linearly divided by turbine-maintenance roads paralleling the ridge. Such internal fragmentation may subdivide populations of some species (Goosem 1997); the magnitude and importance of these effects are influenced by the natural history of the individual taxa and the scale of the fragmentation. The effect of forest roads on aquatic and terrestrial communities has been documented and synthesized elsewhere (Trombulak and Frissell 2000; Forman et al. 2003; NRC 2004, 2005). Trombulak and Frissell summarize seven general effects:

- Direct mortality can result from road construction. The effect is most significant for sessile or slow-moving organisms. Coupled with increased compaction, increased soil temperature beneath the road can adversely affect communities of soil organisms.
- Mortality from collision with vehicles using roads may be significant on large, frequently traveled roads. Because vehicular traffic on wind-energy sites typically is infrequent, it is unlikely that collision with vehicles will be a significant source of mortality resulting from wind-energy development at most sites, including the Mid-Atlantic Highlands.
- Forest roads may result in a modification of animal behavior. Some species (e.g., black bears [*Ursus americanus*]) avoid roads of high traffic volume, and forest roads in areas where they are hunted (Brody and Pelton 1989), while turkey vultures (*Cathartes aura*) are common along forest roads. Typically the roads and the surrounding surfaces at wind-energy facilities are maintained to 15-20 m wide, and are usually lightly traveled. However, roads prove to be barriers for such diverse taxa as land snails (even roads that are unpaved and < 3 m in width) and small mammals (Merriam et al. 1989; Baur and Baur 1990). Moreover, forest roads as small as 5-8 m in width can be barriers to salamander dispersal and gene flow (deMaynadier and Hunter 2000; Marsh and Beckman 2004; Marsh et al. 2005). Such effects are exacerbated by the grade of road verges. Steeper verges tend to decrease the dispersal ability of salamanders (Marsh et al. 2005). In contrast, some species use linear features such as roads as travel corridors or feeding habitat. For example, some species of bats forage along linear landscapes created by road cuts in forested habitats, where they forage mostly on aerial insects (Krusic et al. 1996; Menzel et al. 2002). Even species such as black bears that may avoid roads with high traffic may use forest roads with low traffic as travel lanes (Brody and Pelton 1989).
- Forest roads disrupt the physical environment of the road bed as well as the adjacent edge. Soil density, even on closed roads, increases over time and can persist for periods in excess of 40 years. In addition to soil density, road-induced transformations can include changes in temperature, soil water content, light, dust, surface water flow, pattern of run-off, and sedimentation of downslope aquatic habitats, although sedimentation should be avoided through following the requirements of each facility's National Pollutant Discharge Elimination System (NPDES) permit (EPA 2006d).
- Forest roads can alter the chemical environment of the road bed and adjacent edge habitats. Edges along roads serve as concentrators of both nutrients (nitrogen compounds) and pollutants (sulfur compounds) (Weathers et al. 2001). This in turn can alter basic trophic processes such as food-web relationships between plants, insects, and the predators of insects (Valladares et al. 2006).

• The presence of forest roads increases the spread of invasive species. Three mechanisms have been proposed for the establishment of invasives: the presence of altered habitat, increased stress to or removal of native species, and easier access to disturbed habitats by wild or human vectors (Turton and Freiburger 1997). In addition, poor reclamation practices may lead to lack of germination of desirable plants leaving the unvegetated disturbed site available for the establishment of invasives.

• Forest roads can change humans' use of land and water by increasing access to those resources, or by providing access where none previously was available, allowing increased hunting, fishing, recreational driving, and other activities (e.g., NRC 2003, 2005).

In summary, maintenance roads and areas cleared for turbine installation may result in a diversity of influences on forest-dwelling species. Unfortunately, there are no empirical studies that have investigated impacts of roads associated with wind-energy facilities on ecological processes in the area, and relatively few studies have examined ecological impacts of roads in the central Appalachian Highlands. As a result, the extent to which these impacts are manifested at any particular site are not known, and the population-level consequences also are uncertain.

Influences of Habitat Alteration on Birds

Effects of wind-energy development on habitats used by birds can be divided into two general categories: loss of habitat (including avoidance of disturbed and adjacent areas), and fragmentation effects to remaining habitat. Moreover, for a complete understanding of impacts, effects must be assessed relative to the state of the habitat suitable for individual species prior to the construction of a wind-energy facility. For example, a project located on a reclaimed surface mine would not have the same impact on forest birds as one located in a forest 100 times larger. In general, aerial photographs (e.g., Figure 3-7) indicate that the disturbance caused by wind-energy projects is linear along ridgelines, and that habitat for forest-dependent birds has been removed. Habitat loss has large and consistently negative effects on biodiversity (Fahrig 2003). In addition, many forest-dependent bird species respond to direct habitat loss and to changes in the configuration of habitat (fragmentation) resulting from that forest loss (Villard et al. 1999). Thus, assessments of the effects of wind-energy facilities on bird habitat should not be confined to simple measurement of the area of vegetation removed, but also should include analysis of habitat fragmentation and edge effects.

Impacts of wind-energy facilities on habitat are considered to be greater than collision-related fatalities on birds in Europe (Gill et al. 1996). Studies of both onshore and offshore wind-energy facilities in Europe have reported disturbance effects ranging from 75 m to as far as 800 m from turbines for waterfowl, shorebirds, waders, and passerines (Peterson and Nohr 1989; Winkelman 1989, 1990, 1992a; Vauk 1990; Pedersen and Poulsen 1991; Larsen and Madsen 2000). Avoidance of wind-energy facilities varies among species and depends on site, season, tide, and whether the facility was in operation. Disturbance tends to be greatest for migrating birds while feeding and resting (Crockford 1992); disturbance to breeding birds appears to be negligible and was documented only in one study (Pedersen and Poulsen 1991). In terms of the layout of turbines at wind-energy facilities, Larsen and Madsen (2000) found that in the case of wintering pink-footed geese (*Anser brachyrhynchus*), avoidance distances from wind turbines that are constructed in lines were 100 m; they were 200 m when the turbines were clustered. For other bird groups or species at other European wind-energy facilities, no displacement effects were observed (Karlsson 1983; Winkelman 1989, 1990; Phillips 1994). It is likely that there is a gradient of avoidance, with extent of impact being a function of distance from the facility, although Winkelman (1995) reported reductions in use of up to 95% out to 500 m away from turbines. A recent radar study of bird movements at a wind-energy development off the coast of Denmark (Desholm and Kahlert 2005) found that the percentage of flocks of common eiders (*Somateria mollissima*) and geese entering an offshore wind-energy facility area decreased by a factor of 4 from pre-construction to initial operation. At night, migrating flocks were more prone to enter the wind-energy facility but counteracted the higher risk of collision in the dark by increasing their distance from individual turbines and flying in the corridors between turbines. Desholm and Kahlert (2005) estimated that less than 1% of the ducks and geese migrated close enough to the turbines to be at any risk of collision. However, there is no assessment of the issue of potential interference from turbines on the radar signal, potentially biasing study results.

Bird displacement associated with wind-energy development has received little attention in the United States. Howell and Noone (1992) found similar numbers of raptor nests before and after construction of Phase 1 of the Montezuma Hills, California wind-energy facility. A pair of golden eagles successfully nested 0.8 km from the FCR, Wyoming wind-energy plant for three different years after it became operational (Johnson et al. 2000a), and a Swainson's hawk nested within 0.8 km of a small wind-energy plant in Oregon (Johnson et al. 2003b). Anecdotal evidence indicates that raptor use of the APWRA in California may have increased since installation of wind turbines (Orloff and Flannery 1992; AWEA 1995). Results of more than two years of raptor nest monitoring at the Stateline Wind Project showed no measurable change in raptor-nest density within two miles of the facilities. In a survey of breeding golden eagle territories in the APWRA, Hunt and Hunt (2006) found that within a sample of 58 territories sampled, all territories occupied by eagle pairs in 2000 were also occupied in 2005.

The only case interpreted as avoidance of wind-energy plants by raptors occurred at the Buffalo Ridge facility, Minnesota, where raptor-nest density on 261 km^2 of land surrounding the facility was 5.94/100 km^2, yet no nests were present in the 32 km^2 facility, even though habitat was similar (Usgaard et al. 1997). However, more information would be needed to conclude with confidence that the observed distribution of nests was due to raptor avoidance of turbines, and not due to chance or other factors. Osborn et al. (1998) reported that fewer birds and fewer species were within the Buffalo Ridge wind-energy facility in turbine plots than at reference plots, and concluded that birds avoided flying in areas with turbines. Also at the Buffalo Ridge facility, Leddy et al. (1999), using the impact gradient sampling design and linear regression methods, found that species-specific densities of male songbirds were significantly lower within 180 m of turbine locations in CRP grasslands than in CRP grasslands without turbines. Grasslands without turbines, as well as portions of grasslands located at least 180 m from turbines, had bird densities four times greater than grasslands located near turbines. In a 4-year study designed to evaluate displacement of breeding birds at the Buffalo Ridge site, Johnson et al. (2000b) used a BACI sampling design and linear regression models to assess displacement impacts. Their results indicated that the facility of 354 wind turbines displaced some groups and species of birds, and that the area of displacement was limited primarily to areas ≤ 100 m from turbines.

While similar avoidance of wind turbines has not been documented for other prairie species of conservation concern, such as many prairie-grouse species, studies of the impacts of other human disturbances on prairie chickens and sage grouse indicate that birds do avoid disturbed areas. It is likely that these species will be displaced by wind-power development, although the magnitude of the displacement is unknown. The relationship between wind-energy development and the habitats used by birds in the Mid-Atlantic Highlands has not been investigated, and information from other geographic locations and non-forest vegetation associations provide limited insight into how forest-dwelling birds respond to such habitat perturbation. However, the response of bird species to habitat alterations caused by changes in vegetation associated with timber management, mining, and insect outbreaks have been widely studied in the Mid-Atlantic Highlands (e.g., Duguay 1997; Bell and Whitmore 2000; Duguay et al. 2000, 2001; Hagan and Meehan 2002; Weakland and Wood 2005; Wood et al. 2005, 2006) and these studies provide some insight to the potential effects of wind-energy development. While changes in forest cover from a single wind-energy facility may not be of the same magnitude as those from timber management or an insect outbreak, the total area disturbed by a wind-energy project, including roads and ancillary structures, as well as the depth of edge influence, would likely cover hundreds of hectares.

The response of birds to changes in vegetation structure varies with species, and changes that adversely affect some species may be positive for others. For example, in the Mid-Atlantic Highlands, removal of the forest canopy and subsequent understory release can benefit shrub-nesting species such as the eastern towhee (*Pipilo erythrophthalmus*), which responds positively in both gypsy-moth-defoliated forest tracts (Bell and Whitmore 1997) and timber-managed tracts (Duguay 1997; Duguay et al. 2000, 2001). Conversely, habitat for ovenbirds (*Seiurus aurocapillus*) and Blackburnian warblers (*Dendroica fusca*) is negatively correlated with understory density and positively correlated with the size and density of hardwood trees (Hagan and Meehan 2002). Moreover, data from Breeding Bird Surveys indicate that populations of edge species such as eastern towhee, indigo bunting (*Passerina cyanea*), and song sparrow (*Melospiza melodea*) generally are increasing within the Mid-Atlantic Highlands (Sauer et al. 2005). However, forest-interior species, including ovenbirds, Kentucky warblers (*Oporornis formosus*), and worm-eating warblers (*Helmitheros vermivorus*), are declining (Freemark and Collins 1992; Wenny et al. 1993).

In the Mid-Atlantic Highlands, three species of warbler—cerulean warbler, worm-eating warbler and ovenbird—are of conservation concern and thus are of particular interest with respect to wind-energy development in this region (USFWS 2002b). For example, the cerulean warbler appears to be declining precipitously (Robbins et al. 1992), and is experiencing approximately a 3% annual decrease in abundance (Link and Sauer 2002; Wood et al. 2006). This rate of decline, however, needs to be re-evaluated because cerulean warblers extensively use ridge tops in some areas of the Mid Atlantic Highlands, and these areas are not sampled as much as mid-slopes or valley floors (Wood et al. 2006); as a result, estimates of declines may be biased. Mid-Atlantic Highlands populations of worm-eating warblers are likewise declining, showing a 20% drop between 1996 and 2001 in the Monongahela and George Washington National Forests (Cooper et al. 2005a).

Ovenbirds are declining in eastern forests (Robbins et al. 1989; Sauer et al. 2005) and appear to be particularly sensitive to forest fragmentation, showing decreases in density adjacent to narrow, unpaved, interior forest roads and trails (Ortega and Capen 1999, 2002). Factors implicated in this decline are loss of insect-prey biomass in small forest fragments (Burke and Nol 1998), increased predation (Mattsson and Niemi 2006), and brood parasitism (Lloyd et al. 2005). In addition, both density and fecundity of ovenbirds were lower in large (> 2,000 ha) habitat patches than in unfragmented reference plots (located in > 2 million ha) (Porneluzi and Faaborg 1999). Small forest fragments may act as population sinks that rely on continual re-supply from adjacent large forest tracts for ovenbirds (Nol et al. 2005). Nesting ovenbirds and five other species have recently been reported to decline in habitats altered by a wind-energy project near Searsburg, Vermont (Kerlinger 2002). Openings created for turbines and roads were hypothesized to be the likely cause of this decline (Kerlinger 2002). These are the only before and after data for a wind-energy development in forested habitats in the eastern United States.

Several additional bird species of concern have statutory protection and may occur in habitats impacted by wind-energy development (Table C-6 of Appendix C). All states in the Mid-Atlantic Highlands except West Virginia have State Endangered, Threatened, or Species of Conservation Concern legislation and have published lists of protected species, in addition to those protected under the U.S. Endangered Species Act (ESA). Most of these state-listed species occur at peripheral locations in their historic range (e.g., mourning warbler [*Oporornis philadelphia*]) and may not be at risk from a global perspective. Nonetheless, they do have protected status at the state level and need to be considered in siting assessments.

Long-term trend analysis by Sauer et al. (2005) using Breeding Bird Survey data for North American bird species that winter in the tropics (neotropical migrants) shows that populations of 45 species are declining (Appendix C, Table C-5). Most of these species either nest in Mid-Atlantic Highland habitats or migrate through the region seasonally. All of these species are protected under the Migratory Bird Treaty Reform Act of 2005 and should be included in siting studies as well as in long-term monitoring of existing wind-energy facilities.

Although habitat alteration resulting from wind-energy development often occurs at a relatively small scale, the cumulative effects of wind-energy development, in conjunction with changes in habitat from a variety of other past and present anthropogenic activities, could result in negative impacts on bird populations.

Influences of Habitat Alteration on Bats

Changes in habitat associated with wind-energy facilities can be relatively minor in some situations, such as may be the case in agricultural settings. In forested environments, however, habitat alteration at wind-energy facilities may be considerable. In addition to changes resulting from presence of the turbine itself, alteration of bat habitat results from road construction and maintenance, buildings and structures associated with turbines, and power lines associated with wind-energy facilities. Manipulation of vegetation, including creating and maintaining clearings around turbines, along roadsides, and along power line rights-of-way probably are the most important form of bat habitat alteration associated with wind-energy facilities—alteration that may increase the activity of bats at these sites.

Alteration of vegetation associated with wind-energy facilities could influence bats in two ways. First, changes in vegetation associated with wind-energy facilities could influence the quality of habitat for bats, thereby influencing carrying capacity of the area, and ultimately influencing population abundance. Alternatively, changes in vegetation could alter the behavior of bats, thereby changing the risk of collision with turbine blades. The overall influence of habitat alteration on bats (and birds) at wind-energy facilities is thus a function of the relative influences of changes in population abundance and behavior (Figure 3-8).

Although some studies are under way to evaluate the influence of wind-energy facilities on bats, no studies have been published that directly examine influences of vegetation change associated with wind-energy facilities on bats. However, inference from studies that have examined the ecology and the influences of forest management practices on forest-dwelling bats can provide insight into potential influences of wind-energy facilities. Here we summarize likely influences of vegetation alteration associated with wind-energy facilities on roosts and roosting ecology, habitat use, and vertical patterns of activity of bats.

	Habitat alteration decreases use of area within turbine sweep zone	Habitat alteration does not influence use of area within turbine sweep zone	Habitat alteration increases use of area within turbine sweep zone
Habitat alteration decreases abundance	Dependent on magnitude of influences	Negative impact	Negative impact
	Positive impact	No influence	Negative impact
Habitat alteration increases abundance	Positive impact	Positive impact	Dependent on magnitude of influences

FIGURE 3-8 The influence of habitat alteration associated with wind-energy facilities on bats is a function of the combined influences of the ways that habitat alteration influences abundance and risk of collision with turbine blades.

Influences of Habitat Alteration on Roosts and Roosting of Bats

Bats use roosts as sites for resting, protection from weather and predators, rearing young, hibernation, digestion of food, mating, and social interactions (Kunz 1982a,b,c; Kunz and Lumsden 2003). Roosts have been postulated as limiting factors that influence distribution and abundance of bats (Humphrey 1975; Ports and Bradley 1996; West and Swain 1999). Bats use a variety of structures for roosting, including buildings, caves, bridges, hollow logs, foliage, leaf litter, and hollows, cavities, and crevices in trees, snags, and rock crevices. Of these, wind-energy development in the Mid-Atlantic Highlands and in other forested regions is most likely to influence availability of roosts in trees and snags. The geographic distribution of bats is also influenced by elevation, with males of several species being more common at higher elevations, especially in western states (Cryan 2003).

Large-diameter living and dead trees provide important roosts for many species of forest-dwelling bats (Kunz and Lumsden 2003; Barclay and Kurta 2007). The roosting ecology of the Indiana bat is of particular concern throughout its range in the eastern United States, as this species is listed as endangered by the U.S. Fish and Wildlife Service; Indiana bats roost in cavities and crevices beneath the exfoliating bark of living and dead hardwoods and conifers during summer months (Kurta et al. 1996, 2002; Callahan et al. 1997; Gumbert et al. 2002). Indiana bats also have been reported to roost in buildings (Butchkoski and Hassinger 2002). The roosting ecology of bats of the genus *Lasiurus* also is of interest, as these bats appear to be particularly vulnerable to fatalities at wind-energy facilities. Eastern red bats and hoary bats generally roost in the foliage of several different species of trees and shrubs during the spring, summer, and fall (Constantine 1966; Menzel et al. 1995, 1998; Carter et al. 2003). The silver-haired bat typically roosts in tree cavities (Betts 1996; Vonhof 1996).

Clearing forests at and around wind-energy facilities could result in removal of actual or potential roost sites for Indiana bats, eastern red bats, hoary bats, and silver-haired bats, and several other species that occur in or migrate through the Mid-Atlantic Highlands. In Pennsylvania, the typical foraging habitat of Indiana bats is in upland forests (Butchkoski and Hassinger 2002). Moreover, removing dead trees that are adjacent to roadways developed for wind-energy facilities because of their potential hazards to safety or their risk of obstructing roadways can reduce the number of potential roosts for several species of bats.

Use and quality of roosts also may be influenced by the microclimatic changes resulting from habitat alteration. Microclimate appears to play an important role in determining quality and use of roosts in forest settings (Hayes 2003; Kunz and Lumsden 2003; Barclay and Kurta 2007; Hayes and Loeb 2007). For example, although the primary roosts of Indiana bats are mostly in wooded riparian habitats that receive considerable solar radiation (Humphrey et al. 1977; Callahan et al. 1997; Britzke et al. 2003), more recent evidence suggests that some roost in forested areas (Kurta and Kennedy 2002). Thermal environment also is thought to influence use of roosts by foliage-roosting bats, although less is known about the influences of temperature on foliage-roosting bats or the scale at which it operates. In Kentucky, eastern red bats selected roosts in foliage with lower temperatures than in other points in the same tree (Hutchinson and Lacki 2001), possibly to minimize heat stress during high summer temperatures or to conserve energy by entering daily torpor.

Changes in forest structure and creation of openings are likely to alter microclimatic conditions in forested regions used by roosting bats (Kunz and Lumsden 2003). In general, these changes should increase roost temperatures in the affected area. When these changes are important enough, they may improve roosting conditions for crevice- and cavity-roosting species; however, these influences are difficult to predict with any degree of certainty, are likely to be site-specific, and may differ among species and at different times of the year.

Several species of bats also regularly roost in human-made structures (Kunz 1982a,b,c, 2004). However, we are unaware of records of bats roosting in structures associated with wind-energy facilities in the United States, although bats have gained access to and roosted in the nacelle in Europe (Hensen 2004). Nonetheless, bat species that appear to be most at risk of being killed by wind turbines in the Mid-Atlantic Highlands include eastern red bats, hoary bats, silver-haired bats, and eastern pipistrelles. The latter species typically roosts in foliage during the summer months (Veilleux and Veilleux 2004; Veilleux et al. 2004), although it also is known to roost in buildings (Fujita and Kunz 1984; Hoying and Kunz 1998; Whitaker 1998).

Establishment of artificial roosts (e.g., Burke 1999; Arnett and Hayes 2000; Brittingham and Williams 2000; Chambers et al. 2002; Kunz 2003) is sometimes proposed to mitigate loss of roosts resulting from changes in land-use practices. However, encouraging increased roosting sites at or near wind-energy facilities could increase use of areas and increase risk of fatalities by collisions with turbines. Thus, mitigating loss of natural roosts at or near wind-energy facilities by constructing artificial roosts at these sites may not be effective.

Influences of Habitat Alteration on Habitat Use by Bats

Construction of roadways, management of vegetation, and the selective clearing of forests associated with the development of some wind-energy facilities can influence use of the area by bats. These influences could be manifested as changes in carrying capacity of an area or through influences of patterns of habitat use on risk of collision with turbines.

Many species of bats commonly use edges between forested and non-forested habitat and small forest gaps for commuting and foraging (Furlonger et al. 1987; Clark et al. 1993; Krusic et al. 1996; Walsh and Harris 1996; Wethington et al. 1996; Grindal and Brigham 1999; Zimmerman and Glanz 2000; Hogberg et al. 2002). For example, bat activity was greater along forest-clearcut edges than within clearcuts or uncut forests in British Columbia (Grindal and Brigham 1999), greater in forest clearings ranging from 0.5 to 1.5 ha in size than in intact forests in British Columbia (Grindal and Brigham 1998), greater along logging roads than in intact forest in South Carolina (Menzel et al. 2002), and greater along forest trails than in interior forests in New Hampshire (Krusic et al. 1996). Increased use of gaps, edges, and roadways is likely a consequence of reduced clutter (the number of obstacles a bat must detect and avoid in a given area [Fenton 1990]) along edges, increased availability of prey, or a combination of these factors. It is quite likely that construction of roads and clearings at wind-energy facilities in forested regions improves foraging habitats for several species of bats in the Mid-Atlantic Highlands, and elsewhere where similar habitat exists.

All bat species known to occur in the eastern United States, including the Mid-Atlantic Highlands, are insectivorous. These bats consume large quantities of nocturnal insects (Aubrey et al. 2003); both empirical evidence and anecdotal observations support the hypotheses that bats respond to prey availability and that prey availability is influenced by vegetation structure and to habitat alteration (e.g., agriculture). However, determining the relationship of distribution and abundance of insects to habitat use or population abundance of bats has been hampered by difficulties in determining abundance and availability of insects at appropriate spatial scales (Kunz 1988; Kunz and Lumsden 2003; Hayes and Loeb 2007). Thus, challenges lie ahead in estimating the influences of habitat changes on the prey base for insectivorous bats at wind-energy facilities. Changes that increase actual or relative abundance of insects preyed on by bats, or the vulnerability of insects to predation by bats at altitudes within the rotor-swept area of turbines could influence risk of bats to collisions with turbines. Clearly, large numbers of insects often are present in the vicinity of wind-turbine rotors, judging from insects that are known to accumulate on turbine blades in some regions (Corten and Veldkamp 2001).

Most of the studies of habitat use by bats have been conducted using recording devices. Only a few studies have evaluated vertical patterns of habitat use by insectivorous bats (e.g., Kurta 1982; Kalcounis et al. 1999; Hayes and Gruver 2000; Kunz 2004). Risk of collision with wind turbines is strongly influenced by vertical patterns of habitat use by bats, and is at least partially a function of the altitudes at which bats commute, forage, and migrate. Some of the species-specific differences in fatalities at wind turbines could be related to variation in vertical patterns of nightly foraging or migratory activity, possibly in response to prey resources, although currently there are no data available to test this hypothesis. It is unclear if or how habitat alteration at wind-energy facilities influences vertical patterns of habitat use by bats, but changes in vertical activity in response to habitat alteration and insect resources at wind-energy facilities could strongly influence fatality risks to bats. Vertical activity of bats could be influenced by the vertical distribution and abundance of aerial insects. Typically, insects rise to high altitudes above the ground on daily thermals, and then drop to lower altitudes as the lower atmosphere cools throughout the night (Figure 3-9).

Although habitat alteration resulting from wind-energy development often occurs at a relatively small scale, it is likely that the cumulative effects of wind-energy development, in conjunction with changes in habitat from a variety of other activities, will result in negative impacts on bat populations. Given the distances that bats travel nightly and during migration, contribu-

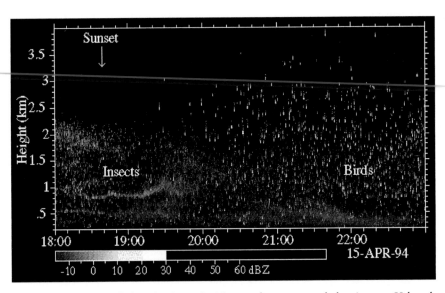

FIGURE 3-9 Vertical distribution of airborne fauna, recorded using an X-band vertically pointing radar on April 15, 1994. Note that insect targets drop markedly in elevation from before sunset until 2400 h. Most of the larger targets (assumed to be migrating birds and bats) occur at higher altitudes.
SOURCE: McGill University 2000. Reprinted with permission; copyright 2000, McGill University.

tions of wind-energy development to changes in landscape characteristics could influence bat populations. Unfortunately, the influences of habitat characteristics on bats at large spatial scales are poorly understood. Some bats have been shown to respond negatively to forest fragmentation in a number of areas (e.g., Pavey 1998; Law et al. 1999; Schulze et al. 2000; Estrada and Coates-Estrada 2002), but there is little information available about responses of bats to characteristics at the landscape scale in North America (Hayes and Loeb 2007). Lack of information on influences of landscape-scale patterns on bats precludes assessment of the likely impacts of habitat alterations at wind-energy facilities at broad spatial scales.

The combined influences of changes in availability of roosts, microclimatic conditions at roosts, availability of prey, vertical patterns of use, and landscape structure on bat populations in the Mid-Atlantic Highlands are difficult to predict with any precision. Moreover, the magnitude of influence of these factors may be site-specific and depend on site characteristics prior to construction of wind-energy facilities and associated infrastructure. If these changes were considered in the absence of direct influences of turbines on fatalities of bats, it is likely that we would conclude that impacts were not significantly negative in light of other threats to bats in the region and habitat changes resulting from other land uses. However, even this provisional conclusion must be tempered by the scale of habitat alteration; broad-scale proliferation of wind-energy facilities in the Mid-Atlantic Highlands and in other regions of the United States could result in significant consequences for habitat for bats and other species. For bats, the interaction among habitat alteration, influences on bat activity patterns, and risk of collision with wind turbines could be an important factor in bat fatalities in the Mid-Atlantic Highlands. Gaining increased understanding of these interactions could help inform in pre-siting risk assessments for bats.

Influences of Habitat Alteration on Terrestrial Mammals

Historically, higher elevation ridges of the Mid-Atlantic Highlands consisted of forest stands dominated by red spruce (*Picea rubens*). Late 19th- and early 20th-century logging operations reduced these stands to scattered remnants of mixed hardwood and spruce composition (Brooks 1965; Mielke et al. 1986). The federally listed (endangered) subspecies of the northern flying squirrel, the West Virginia northern flying squirrel (*Glaucomys sabrinus fuscus*), sometimes referred to as the Virginia northern flying squirrel, is closely associated with this spruce habitat. Genetically distinct from other populations of the species (Arbogast et al. 2005), this subspecies has been found at more than 100 separate sites along the ridge tops of the Mid-Atlantic Highlands (USFWS 2006). Current populations of the squirrel can be found in mixed stands of red spruce, cherry (*Prunus* serotina), and yellow poplar (*Liriodendron tulipifera*), although spruce is preferred (Menzel 2003; Menzel et al. 2006). Populations are locally expanding due to second-growth regeneration of upper-elevation forest tracts (USFWS 2006). The West Virginia northern flying squirrel is unique among squirrels in being active year-round and subsisting primarily on lichens, mushrooms, and mycorrhizal fungi, the latter of which are located by olfaction (Loeb et al. 2000; Mitchell 2001). There is an apparent symbiotic relationship between the squirrels and mycorrhizal fungi. The squirrels depend on fungi for food, while the fungi depend on the squirrels to disperse their spores as well as nitrogen-fixing bacteria, which are essential to the growth of red spruce (Mitchell 2001; USFWS 2006). Moreover, the overall condition of red-spruce forests appears to be strongly influenced by the presence of the squirrels (Mitchell 2001; USFWS 2006). Construction of wind turbines and associated roads can result in loss of mixed spruce/hardwood forest habitat and could lead to concomitant drops in squirrel population densities. The lack of quantitative data pertaining to the loss of spruce forest and squirrel habitat at wind-energy facilities limits our understanding of the potential impacts of wind-energy development.

Also of conservation interest is the Allegheny woodrat (*Neotoma magister*). Although not listed under the federal Endangered Species Act, this species is identified as endangered on state lists in New York, New Jersey, and Maryland; threatened in Pennsylvania; species of concern in North Carolina and Virginia; and a species "somewhat vulnerable to extirpation" in West Virginia. It is believed to be extinct in New York, New Jersey, and Connecticut. It is patchily distributed throughout the Mid-Atlantic Highlands in cliff lines and rock outcroppings, which provide their required nest locations (Castleberry 2000). Recent population declines have been dramatic and potential causal factors include anthropogenic disturbance near nest locations, increased predation by great horned owls (*Bubo virginianus*) and raccoons (*Procyon lotor*) directly linked to forest fragmentation, increased incidence of the parasitic raccoon roundworm (*Baylisascaris procyonis*), and diminished colonization of new locations because they need rock-outcrop habitats (Balcom and Yahner 1996; Castleberry et al. 2001, 2002; LoGiudice 2003; Hassinger 2005). A recent study based on 735 defined Allegheny woodrat "habitat sites" in higher-elevation forests in Pennsylvania showed that the occupancy rate of these sites increased with distance to non-forest edge (Hassinger et al. 2005). Moreover, habitat sites >2 km from a forest edge were 1.7-11.1 times more likely to be occupied than habitat sites within 1 km of a forest edge. Similarly, habitat sites 1-2 km from a forest edge were 1.7-3.8 times more likely to be occupied (Diefenbach et al. 2005). The lack of quantitative data pertaining to the loss of potential Allegheny woodrat habitat in the Mid-Atlantic Highlands is a data gap in the development of wind-energy projects.

Another mammalian species with unique habitat requirements in the Mid-Atlantic Highlands region is the snowshoe hare (*Lepus americanus*). Cyclically abundant in more northern habitats, this species reaches its southernmost distribution along the high ridges of Pennsylvania, Virginia, Maryland, and West Virginia (Brooks 1965). While this species is not protected under the U.S. Endangered Species Act, it is listed as "endangered/extirpated" in Maryland (MDDNR 2003) and "extremely rare" in Virginia (Roble 2006). This species is legally hunted in West Virginia. Populations of snowshoe hares occupy boreal forests at the northern end of their range while "southern populations occur primarily in insular patches of suitable habitat set amidst less-preferred areas" (Wirsing et al. 2002, p. 170). Brushy undergrowth and tree saplings, often aspen (*Populus tremuloides*), cottonwood (*P. deltoides*), or birch (*Betula spp.*) are the preferred habitat in the Mid-Atlantic Highlands. Tree removals in conjunction with wind-energy development could alter habitat for hares, and given their protected status in Maryland and Virginia, accurate pre-siting surveys should be conducted. The isolated population in Garrett County, Maryland, occurs in a location suitable for wind-energy development.

In the Mid-Atlantic Highlands, managed populations of large game mammals include the black bear and white-tailed deer, while managed furbearers include raccoon, beaver (*Castor canadensis*), red fox (*Vulpes vulpes*), gray fox (*Urocyon cinereoargenteus*), mink (*Mustela vison*), and fisher (*Martes pennanti*). Generally, trading the forested habitats of these species for gravel roads and foundation pads is unlikely to be beneficial. For example, black bears rely on forest habitats for food, cover, and denning sites (Brody and Pelton 1989). Because their selected habitats include a variety of interspersed vegetation types ranging from dense old-growth forests to forest openings rich in berries, bears have been referred to as "landscape species" (Gaines et al. 2005). Thus, analysis of any one vegetation type may be inconclusive and broad spatial analysis of the cumulative effects of human activity are required for effective habitat management (Gaines et al. 2005). However, forest-management practices in the region, such as thinning, clearcutting, and the construction of forest roads generally increase the amount of available soft mast (berries, shrub, and regenerating tree saplings) but also decrease the amount of hard mast (acorns and other nuts) available to black bears (Mitchell and Powell 2003). Soft mast would be reduced by maintenance of wind-energy facility roads and tower pads in a gravel state. Moreover, black bears avoid high-traffic roads, such as interstate highways and other divided highways, as well as low-traffic forest roads that provide access to hunters and their dogs (Brody and Pelton 1989). However, bears can learn to use low-traffic roads to move within their home range (Brody and Pelton 1989). In summary, the effects of wind-energy development in the Mid-Atlantic Highlands on black bears needs to be assessed at the landscape level and in conjunction with the cumulative aspects of all anthropogenic changes in forest structure. The relationship between wind-energy development and furbearer population biology also is unstudied at this time.

Small-mammal (e.g., *Peromyscus sp.*, *Microtus sp.*, and *Blarina sp.*) populations probably would not be affected by wind-energy development. Small-mammal populations may sometimes form demographic metapopulations under some conditions (Merriam et al. 1989). Even narrow (< 3 m), gravel roads can act as barriers to movements of prairie voles (*Microtus ochrogaster*) and white-footed mice (*Peromyscus leucopus*), and thus may isolate some populations genetically (Swihart and Slade 1984; Merriam et al. 1989). It is unclear what, if any, effect this isolation might have on small-mammal populations in the Mid-Atlantic Highlands. The lack of information on the effects of isolation is identified as a data gap in assessment of the ecological consequences of wind-energy development in the region.

Influences of Habitat Alternation on Amphibians and Reptiles

Amphibians play important roles in the functioning of forested ecosystems in the central Appalachians (Burton and Likens 1975a; Wyman 1998). It has been estimated that salamander biomass in eastern deciduous forests is 24 times that of birds (Greenberg 2001) and that it exceeds that of birds and mammals combined (Burton and Likens 1975b; Hairston 1987). Moreover, amphibians often are more sensitive to habitat alteration than birds and mammals (Marsh and Beckman 2004). Amphibians native to Mid-Atlantic Highland forest environments require aquatic or moist terrestrial habitats to complete their life cycles. Populations of both groups are influenced by the microclimate of forest floor habitats, specifically soil moisture and temperature, and species that lay eggs in aquatic systems also rely on free-standing water, even if it is ephemeral. Even without grading and construction of roads, slight removal of canopy vegetation may result in significant reduction of the amphibian fauna from forest tracts in some situations (Petranka et al. 1993; Ash 1997; Knapp et al. 2003). Knapp et al. (2003), for example, detected significant reduction in densities in *Plethodon* and *Desgmognathus* salamanders as a result of removal of canopy vegetation and almost all salamander taxa were adversely affected by timber removal (Petranka et al. 1993).

Amphibian species that require vernal pools for mating and egg-laying may be attracted to roadside ditches and ruts in maintenance roads by the presence of temporary water. However, if they become dry before the larvae become independent of water, such features may be "attractive sinks" (Delibes et al. 2001; Battin 2004), because animals that use them have reduced reproductive output that could contribute to the decline or loss of local populations. In a forest study of anthropogenic and natural

pools, both larval wood frogs (*Rana sylvatica*) and larval spotted salamanders (*Ambystoma maculatum*) suffered high mortality from premature
drying in the anthropogenic pools (DiMauro and Hunter 2002). During
"wet years" the larvae that metamorphosed were significantly smaller in
anthropogenic ponds than in natural ones; the anthropogenic pools were
subject to increased solar radiation and a more porous substrate, which
resulted in elevated water temperatures and faster drying rates (DiMauro
and Hunter 2002).

One species of amphibian in the Mid-Atlantic Highlands has is listed
as threatened under the ESA. Cheat Mountain salamanders (*Plethodon
nettingi*) occur in high forested landscapes in five West Virginia counties:
Pocahontas, Pendleton, Grant, Tucker, and Randolph (Green and Pauley
1987, T. Pauley, Marshall University, personal communication 2006). The
species was originally thought to occur only in spruce forests, but now is
known also to occur in high mixed hardwood/conifer tracks (Pauley 1981).
Removal of mixed hardwood/spruce trees and replacement with gravel
roads and tower pads could be detrimental to this species.

Ecology and natural history of reptiles are poorly studied in forest
communities potentially modified by wind-energy development in the Mid-
Atlantic Highlands. Generally, reptiles respond differently to the creation
of edge habitats than amphibians. Reptiles are more mobile than most
amphibians and certain species patrol forest edges in search of prey. In addition, since reptiles are typically associated with warmer, drier environments
than amphibians are, they may gain a positive thermoregulatory advantage
by taking advantage of increased solar radiation associated with forest
clearings (Greenberg 2001). One reptilian species of concern is the timber
rattlesnake (*Crotalus horridus*), which has been extirpated from most of its
historic range (Clark et al. 2003) and survives in isolated patches of forests,
including locations on or near ridge tops in the central Appalachians (Green
and Pauley 1987, F. Jernajic, West Virginia Division of Natural Resources,
personal communication 2006). Winter dens also occur along Appalachian
ridges and are shared by rattlesnakes, copperheads (*Agkistrodon contortix*),
and black rat snakes (*Elaphe obsoleta*). Timber rattlesnakes are of conservation importance because they have low fecundity, long reproductive cycles
(Brown 1993; Martin 1993), and are heavily persecuted by humans (Clark
et al. 2003). Alteration of habitat related to wind-energy development could
influence habitat suitability for this species, but we are unaware of any studies at wind-energy developments that have examined these effects.

Influences of Habitat Alteration on Fish and Other Aquatic Organisms

Aquatic habitats are not common along Mid-Atlantic Highland ridges.
By the very nature of the terrain, establishment of permanent bodies of

water and associated wetland habitat is reduced when compared with nearby downstream valleys. Uncontrolled erosion caused by anthropogenic activities at wind-energy facilities could have far-reaching consequences for aquatic habitats. Since wind-energy facilities in the Mid-Atlantic Highlands are at or near the top of mountain ridges, and hence they are in areas that receive large amounts of rain (> 125 cm per year, see CPC 2004), the potential exists for run-off and erosion. Erosion and sedimentation are avoided through following the requirements of each wind-energy facility's NPDES permit (EPA 2006d).

PROJECTED CUMULATIVE IMPACTS OF BIRD AND BAT FATALITIES: A WORKING HYPOTHESIS

Because we lack extensive data on the ecological influences of wind-energy facilities, projection of likely impacts in the Mid-Atlantic Highlands is challenging. Among the uncertainties that restrict our ability to assess impacts accurately are uncertainties in magnitude and pattern of future wind-energy development in the region, and lack of spatial and temporal replication in fatality assessments in the region. Nonetheless, it is valuable to prepare a preliminary assessment of potential cumulative impacts based on the limited information that is currently available. Here we estimate expected cumulative impacts on bats and birds based on current estimates of fatalities and projections of installed capacity of wind-energy facilities in the Mid-Atlantic Highlands.

Assumptions

Future development of wind-energy facilities in the Mid-Atlantic Highlands region, and elsewhere, depends on complex interactions among economic factors, technological development, regulatory changes, political forces, and other factors that cannot be predicted easily or accurately (Chapter 2). Here we provide a range of estimates of potential impacts for both birds and bats under the assumption that the National Renewable Energy Laboratory (NREL) Wind Energy Deployment System (WinDS) model and the PJM Interconnection queue (Table 3-5) estimates of projected installed capacity represent the range of potential wind-energy development that will occur in the Mid-Atlantic Highlands. The projections provide an upper and lower boundary, based on estimates of 2020 installed capacity (Table 3-5), and thus provide important hypotheses for testing. While it is conceivable that radically different fatality rates could occur in other locations in the eastern United States, using the information available from the few sites surveyed in the eastern United States to date (Tables 3-2 to 3-4) is the most realistic approach for evaluating potential cumulative impacts at this time.

TABLE 3-5 Estimates of Existing and Projected Installed Capacity for
Wind-Energy Facilities in the Mid-Atlantic Highlands by 2020, and the
Equivalent Number of 1.5 MW Wind Turbines That Would Generate
This Capacity

Basis for Estimate	Capacity (MW)	Equivalent Number of 1.5 MW Turbines
NREL estimate of total technical capacity[a]	8015	5344
NREL WinDS model reference case projection for 2020[b]	2158	1439
In-service, or approved by state regulatory authority[c]	1144	763
PJM (electricity grid operator) interconnection queue[d]	3856	2571

[a]Wind-capacity potential for MD, PA, VA, and WV provided on March 16, 2006, by
National Renewable Energy Laboratory (NREL), Golden, CO. Estimate limited to Class 3
and better wind areas above 1,000 feet elevation. Standard exclusions applied by NREL for
defining available wind resource, including environmental, land-use, and other criteria. See
Appendix B for description of the wind resource database and exclusion criteria.

[b]Modeled onshore capacity totals for MD, PA, VA, and WV provided on March 16, 2006,
by NREL, Golden, CO, based on application of the Wind Deployment System (WinDS) model
(for model information see NREL 2006b). As indicated in Table 2-3, the WinDS projections for U.S. wind-energy development are much larger than those provided by the Energy
Information Agency (EIA 2006a). EIA projections for MAH development, however, are not
available.

[c]Based on assembled information for in-service wind projects and wind projects with state
or local-level approval listed in the PJM interconnection queue (Boone 2006).

[d]Based on assembled information for wind-energy projects listed in the PJM Interconnection
queue in addition to in-service projects and projects with state or local-level approval (Boone
2006).

We base our estimation of fatalities on the information available in
the eastern United States for birds (Table 3-2) and for the Mid-Atlantic
Highlands for bats (Table 3-4). Our estimates for the lowest and highest
fatality rates reported for the Mid-Atlantic Highlands (Tables 3-1 to 3-4)
are based on only two studies selected as bounds; thus they may not bracket
the true extremes that might occur and thus provide estimates of cumulative
impacts to be expected in 2020, based on stated assumptions. These assumptions are: (1) reported fatality estimates are representative of the range
that could be expected (i.e., estimates based on more sites and improved
bias corrections are not likely to increase the range of the numbers of birds
and bats killed by wind turbines); (2) observed variation in fatality rates
are representative of the Mid-Atlantic Highlands (i.e., as more wind-energy
facilities are developed, minimum and maximum fatalities may change); (3)
there will be no significant technological changes that reduce or increase
fatalities (i.e., more and larger wind turbines than NREL- or PJM-based

projections will not be installed; (4) the numbers of resident and migrating bird and bat species will remain constant (i.e., no decline in populations from wind-turbine-related fatalities or other factors is expected); and (5) the relationship of installed capacity to operational hours and rotor-swept area will not change. Because our estimates are specific to the Mid-Atlantic Highlands, the number of reported fatalities and assumption might differ significantly for other geographic regions and should not be applied to them without additional study (Kunz et al. 2007).

Projected Cumulative Impacts

Based on the assumptions noted above for wind-energy development in the Mid-Atlantic Highlands, at projected levels of development by the NREL WinDS model for 2020 and the best available information (lowest and highest mean fatality rates; Table 3-2), we estimate that the projected avian fatalities in the mid-Atlantic regions could range from a mean minimum of approximately 5,805 birds per year (based on the fatality rate at the Mountaineer Wind Energy Center, West Virginia), to a maximum of approximately 25,183 birds per year (based on the fatality rate estimated for the Buffalo Mountain Wind Park in Tennessee). Using similar logic and the PJM-based projections for development, the projected range of avian fatalities increases to approximately 10,372 to 44,999 per year.

Under the assumption that the species composition of fatalities will be similar to the data presented above (Figure 3-1), we predict that these fatalities will primarily consist of passerines (Table 3-6). In the existing studies in this region at Mountaineer (Kerns and Kerlinger 2004) and Buffalo Mountain (Nicholson 2003), most individual passerine species made up a relatively small percentage of the passerine fatalities, up to 5%, resulting in the potential for approximately 200 to 1,000 individuals of any one species being killed per year using data from the NREL WinDS model projections and 400 to 1,800 killed per year using data from the PJM-based projections. However, at the Mountaineer site approximately 35% of the passerines killed were of the same species (red-eyed vireo, *Vireo olivaceus*). Thus, it is possible that from 1,600 to 7,000 individuals of a single species could be killed per year using NREL WinDS model projections and 2,900 to 12,700 per year using PJM-based projections.

The biological importance of these fatalities depends on the number of passerines in the affected population and whether the birds killed were migrant or resident in the areas of impact. Based on the existing data, it appears that approximately 50% of the passerines are migrant and losses to migrating and resident populations of passerines in this region would be approximately 2,400 to 10,000 each per year using NREL WinDS model projections and 4,200 to 18,000 per year using PJM-based projec-

TABLE 3-6 Projected Annual Number of Bird Fatalities from Wind Turbines Expected in 2020. Based on Estimates of Current Proportional Fatality Rates and Available Estimates of Installed Capacity for the Mid-Atlantic Highlands Region

	Proportion of Total Fatalities[b]	Projections Based on the NREL WinDS Model of Installed Capacity[a]	
		Minimum Projected Number of Bird Fatalities[c]	Maximum Projected Number of Bird Fatalities[d]
Total		5,805 (6,000)	25,183
Species Group[b]			
Doves/pigeons	.02	116	503
Gamebirds	.02	116	503
Other birds	.06	348	1,510
Passerines	.81	4,702	20,398
Rails/coots	.02	116	503
Raptors/vultures	.03	174	755
Unidentified birds	.02	116	503
Waterfowl	.02	116	503

	Proportion of Fatalities	Projections Based on the PJM Grid-Operator Queue[e]	
		Minimum Projected Number of Bird Fatalities[f]	Maximum Projected Number of Bird Fatalities[g]
Total		10,372	44,999
Species Group[b]			
Doves/pigeons	.02	207	899
Gamebirds	.02	207	899
Other birds	.06	622	2,699
Passerines	.81	8401	36,449
Rails/coots	.02	207	899
Raptors/vultures	.03	311	1,349
Unidentified birds	.02	207	899
Waterfowl	.02	207	899

[a]Estimated installed capacity of 2,158 MW based on National Renewable Energy Laboratory (NREL) WinDS Model for the Mid-Atlantic Highlands for the year 2020 (NREL 2006b).

[b]Estimated species-specific fatality rates are based on data collected in the eastern United States (Figure 3-1).

[c]Minimum projected number of fatalities in 2020 is based on the product of 2.69 bird fatalities/MW reported from the Mountaineer Wind Energy Center, WV (from Table 3-2), and the estimated installed capacity (2,158 MW) = 5,805. The species group-specific annual minimum number of projected bird fatalities is the product of the minimum number of projected fatalities and the species group-specific proportional fatality rates (column 2).

[d]Maximum projected number of fatalities in 2020 is based on the product of 11.67 bird fatalities/MW reported from the Buffalo Mountain Wind Energy Center, TN (from Table 3-2), and the estimated installed capacity (2,158 MW) = 25,183. The species group-specific annual

continued

TABLE 3-6 Continued

maximum number of projected fatalities is the product of the maximum number of projected fatalities and the species group-specific proportional fatality rates (column 2).

[e]Estimated installed capacity of 3,856 MW based on PJM (electricity grid operator interconnection queue) for the Mid-Atlantic Highlands for the year 2020 (Boone 2006).

[f]Minimum projected number of fatalities in 2020 is based on the product of 2.69 bird fatalities/MW reported from the Mountaineer Wind Energy Center, WV (from Table 3-2), and the estimated installed capacity (3,856 MW) = 10,372 (10,500). The species group-specific annual minimum number of projected bird fatalities is the product of the minimum number of projected fatalities and the species group-specific proportional fatality rates (column 2).

[g]Maximum projected number of fatalities in 2020 is based on the product of 11.67 bird fatalities/MW reported from the Buffalo Mountain Wind Energy Center, TN (from Table 3-2), and the estimated installed capacity (3,856 MW) = 44,999. The species group-specific annual maximum number of projected fatalities is the product of the maximum number of projected fatalities and the species group-specific proportional fatality rates (column 2).

tions. Estimating the fatalities for local populations based on projections for the year 2020 requires the assumption that several local populations are affected. On the assumption that the Mountaineer facility represents a typical development for the future (66 MW) in the region, and that a total of 2,158 to 3,856 MW of capacity will be installed by then, there would be 33 to 58 wind-energy facilities. Furthermore, the upper end of the range of projected fatalities for the two development scenarios would result in approximately 300 passerines killed per facility per year. Thus, if up to 5% of the birds killed locally are of the same species, one could expect that most local populations would suffer the loss of approximately 15 birds per year. Under the assumption that an individual species could be much more vulnerable than the average to collisions, and using the red-eyed vireo as an example, up to 35% of the birds killed locally could be of one species (105 birds per year) and presumably be from one local population.

Local populations of raptors and vultures are much smaller than passerine populations and thus potentially more at risk for population effects of fatalities from wind-energy generation. Using the same logic and data sources for raptors and vultures as were used for passerines, approximately 9-23 individuals per year of these species are projected to be killed at each of these sites using the lowest and highest range of projected wind-energy development. Some of the birds would be resident and some migrant.

Based on currently available information on bat fatalities in the eastern United States, projected cumulative impacts using estimates of installed capacity for the Mid-Atlantic Highlands in the year 2020, along with supporting data, assumptions, and calculations, are in Table 3-7. Minimum and maximum estimates of installed capacity for this region range from 2,158

TABLE 3-7 Projected Annual Number of Bat Fatalities from Wind Turbines Expected in 2020. Based on Projections of Installed Capacity for This Region and Current Proportional Fatality Rates Available from the Eastern United States

Species[b]	Fatality Rate[c]	Projections Based on the NREL WinDS Model of Installed Capacity[a]	
		Minimum[d]	Maximum[e]
Hoary bat	0.289	9,542	17,899
Eastern red bat	0.344	11,358	21,306
Silver-haired bat	0.052	1,717	3,221
Eastern pipistrelle	0.185	6,108	11,458
Little brown *myotis*	0.087	2,873	5,388
Northern long-eared *myotis*	0.006	198	372
Big brown bat	0.025	825	1,548
Unknown/other	0.012	396	743
TOTAL		33,017	61,935

Species[b]	Fatality Rate[c]	Projections Based on the PJM Grid Operator Interconnection Queue[f]	
		Minimum[g]	Maximum[h]
Hoary bat	0.289	17,050	31,983
Eastern red bat	0.344	20,295	38,069
Silver-haired bat	0.052	3,068	5,756
Eastern pipistrelle	0.185	10,914	20,473
Little brown *myotis*	0.087	5,133	9,628
Northern long-eared *myotis*	0.006	354	664
Big brown bat	0.025	1,475	2,767
Unknown/other	0.012	708	1,328
TOTAL		58,997	110,667

[a]Estimated installed capacity of 2,158 MW based on National Renewable Energy Laboratory (NREL) WinDS Model for the Mid-Atlantic Highlands for the year 2020 (Table 3-5).

[b]Eastern red bats, hoary bats, and silver-haired bats are the only species in the eastern United States known to undertake long-distance migrations (Barbour and Davis 1969).

[c]Estimated species-specific fatality rates are based on data collected in the eastern United States (Table 3-4).

[d]Minimum projected number of fatalities in 2020 is based on the product of 15.3 bat fatalities/MW/year reported from the Meyersdale Wind Energy Center, PA (from Table 3-4), and the estimated installed capacity (2,158 MW) = 33,017. The species-specific annual minimum number of projected bat fatalities is the product of the species-specific fatality rates (column 2) and the minimum total number of fatalities (e.g., for the hoary bat, 0.289 * 33,017 = 9,542).

[e]Maximum projected number of fatalities in 2020 is based on the product of 28.7 bat fatalities/MW/year (average for 2003 and 2004) reported from the Mountaineer Wind Energy Center, WV (from Table 3-4), and the projected installed capacity (2,158 MW) = 61,935. The

continued

TABLE 3-7 Continued

species-specific annual maximum number of projected bat fatalities is the product of the species-specific fatality rates (column 2) and the total maximum number of fatalities.

[f]Estimated installed capacity of 3,856 MW based on PJM (electricity grid operator interconnection queue) for the Mid-Atlantic Highlands for the year 2020 (Table 3-5).

[g]Minimum projected number of fatalities in 2020 is based on the product of 15.3 bat fatalities/MW/year reported from the Meyersdale Wind Energy Center, PA (from Table 3-4), and the projected installed capacity (3,856 MW) = 58,997. The species-specific annual minimum number of projected bat fatalities is the product of the species-specific fatality rates (column 2) and the total minimum projected number of fatalities.

[h]Maximum projected number of fatalities in 2020 is based on the product of 28.7 bat fatalities/MW/year (average of year 2003 and 2004) reported from the Mountaineer Wind Energy Center, WV (from Table 3-4), and the projected installed capacity (3,856 MW) = 110,667. The species-specific annual maximum number of projected bat fatalities is the product of the species-specific fatality rates (column 2) and the total maximum projected number of fatalities.
SOURCE: Kunz et al. 2007.

MW (based on the NREL WinDS model) to 3,856 MW (from the PJM Interconnection queue), as was the case for the bird-fatality projections.

These cumulative fatality projections for bats based on fatality rates determined for this region should be regarded as provisional (Table 3-4). Although some of the empirical data for this region were not collected consistently, the data summarized in Table 3-3 are the best available data for assessing cumulative impacts.

Based on estimates of installed capacity and the limitations and assumptions regarding fatality rates noted above, the minimum and maximum projected fatalities of bats presented in Tables 3-4, 3-5, and 3-7 would range from 33,017 to 61,935 per year based on the NREL's WinDS model and 58,997 to 110,667 per year based on the PJM Interconnection queue. These projected cumulative impacts in 2020 based on the WinDS model and PJM Interconnection queue would cause annual fatalities of 9,542 to 31,983 hoary bats, 11,358 to 38,069 eastern red bats, 1,717 to 5,755 silver-haired bats, and 6,108 to 20,473 eastern pipistrelles in the mid-Atlantic region. These projections should be considered as hypotheses, until improved estimates (or enumerations) of installed capacity and bat fatalities become available for this region (Kunz et al. 2007).

No projections were made for the endangered Indiana bat, Rafinesque's big-eared bat (*Corynorhinus rafinesquii*), or the regionally listed small-footed myotis (*Myotis leibii*), because no fatalities for these three species have been reported at wind turbines in the Mid-Atlantic Highlands. This should not be interpreted as reflecting a judgment that no members of those species will be killed. It is possible that their behavior and distribution prevent them from coming into contact with turbines, or it is possible that their rarity has not yet led to a recorded fatality of any of those species.

Ecological Implications of Projected Cumulative Impacts

These projections of cumulative bat and bird fatalities for the Mid-Atlantic Highlands by the year 2020 assume that bat and bird populations living in or migrating through the region each year would be constant. The latter assumption is likely to be violated given assorted caveats about expected inter-annual variability; however, given that we have presented both worst-case (maximum number of fatalities/year) and best-case (minimum number of fatalities/year) scenarios, our projected fatality rates in the Mid-Atlantic Highlands bracket expected extremes. These projected fatalities can best be considered as hypotheses to be tested with future data on fatalities from the Mid-Atlantic Highlands and other regions where bird and bat fatalities have been reported, and by adjusting monitoring protocols to minimize potentially confounding assumptions (Kunz et al. 2007).

A question that arises from these projections is whether they are of biological importance to bat and bird populations. The answer differs for birds and bats and for migratory and local populations. For birds, it is unlikely that this predicted level of fatalities would result in measurable impacts to migratory populations of most species. However, for rare species and local populations, the impacts, when combined with other sources of mortality such as large weather-related bird kills, could affect viability, and thereby affect overall risks to populations. A definitive conclusion on these predicted impacts requires more information on the demographics of rare and local populations of birds than is currently available.

For bats, the question draws attention to the almost complete lack of data for population estimates of any species considered here, either on a regional or continental scale (Kunz et al. 2007). A risk assessment of biological impacts typically requires knowledge of baseline populations. Nonetheless, the numbers of fatalities projected above for bats in the Mid-Atlantic Highlands suggest that bat populations might be at risk, because they reflect fatality rates as high as or higher than fatality rates that have been reported for bats from other measurable anthropogenic sources (Kunz et al. 2007).

CONCLUSIONS AND RECOMMENDATIONS

Our understanding of the ecological effects of wind-energy development in the Mid-Atlantic Highlands region and elsewhere is limited by minimal monitoring efforts at existing wind-energy facilities and by poor understanding of key aspects of species ecology, of causal mechanisms underlying fatalities at wind-energy facilities, and of the reliability of our projections of fatalities at wind-energy facilities. This section contains the committee's conclusions about the known and potential ecological effects of wind-energy projects, identification of information needs, and recommendations for research and monitoring.

Ecological Effects of Wind-Energy Projects

• While research and monitoring studies admittedly are limited, a synthesis of the existing studies indicates that adverse effects of wind-energy facilities on ecosystem structure and functioning have occurred. This knowledge should be used to guide decisions on planning, siting, and operation.

• Wind turbines cause fatalities of birds and bats through collision, most likely with the turbine blades.

• Species differ in their vulnerability to collision. The probability of fatality is most likely a function of abundance, local concentrations, and the behavioral characteristics of species.

• Migratory tree-roosting bat species appear to be most susceptible to direct impacts. To date, the highest fatality rates have been reported in the Mid-Atlantic Highlands, although recent evidence suggests that bats from grassland and agricultural landscapes may also experience high fatality rates. Migratory tree bats constitute over 78% of all fatalities reported at wind-energy facilities, and thus appear to be killed disproportionately to highly colonial species. To date, no endangered species have been reported being killed at existing wind-energy facilities, although only a few sites have been monitored. Increased risks are expected as more wind-energy facilities are developed. Risks of fatalities to bats in the southwestern United States, especially in Texas, where large wind-energy facilities exist and have been proposed, are largely unknown because data have not been reported for most of these facilities.

• Abundance interacts with behavior to influence exposure of breeding passerines, raptors, and bats to the risk of collisions. Raptors appear to be the most vulnerable to collisions. On average raptors constitute 6% of the reported fatalities at wind-energy facilities, yet they are far less abundant than most other groups of birds (e.g., passerines). By contrast, crows, ravens, and vultures are among the most common species seen flying within the rotor-swept area of turbines, yet they are seldom found during carcass surveys. Nocturnally migrating passerines are the most abundant species at most wind-energy facilities and are the most commonly reported fatalities. Nonetheless, fatalities among passerines vary more than can be explained by abundance alone.

• Species differ in the extent to which their fatalities are discovered and publicized. Small birds and bats are more difficult to find than others during planned searches and incidentally. Large birds such as raptors are more easily seen, and are often more publicized because of their charismatic status and perceived importance in the environment.

• The location of wind-energy facilities on the landscape (e.g., agricultural lands, ridge tops, canyons, grasslands) influences bird and bat fatalities. Available evidence suggests that fatalities are positively correlated with bird abundance. Landscape features influence density by concentrating prey or through providing favorable conditions for other activities such as nesting, feeding, and flying (e.g., updrafts for raptor soaring and linear landscapes for bats).

• The characteristics (e.g., rotor-swept area, height, support structure, lighting, number of turbines) of wind-energy facilities may act synergistically to cause bird and bat fatalities. Newer, larger turbines installed on monopoles may cause fewer bird fatalities per MW than the smaller, older, lattice-style turbines, but the ability to determine the significance of these characteristics is limited by sparse data; in addition, other factors such as the local and regional abundances of birds and bats and landscape variation confound understanding of the effects of turbine characteristics noted above.

• The lack of estimates of population sizes and other population parameters for birds and bats and the lack of multiyear studies at most existing wind-energy facilities make it difficult to draw general conclusions about how wind turbines and population characteristics interact to influence mortality of birds and bats. In addition, lack of replication of studies among facilities and years makes it impossible to evaluate natural variability, in particular unusual episodic events, in relation to fatalities and to predict the potential for future population effects. It is essential that the potential for population effects be evaluated as wind-energy facilities become more numerous.

• Fatality rates of migratory tree bats appear to be high in some landscapes (e.g., forested ridge tops), although almost nothing is known about the population status of these species, and the biological significance of reported fatalities. Nonetheless, this lack of data on bat populations points to a critical need to evaluate the status of these and other species that may be at risk, especially as wind-energy facilities proliferate, and a need to evaluate where major cumulative impacts could be expected.

• The construction and maintenance of wind turbines and associated infrastructure (e.g., roads) alters ecosystem structure through vegetation clearing, soil disruption, and potential for erosion and noise.

• Based on similar types of construction and development, it is likely that wind-energy facilities will adversely alter ecosystems indirectly, especially through the following cumulative impacts:

1.   Forest clearing resulting from road construction, transmission lines leading to the grid, and turbine placements represents perhaps the most significant potential change through habitat loss and fragmentation for forest-dependent species. This impact is particularly important in the Mid-Atlantic Highlands, because wind-energy projects there all have been constructed or proposed in forested areas.

2.   Changes in forest structure and the creation of openings may alter microclimate and increase the amount of forest edge.

3.   Plants and animals throughout the ecosystem respond differently to these changes, and particular attention should be paid to species listed under the ESA and species of concern (Appendix C) that are known to have narrow habitat requirements and whose niches are disproportionately altered.

Information Needs

Here we identify information needs related to understanding, predicting, and managing bird and bat fatalities and landscape and habitat alterations. For each of these categories we suggest important information needs that we judge should be given the highest priority for monitoring and research based on our collective understanding of the issues, weighed by tractability and best practices. The following recommendations are not meant to apply to every situation and should be modified given the characteristics of the site being developed, the species of concern, the results of pilot studies, and the amount of information applicable to that site. If wind-energy development continues in a region, research and monitoring protocols should evolve as more becomes known.

Research is needed to develop mitigation approaches for existing facilities and to aid in assessing risk at proposed facilities. The latter is particularly important in landscapes where unusually high bird and bat fatalities have already been reported and in regions where facilities are planned where little is known about migration, foraging, and fatalities associated with wind-energy facilities (e.g., the Mid-Atlantic Highlands and the southwestern United States).

Following accepted scientific protocols, hypotheses should be developed to help address unanswered questions. Testing hypotheses promises to provide science-based answers that will help inform developers, decision makers, policy makers, and other stakeholders concerning actual and expected impacts of wind-energy development on bat and bird population and on landscapes and habitats of other animals that might be altered by construction.

Some of these information needs are beyond the scope of any individual developer (e.g., population status of affected species). Therefore, a collaborative effort by industry and agencies to fund the necessary research to address these overarching questions should be initiated. Other information could be developed as part of the permitting process. Decision makers could require owners and developers to fund research and monitoring studies by qualified researchers at the proposed wind-energy facilities; developers and operators should provide full access (subject to safety and proprietary concerns) to researchers at existing wind-energy facilities. The research should be conducted openly and the protocols and results should be subject to peer review.

1. Follow established scientific principles in conducting monitoring studies and experiments.

2. Follow established research methods and metrics (summarized in Appendix C).

3. Evaluate the efficacy of tools needed to make reliable predictions that would assess measures to reduce the risk of fatalities (e.g., evaluate potential mitigation measures).

4. Develop new quantitative tools to predict fatalities at proposed and existing wind-energy facilities.

    a. Develop estimates of exposure for use in evaluating fatalities and for estimating risk (e.g., radar studies at existing facilities in combination with fatality data to develop stronger risk-assessment tools).

    b. Improve tools and protocols that can discriminate migrating birds from migrating bats, operate in inclement weather, and provide cost-effective estimates of numbers and movements of flying birds and bats.

    c. Develop models to predict risk based on geographic region, topography, season, weather, lunar cycles, and characteristics of different turbines.

    d. Improve methods and metrics to determine the context of the number of fatalities related to the number of birds moving through the airspace (proportionality).

    e. Identify potential biases associated with estimation of fatalities, including necessary search effort (plot size, frequency of search, methods of searching), the probability that a carcass will be detected if present, and the probability that a carcass will be removed so that its detection probability is zero.

5. Encourage and conduct studies to support impact assessments.

    a. Assess effects of changing technologies (e.g., larger turbines) on bird and bat fatalities.

    b. Identify impacts of different types of lighting on bat and bird fatalities.

   c. Assess how different landscape features may affect bird and bat fatalities (mountain ridges, agriculture, grassland, canyons).

   d. Assess how weather fronts influence bat and bird fatalities.

   e. Identify bat and bird migratory patterns over space and time.

   f. Determine whether migratory birds and bats adjust their migratory paths or exhibit other behaviors that may cause them to avoid turbines.

   g. Determine whether fatalities from turbines reduce the breeding or stopover density and reproductive success of birds and bats.

   h. Conduct studies to identify methods of mitigating impacts of wind turbines on bats, birds, and other wildlife.

Hypothesis-Based Research on Bats

Knowledge about bat fatalities at wind-energy plants is very limited, mainly because the large number of bats killed has been recognized only recently. Eleven hypotheses are listed below, as examples, to help address how, when, where, and why bats are being killed at wind-energy facilities (Kunz et al. 2007). These hypotheses are not mutually exclusive, as several postulated factors might act synergistically to produce the high fatalities that have been reported.

• *Linear-Corridor Hypothesis:* Wind-energy facilities constructed along forested ridge tops create clearings with linear landscapes that are attractive to bats. Bats frequently use these linear landscapes during migration and while commuting and foraging (Limpens and Kapteyn 1991; Verboom and Spoelstra 1999; Hensen 2004; Menzel et al. 2005a), and thus may be placed at increased risk of being killed (Dürr and Bach 2004).

• *Roost-Attraction Hypothesis:* Tree-roosting bats commonly seek roosts in tall trees (Pierson 1998; Kunz and Lumsden 2003; Barclay and Kurta 2007) and thus if wind turbines are perceived as potential roosts (Ahlén 2002, 2003; Hensen 2004), their presence could contribute to increased risks of being killed when bats search for night roosts or during migratory stopovers.

• *Landscape-Attraction Hypothesis:* Modifications of landscapes needed to install wind-energy facilities, including the construction of wide power-access corridors and removal of trees to create clearings (usually 0.5-2 ha) around each turbine site, create conditions favorable for insects on which bats feed (Lewis 1970; Grindal and Brigham 1998; Hensen 2004). Thus, bats that are attracted to and feed on insects in these altered landscapes may be at an increased risk of being killed by wind turbines.

• *Low Wind-Velocity Hypothesis:* Fatalities of aerial feeding and migrating bats are highest on nights during periods of low wind velocity (Fiedler 2004; Hensen 2004; Arnett 2006), in part because flying insects are most active under these conditions (Ahlén 2002, 2003).

• *Heat-Attraction Hypothesis:* Flying insects are attracted to the heat produced by nacelles of wind turbines (Corten and Veldkamp 2001; Ahlén 2002, 2003; Hensen 2004). As bats respond to high densities of flying insects near wind turbines, they may be at increased risk of being struck by turbine blades.

• *Acoustic-Attraction Hypothesis:* Bats are attracted to audible and ultrasonic sound produced by wind turbines (Schmidt and Joermann 1986; Ahlén 2002, 2003). Sounds produced by the turbine generator and the swishing sounds of rotating turbine blades may attract bats, thus increasing risks of collision and fatality.

• *Visual-Attraction Hypothesis:* Insects flying at night are visually attracted to wind turbines (von Hensen 2004). Inasmuch as bats may feed on those insects, they become vulnerable to collisions with the turbine blades.

• *Echolocation-Failure Hypothesis:* Migrating and foraging bats fail to detect wind turbines by echolocation, or miscalculate rotor velocity (Ahlén 2002, 2003). If bats are unable to detect the moving turbine blades, they may be struck and killed directly.

• *Electromagnetic-Field Disorientation Hypothesis:* If bats have receptors sensitive to magnetic fields (Buchler and Wasilewski 1985), and wind turbines produce complex electromagnetic fields in the vicinity of the nacelle, the flight behavior of bats may be altered by these fields and thus increases their risk of being killed by rotating turbine blades.

• *Decompression Hypothesis:* Bats flying in the vicinity of turbines may experience rapid decompression (Dürr and Bach 2004; Hensen 2004). Rapid pressure change may cause internal injuries or disorientation, thus increasing risk of death.

• *Thermal-Inversion Hypothesis:* The altitude at which bats migrate and/or feed may be influenced by thermal inversions, forcing them to the altitude of rotor-swept areas (Arnett 2005). The most likely impact of thermal inversions is to create dense fog in cool valleys, possibly concentrating both bats and insects on ridges, and thus encouraging bats to feed over the ridges on those nights, if for no other reason than to avoid the cool air and fog.

Research Recommendations

Research should focus on two general lines of inquiry, including methodological research addressing improved tools and monitoring protocols as necessary, and hypothesis-driven research to provide information that will help inform developers, decision makers, policy makers, and other stakeholders to deal with actual and expected impacts of wind-energy development on populations and ecosystems.

At a national scale, it would be appropriate to identify multiyear research goals that place the impacts of wind-energy development into a broad environmental perspective. Research initiatives should be encouraged to identify biological impacts of wind-energy development, and compare these impacts and risks with those of competing power-generating technologies.

Research should focus on regions and sites where existing and new information suggest the greatest potential for biologically significant adverse impacts on birds and bats at proposed and existing wind-energy facilities. For example, while current evidence suggests that bat fatalities have been the highest at wind-energy facilities in forested mounted ridge tops in the Mid-Atlantic Highlands, recent monitoring studies in agricultural landscape in the Midwest and at wind-energy facilities in southwestern Alberta, Canada, suggest that fatality rates of migratory tree bats may be as high as those reported for the Mid-Atlantic Highlands. We also expect that high bat fatalities are occurring or will occur in the southwestern United States, where large numbers of Brazilian free-tailed bats form maternity colonies (McCracken 2003), and where there is high bat-species richness (O'Shea and Bogan 2003). However, to date, no appropriately designed fatality surveys have been reported at wind-energy facilities in this region. Given the observed geographic variation in fatality rates of both birds and bats, research is needed to evaluate where the risks or fatalities are high so that similar areas can be avoided. Improved assessments, with a focus on evaluation of causes and cumulative impacts, should be an urgent research priority. Proceeding with large-scale development of wind-energy facilities before identifying risks likely threatens both bats and the public acceptance of wind energy as an environmentally friendly form of energy (Kunz et al. 2007). Thus, the initial developments should be used as an opportunity to understand the risks before the full wind-energy potential of the Mid-Atlantic Highlands is developed.

The highest priority for avian habitat is the quantification and prediction of habitat impacts, including loss because of the spatial demands of wind-energy facilities (e.g., roads and turbine pads) and displacement impacts because of behavioral response or habitat degradation, particularly on forest-dwelling and shrub-steppe and grassland birds. In addition, the role of wind in large-scale fragmentation of habitat for species dependent on forests should be evaluated. Finally, the impact of habitat loss or modification should be evaluated in terms of the potential for demographic impacts on ground-nesting birds.

Clearly defined pre- and post-construction studies are needed to inform decision makers about the feasibility of constructing a new project and mitigating the adverse effects of existing facilities. The studies should be replicable and compared with other studies conducted in areas with similar topography and habitat. Where appropriate, pre- and post-construction studies should be conducted as recommended below.

- Pre-siting Studies
  1. Conduct pre-siting studies that allow the comparison of multiple sites when making decisions about where to develop wind energy.
  2. Identify species of special concern and their habitat needs; these include species listed under the federal ESA, such as the West Virginia northern flying squirrel, as well as species listed by the appropriate state, such as the Allegheny woodrat.
- Pre-construction Studies
  1. Conduct regional assessments to identify species of concern, including those vulnerable to direct impacts and those vulnerable to habitat loss.
  2. Develop pre-construction estimates of potential biological significance of fatalities based on estimated fatality rates and demographics of the species of concern.
  3. Conduct multiyear studies when appropriate to assess daily, seasonal and interannual variability of bird and bat populations.
  4. Establish species-specific abundance, periods of use (both seasonally and within a day), and behavior in relation to proposed turbines placement locally, regionally, and nationally.
  5. Identify habitat characteristics for birds, bats, and other animals, such as topography and types of vegetation at each proposed sites.
- Post-construction Studies
  1. Conduct full-season, multiyear, post-construction studies where appropriate to assess variability of bird and bat fatalities.
  2. Identify number, species composition, and timing of fatalities.
  3. Estimate the biological significance of bird and bat fatalities.
  4. Clarify the relationship of small-scale (e.g., habitat disturbance and species displacement) versus large-scale impacts (e.g., landscape alteration and fragmentation) of development on bird and bat populations.
  5. Conduct experiments to test alternative mitigation procedures (strategic shutdowns, feathering, blade painting and other potential deterrents, and lighting) that could avoid or reduce current fatality rates—independent of a meta-analysis to assess biological significance and adverse cumulative impacts.

- General

    1. Develop predictive and risk-assessment models of potential cumulative impacts of proposed wind-energy facilities, based on monitoring studies and hypothesis-based research.

Summary

More information is needed on the characteristics of bird and bat fatalities at wind facilities in all regions of the county, and in particular areas that are relatively unstudied such as the Mid-Atlantic Highlands, the arid southwest, and coastal areas. Turbine characteristics, turbine siting, and abundance appear to be important factors in determining the risk of raptor fatalities at wind-energy facilities. Compared to relatively high raptor fatalities at some older facilities in California, direct impacts of wind-energy development on passerines at the current level of development appear to be minimal. At current levels of development existing data suggest that new-generation turbines (e.g., fewer turbines mounted on monopoles with greater rotor-swept zones) may cause lower bird fatalities in agricultural and grassland areas than older smaller turbines have caused in California. Data on bird fatalities are absent for many existing wind-energy facilities, particularly in Texas and the southwestern United States. Additionally, new areas are being proposed for development where no previous data on bird and bat fatalities exist. It is important to assess impacts in existing and new areas to determine if trends are consistent with existing information. In particular, only two short-term post-construction studies have been conducted in the Mid-Atlantic Highlands and any new facilities should be used as learning opportunities.

Additional information also is needed to characterize bat fatalities in all regions of the country where wind-energy development has occurred or where it is expected. Most wind-turbine-related bat fatalities in the United States have been of migratory species. To date, no fatalities of federally listed bat species have been documented, although as wind-energy development increases geographically, some threatened and endangered species could be at risk. Among the studies that have been conducted, the highest bat fatality rates appear to occur episodically in late summer and early autumn during periods of relatively low wind speeds (< 6 m/sec), at times when wind-energy generation is low, especially following passing weather fronts. To date, few studies have evaluated fatalities during spring migration or during the summer maternity period. Moreover, among fatality surveys that have been conducted, few have consistently corrected results for observer bias and scavenger removal, protocols that are needed to provide reliable data on fatalities. While current evidence suggests that the highest fatality rates are of migratory tree-roosting species along ridge tops in eastern deciduous forests, recent evidence suggests that similar fatality rates may occur in some agricultural and grassland regions. Bats in other regions of the country that have high wind capacity and are currently undergoing rapid wind-energy development (e.g., southwestern United States), where some of the largest bat colonies in North America are known, may be at considerable risk from wind-energy development during both migratory and maternity periods. Projected development of wind-energy facilities throughout the United States should be evaluated for cumulative impacts on different species considered at risk.

4

Impacts of Wind-Energy Development on Humans

INTRODUCTION

Although they have some unusual characteristics, such as visibility at a distance, wind-energy projects are not unique in their impacts on people. They share many characteristics with other projects—not only energy-production projects but also landfills, waste incinerators, etc.—that create both benefits and burdens. In considering how to undertake local interactions and how to temper negative socioeconomic impacts while enhancing benefits, much can be learned from past experiences with other potentially controversial issues.

One important lesson—and an important prelude to this chapter—is that concern about visual, auditory, and other impacts is a natural reaction, especially when the source of the impacts is or will be close to one's home. The project's potential for negative impacts as well as benefits, and the fact that different people have different values as well as different levels of sensitivity, are important aspects of impact assessment.

This chapter addresses some key potential human impacts, positive and negative, of wind-energy projects on people in surrounding areas. The impacts discussed here include aesthetic impacts; impacts on cultural resources such as historic and archeological sites and recreation sites; impacts on human health and well-being, specifically, from noise and from shadow flicker; economic and fiscal impacts; and the potential for electromagnetic interference with television and radio broadcasting, cellular phones, and radar.

The topics covered in this chapter do not represent an exhaustive list of all possible human impacts from wind-energy projects. For example, we have not addressed potentially significant social impacts on community cohesion, sometimes exacerbated by differences in community make-up (e.g., differences in values and in amounts and sources of wealth between newcomers and long-time residents). Also not covered are psychological impacts—positive as well as negative—that can arise in confronting a controversial project (Gramling and Freudenburg 1992; NRC 2003). We have not focused on these matters because they can vary greatly from one local region or project site to another; and also as a function of population density and local and regional economic, social, and economic conditions; and in other ways. As a result, it is very difficult to generalize about them. In addition, not covered in this chapter but discussed elsewhere in this report (see especially Chapter 2) are diffuse health and economic effects of wind-energy projects. The topics covered in this chapter are, however, the chief local environmental impacts that have been recognized to date.

Thus far, there has been relatively little dispassionate analysis of the human impacts of wind-energy projects. Much that has been written has been from the vantage points of either proponents or opponents. There also are few data that have been systematically gathered on these impacts. In the absence of extensive data, this chapter is focused mainly on appropriate methods for analysis and assessment and on recommended practices in the face of uncertainty. Several of the methods discussed follow general principles and practice in socioeconomic impact assessments conducted as part of environmental impact statements; nevertheless, the chapter is tailored to the potential local human impacts of wind-energy projects and to their predominantly rural settings.

Wind-energy projects, like other potentially controversial developments, vary in their social context and thus in their social complexity. In this chapter, comments and methodological recommendations are directed toward relatively complex wind-energy facilities such as those being proposed for the Mid-Atlantic Highlands. While still applicable to smaller, less controversial installations, recommended methods should be simplified accordingly.

AESTHETIC IMPACTS

Aesthetics is often a primary reason for expressed concern about wind-energy projects (Figure 4-1). Unfortunately, few regulatory review processes adequately address aesthetic issues, and far fewer address the unique aesthetic issues associated with wind-energy projects in a rational manner. This section begins by describing some of the aesthetic issues associated with wind-energy projects. It then discusses existing methods for identifying visual resources and evaluating visual impacts in general, and it

FIGURE 4-1 View of Mountaineer Project from .5 mile. The project includes a total of 44 wind turbines.
SOURCE: Photograph by Jean Vissering.

provides recommendations for adapting those methods to the assessment of visual impacts associated with wind-energy projects. Finally, the section briefly examines the potential for developing guidelines to protect scenic resources when planning for, siting, and evaluating prospective wind-energy projects.

Visual impacts are the focus of this discussion of aesthetic impacts, but noise is considered to the extent that it is related to the overall character of a particular landscape. Noise and shadow flicker are discussed further in this chapter, under the section addressing potential impacts on human health and well-being associated with wind-energy projects.

Aesthetic Issues

The essence of aesthetics is that humans experience their surroundings with multiple senses. We often have a strong attachment to place and an inherent tendency to protect our "nest." Concern over changes in our personal landscapes is a universal phenomenon; it is not limited to the United States or to the present day. Public perceptions of wind-energy projects vary widely. To some, wind turbines appear visually pleasing, while others view them as intrusive industrial machines. Unlike some forms of development (e.g., cell towers), there are many people who find wind turbines to be beautiful. Nevertheless, even beautiful objects may not be desirable in one's current surroundings. Research has shown strong support for wind energy generally but substantially less support for projects close to one's home (Thayer and Hansen 1989; Wolsink 1990; Gipe 2002).

There are a number of reasons why proposed wind-energy projects evoke strong emotional reactions. Modern wind turbines are relatively new to the United States. Some of the early projects were built in remote areas, but increasingly, they are being built in or proposed for areas that are close to residential and recreational uses, and often in areas never before considered for industrial land uses. They must be sited where wind resources, transmission lines, and access exist; in some cases, particularly in the eastern United States, these sites are relatively high in elevation (e.g., mountain ridgelines) and highly visible. Some projects extend over fairly extensive land areas, though only small portions of the area are occupied by the turbines themselves. The turbines[1] often are taller than any local zoning ordinance ever envisioned, and they are impossible to screen from view. The movement of the blades makes it more likely that they will draw attention (Thayer and Hanson 1988; Gipe 2002).

Federal Aviation Administration obstruction lighting (pulsing red or white lights at night) is another aesthetic issue, and one that may result in some of the greatest aesthetic concerns (Hecklau 2005). In addition, wind turbines may produce noise, and the movement of the blades can result in shadow flicker from certain vantage points. Both the noise and the shadow flicker can be aesthetically troubling for some people who live nearby. While less concern has been raised about other project infrastructure such as meteorological towers, roads, power lines, and substations along with their associated site clearing and regrading, these can also result in negative visual impacts. Finally, a lack of regulatory guidance and stakeholder participation can contribute to fears of cumulative impacts if numerous projects are within a single viewshed.

Based on the few studies that have been conducted, it appears that despite low public acceptance during the project-proposal phase, acceptance levels generally have increased following construction (Thayer and Hanson 1989; Wolsink 1990; Palmer 1997). It is possible to find communities that identify their local wind projects as tourist attractions. Part of the positive image many people hold is linked to wind energy's "green image" and spe-

---

[1]Currently (late 2006), the most common commercial turbines being installed in the United States are 1.5 MW machines, usually 65-80 meters tall to the center of the rotor with rotor diameters of around 70 meters. The material in this chapter applies to turbines of this size. At several sites in the United States, 2.5 MW turbines are being used but are not yet in widespread use.

cifically to its potential for replacing CO_2-emitting electricity sources, with the hopeful prospect of reducing air pollution and global warming.

When evaluating the visual impacts of wind-energy projects, the essential question is not whether people will find them beautiful or not, but instead to what degree they may affect the important visual resources in the surrounding area. It is impossible to predict how any one individual will react to a wind-energy project. It is, however, possible to identify the visual character and scenic resources of a particular site and region. Evaluating the aesthetic impacts of wind-energy projects needs to focus on the relationship of the proposed project to the scenic landscape features of the site and its surrounding context. The factors that contribute to scenic quality can be identified and described with reasonable accuracy (Appleton 1975; Zube and Mills 1976; Litton 1979). This is especially true when viewing natural landscapes. Preferences are harder to predict for altered landscapes, although particular qualities of such landscapes have been identified in research of human preferences (Palmer 1983; Smardon et al. 1986). Nevertheless, we know enough to develop meaningful processes for reviewing aesthetic impacts. Despite the tremendous importance of a wind-energy project's aesthetic impacts, especially on nearby residents, this issue is too often inadequately addressed.

Current Information

There is a growing body of information concerning the aesthetic impacts of wind-energy projects. The National Wind Coordinating Committee (NWCC) provides general outlines of aesthetic issues and some examples of local ordinances addressing wind-energy projects. The latter are very basic and do not address the broader issues of protecting particular landscape values. More comprehensive are the Proceedings of the NWCC Siting Technical Meeting (December 2005), which cover a range of relevant topics and provide a useful bibliography. The visual issues are addressed at length by Pasqualetti et al. (2002). While providing an excellent overview, that book predates the use of modern 1.5-3 MW turbines. And while it provides excellent guidance for mitigating impacts, it does not address siting or landscape characteristics. Research on public perceptions of specific wind-energy projects is fairly common in Europe (both pre- and post-construction studies), but there are fewer examples in the United States (Stanton 2005). Of those in the United States, most are focused on western landscapes (Thayer and Hansen 1989), while few are focused on eastern landscapes, including wooded ridgelines. While such studies are useful in understanding public reactions generally, visual impacts are largely site-specific (Pasqualetti 2005). Other available resources include legal and regulatory guidelines for review of wind-energy projects. New York's State Environmental Quality Review Act (SEQRA) is one of the more explicit in the eastern United

States in terms of specifying what applicants need to submit and what will be considered (NYSDEC 2005; NYSERDA 2005a). Maine's Department of Environmental Protection adopted similar language in its environmental-review process (MEDEP 2003). In addition, there are several visual resource methods used for identifying scenic landscapes and for addressing visual impacts. Some important ones are discussed below.

Visual Assessment Methods

Two complementary approaches have been used to identify scenic resources and assess the impacts of proposed development projects. The first often is called a "professional approach" and relies on an individual or group with training in visual-resource and visual-impact assessment. These assessments rely on the research concerning human perceptions of landscapes (USFS 1979; Smardon et al. 1986) and on the adaptation of well-established methods for evaluating scenic landscape quality and for assessing visual impacts on particular landscapes. The second approach involves an assessment of public perceptions, attitudes, and values concerning a proposed project and its visual impacts on scenic resources. Landscapes are complex and imbued with cultural meaning that may not be understood by outside professionals. Techniques for assessing public perceptions, values, and attitudes include surveys, public meetings, interviews, and forums as well as examination of public documents identifying valued scenic resources (Smardon et al. 1986; Priestley 2006).

Among the best known and established methods for evaluating the scenic attributes of landscapes are the Visual Management System (USFS 1974) and the later Scenery Management System (USFS 1995) established by the U.S. Forest Service (USFS). Similarly, the U.S. Bureau of Land Management (BLM) uses a method called Visual Impact Assessment. The USFS and the BLM assessment methods have been used and adapted by numerous state and local agencies either for planning purposes (e.g., identifying scenic landscapes) or for assessing the impacts of proposed projects such as highways, ski areas, power plants, and forest harvesting (MADEM 1982; Smardon et al. 1986; RIDEM 1990).

While these methods are useful starting points, federal agencies such as the USFS usually go further in managing visual impacts on federal lands: they generally have plans in place that identify scenic values and set acceptable thresholds for alterations to the landscape. Even with detailed plans, these methods often fall short of providing meaningful guidance for evaluating the visual impacts of projects such as wind-energy facilities.

Most wind-energy projects are proposed on private land where there is far less guidance, especially with respect to evaluating aesthetic impacts. Many regulatory requirements adopted by states focus only on the tools for understanding the visibility of projects and fail to describe how visual impacts should be evaluated. In other words, most processes are not very successful in addressing questions of what landscape or project characteristics would make a project aesthetically unacceptable or the impacts "undue."

Below we outline a process for evaluating the conditions under which the aesthetic impacts of a proposed wind project might become unacceptable or "undue" in regulatory terms.

An Assessment Process for Evaluating the Visual Impacts of Wind-Energy Projects

The following steps summarize a process for moving from collecting measurable and observable information about visibility and landscape characteristics to analyzing the significance and importance of the visual resources involved and the effects of the proposed project on the landscape character and scenic resources of the surrounding area. Finally and most important, this process helps to inform the regulatory process about whether a proposed project is acceptable as designed, potentially acceptable with appropriate mitigation techniques, or unacceptable. The steps outlined below are described in greater detail in Appendix D.

Project Description

All site alterations that will have potential visual impacts must be identified by the developer in detail. These should include the turbine characteristics (height, rotor diameter, color, rated noise levels, proposed lighting) as well as the number of turbines and their locations; meteorological towers; roads; collector, distribution, and transmission lines; permanent and temporary storage "laydown" areas; substations; and any other structures associated with the project. In addition, all site clearings should be identified, including clearings for turbines, roads, power lines, substations, and laydown areas. All site regrading should be presented in sufficient detail to indicate the amount of cut and fill, locations, and clearing required. This information forms the basis for the visual assessment.

Project Visibility, Appearance, and Landscape Context

Viewshed mapping, photographic and virtual simulations, and field inventories of views are useful tools for determining with reasonable accuracy the visibility of the proposed project and for describing the characteristics of the views as well as identifying distinctive features within views (see Appendix D for more detail). Viewshed maps show areas of potential project visibility based on digital-elevation modeling. The modeling also can be used to determine the number of turbines that would be visible from a particular viewpoint. Actual visibility must be field-verified as trees, buildings, and other objects may restrict views. Field inventories also are necessary to document descriptive characteristics of the view. Inventories normally focus on areas of public use within a 10-mile radius of a project (Box 4-1). These include public roads, recreation areas, trails, wilderness and natural areas, historic sites, village centers, and other important scenic or cultural features identified in planning documents or in public meetings.

Photomontages or simulations provide critical project information for analysis. They should most usefully illustrate visually sensitive viewpoints and a range of perspectives and distances. They should also illustrate "worst-case" conditions to the greatest extent possible (clear weather and leaf-off conditions). Excellent software is available for creating simulations, but the technical requirements for accuracy should be clearly understood and specified (see Appendix D).

Identifying impacts from private residences can be more difficult without entering private property. Viewshed mapping can identify potential visibility. Geographic Information System (GIS) data generally provide additional information concerning existing vegetation and structures along with their primary use (residence, camp, or business). Providing regular notices to residents within a certain distance of the project can offer a means of learning more about visibility from private properties.

BOX 4-1
Area of Assessment: 10-Mile Radius

The size of the area for analysis may vary from location to location depending on the particular geography of the area and on the size of the project being proposed. Modern wind turbines of 1.5-3 MW can be seen in the landscape from 20 miles away or more (barring topographic or vegetative screening), but as one moves away from the project itself, the turbines appear smaller and smaller, and occupy an increasingly small part of the overall view. The most significant impacts are likely to occur within 3 miles of the project, with impacts possible from sensitive viewing areas up to 8 miles from the project. At 10 miles away the project is less likely to result in significant impacts unless it is located in or can be seen from a particularly sensitive site or the project is in an area that might be considered a regional focal point. Thus, a 10-mile radius provides a good basis for analysis including viewshed mapping and field assessment for current turbines. In some landscapes a 15-mile radius may be preferred if highly sensitive viewpoints occur at these distances, the overall scale of the project warrants a broader assessment, or if more than one project is proposed in an area. In the western United States, landscape scale and visibility may require a larger area of assessment.

Scenic Resource Values and Sensitivity Levels

Some landscapes are more visually sensitive than others due to such factors as numbers of viewers, viewer expectations, and identified scenic values. Processes exist for determining the relative visual quality of landscapes, the features that contribute to visual quality, and the sensitivity levels of particular landscape features and their uses. These are outlined in Appendix D and also can be found in methods used by the USFS Visual Management System (USFS 1974) and its later Scenery Management System (USFS 1995). Scenic resources values can also be determined in public planning documents and through public meetings.

Assessment of Visual Impacts

Visual impacts vary considerably depending on the particular characteristics of the project and its landscape context. Visibility of a project is only one of many variables that should be examined. Significant visual impacts generally arise because of the combination of many factors such as proximity of views, sensitivity of views, duration of views, the presence of scenic resources of statewide or national significance, and the scale of the project in relation to its setting (see Appendix D). Some examples of potentially significant impacts might include the following:

• The project is located within a scenic context and is viewed in close proximity, for an extended duration (e.g., broad area or linear miles) from a highly sensitive use area, especially one for which the enjoyment of natural scenery is important, and that is an identified resource of statewide or national significance.
• The project is located on a landform that is an important focal point that is highly visible throughout the region.
• The project is of a scale that would dominate views throughout a region (or 10-mile assessment area) so that few other scenic natural views would be possible without including turbines.

Mitigation Techniques

A well-designed project will incorporate a number of techniques into the planning and design of the project to minimize visual impacts, including sensitive siting and ensuring that project infrastructure is well screened from view. Establishing "Best Practice" Guidelines can help ensure that minimum standards are met before project permit applications are submitted. Nevertheless, a thorough review by interested parties may result in further adjustments. If the visual impacts are deemed unacceptable, additional mitigation techniques can be explored (see Appendix D). In some cases, however, mitigation techniques may not solve inherent concerns, and the project may be found to have "undue aesthetic impacts."

Determination of Unacceptable or Undue Aesthetic Impacts

Guidance on when projects may be found unacceptable tends to be lacking or inadequate in many review processes. The information gathered in the above process can inform this decision by providing a detailed understanding of the particular issues involved in the visual relationship between the project and its surrounding context. Appendix D provides questions that could help determine the degree of visual impact.

Among the factors to consider are:

• Has the applicant provided sufficient information with which to make a decision? These would include detailed information about the visibility of the proposed project and simulations (photomontages) from sensitive viewing areas. New York's SEQRA process offers an example of clearly identifying the information required and the mitigation measures that need to be considered.

• Are scenic resources of local, statewide, or national significance located on or near the project site? Is the surrounding landscape unique in any way? What landscape characteristics are important to the experience and visual integrity of these scenic features?

• Would these scenic resources be significantly degraded by the construction of the proposed project?

• Would the scale of the project interfere with the general enjoyment of scenic landscape features throughout the region? Would the project appear as a dominant feature throughout the region or study area?

• Has the applicant employed reasonable mitigation measures in the overall design and layout of the proposed project so that it fits reasonably well into the character of the area?

• Would the project violate a clear, written community standard intended to protect the scenic or natural beauty of the area? Such standards can be developed at the community, county, region, or state level.

Guidelines for Protecting Scenic Resources

Planning and Siting Guidelines

Siting guidelines that prospectively identify suitable and unsuitable locations for wind-energy projects have been considered in many regions. Problems with such guidelines arise, however. Each site is visually different, local attitudes toward wind-energy development vary, and a wind developer must grapple with several non-aesthetic factors in locating a potentially developable site (e.g., willing property lessors, adequate wind resources, access to transmission lines, and a market for the electricity generated). Several combined approaches may be the most feasible. As discussed in more detail in Chapter 5, they would include the following:

- State and regional guidance providing criteria concerning site conditions that may be inherently suited or unsuited to wind development due to particular scenic values, and/or sensitivity levels that would raise concerns requiring additional detailed study. Policies regarding aesthetic conditions and wind development on state-owned lands would also be appropriate.
- Local and state planning documents that identify valuable scenic, recreational, and cultural assets. Defining particular landscape attributes or other public values that contribute to the resources is helpful when making decisions concerning proposed landscape development proposals.[2] In addition, insofar as a "comprehensive plan" is voted on by the local governing body, the plan may provide guidance to a developer as an expression of the will of the community.
- Statewide policies that address the relationship between the development of wind energy and the protection of valuable scenic resources.

Guidelines for Evaluating Cumulative Aesthetic Impacts

While wind-energy development is relatively new in the United States, the potential for cumulative aesthetic impacts resulting either from several new projects in a particular region or from expansion of existing projects is likely to become an issue that may need to be addressed at local, regional, and state levels. The following questions could help to evaluate the potential for undue cumulative aesthetic impacts:

---

[2]Clear and reasonably objective guidance is more useful than vague statements such as "the ridgelines in our town are valuable to our rural character and no development is allowed." A statement that identifies the resource(s), its particular valued attributes, and appropriate and inappropriate development characteristics provides a clear written community standard. Statements that exclude wind development are generally not appropriate unless clear reasons are provided for this exclusion. For example, "the Town of Jonesville is characterized by the Green Range, which is composed of numerous hills and ridges. Several of the hills stand out because of their distinct shapes, including Mount Grant, Morris Mountain, and Jones Peak. Mount Grant is also valued for a popular hiking trail and the spectacular views looking west…" Such statements provide helpful guidance in decision making. In other words, a project located on another ridge but out of the view from the summit of Mount Grant might be acceptable, whereas a wind project located on Mount Grant probably would not be.

- Are projects at scales appropriate to the landscape context?
- Are turbine types and sizes uniform within the wind resource area and over time?
- How great is the offsite visibility of infrastructure?
- Have areas that are inappropriate for wind projects due to terrain or important scenic, cultural, or recreational values been identified and described?
- If the project is built as proposed, would each region retain undeveloped scenic vistas?
- Would any one region be unduly burdened with wind-energy projects?

Considerations for Improving the Evaluation of Aesthetics and Implementation of Projects

- Accurate and detailed information about the visual appearance of all aspects of a proposed project is extremely important. Incomplete or inaccurate information often results in public mistrust.
- Generally, an area of 10 miles surrounding the project site is adequate for viewshed mapping and field assessment for turbines of a size currently used in the United States. In some landscapes, a 15- to 20-mile radius may be preferred, especially if highly sensitive viewpoints occur at these distances, the overall scale of the project warrants a broader assessment, or more than one project is proposed in an area.
- In evaluating the aesthetic impacts of wind-energy projects, the discussion should focus not on whether people find wind-energy projects attractive but on the characteristics of the landscapes in which the projects will be located; the particular landscape features that contribute to scenic quality; the relative sensitivity of viewing areas; and the degree of degradation that would result to valued scenic resources, especially documented scenic values.
- Computerized viewshed analyses provide useful information about potential project visibility but are best used as the basis for conducting field investigations. Within forested areas, views are likely to be minimal at best. The software allows more detailed analysis of numbers of turbines that can be seen from any one point.
- Photomontages and photo simulations are essential tools in understanding project visibility, and appearance. Accurate representations involve exact technical requirements, such as precise camera focal lengths, Global Positioning System records of the photo location, and digital elevation (GIS-based) software. The technologies are changing, and it is important that simulations are accurately constructed (Stanton 2005). Local planning boards and the general public should be consulted in determining photo-

montage locations. They should illustrate sensitive or scenic viewpoints as well as "worst-case" situations such good weather conditions and the most scenic perspectives.

•   An independent assessment of visual impacts by trained professionals can provide more unbiased information than assessments provided on behalf of either developers or other interested and affected parties, and can provide useful comparisons with those assessments.

•   Meaningful public involvement is essential, and standards for providing information and opportunities for involvement can be helpful (see also Chapter 5).

•   Equally important are perceptions of clear benefits from wind-energy projects. Aesthetic perceptions are linked to our sense of general well-being. This has to do both with financial or material benefits (contributions to local taxes, payments for use of property, offsets such as protection of open space) and with making a real difference in terms of reducing pollution and CO_2 levels (Damborg 2002).

•   Towns, counties, regions, and states can provide helpful guidance to developers and decision makers by identifying landscape resources of value. This process is particularly useful when it is part of formally adopted documents such as comprehensive land-use plans, but it can also be used for developing guidelines.

•   Wind-energy projects will not necessarily conflict with areas of moderate to high scenic quality, and may even appear more attractive in these settings. Problems can arise when the setting is an important regional focal point, or when a project will be seen close to highly sensitive viewing areas where a natural or intact landscape is important.

•   The potential for cumulative impacts either from the location of several projects within a region, or from future expansions of existing projects, could become a problem. Cumulative impacts cannot be addressed at the project or local scale, and so a regional or statewide perspective is needed.

•   Scale is relative. The apparent size of a wind turbine in relation to its surrounding is most relevant. Despite their large sizes, modern wind turbines can fit well in many landscapes. Vertical scale is likely to be an issue primarily if the turbines appear to overwhelm an important ridgeline, focal point, or cultural feature that appears diminished in prominence due to the relative height of the turbines.

•   The number of turbines or horizontal scale of wind projects will be an important determination of reasonable fit within a region. A project that dominates views throughout a region is more likely to have aesthetic impacts judged unacceptable than one that permits other scenic or natural views to remain unimpaired throughout the region. If residences, especially those not directly benefiting from a proposed project, are surrounded by wind turbines, adverse aesthetic impacts are likely to be reported.

- Visual clutter often is adversely perceived and commonly results from the combination of human-made elements in close association that are of differing shapes, colors, forms, patterns, or scales. Generally simple and uniform arrays or groupings of wind turbines are more visually appealing than mixed types and sizes. Screening of associated infrastructure also is important in reducing visual clutter.
- Turbines with rotating blades have been shown to be more visually appealing than those that are still. Maintenance or removal of poorly functioning turbines can be important.
- Turbine noise usually is most critical within a half-mile of a project. Efforts to reduce potential noise impacts on nearby residents therefore may be most important within that distance.
- Decommissioning wind-energy projects appropriately would be considered in initial permit approvals. While some wind-energy projects may have longer life spans than originally anticipated, provisions are needed for removal of site structures that no longer contribute to the project, and for site restoration. Funding provided in escrow for decommissioning is sometimes essential.
- Obstruction lighting required on objects more than 200 feet tall often is an extremely important aesthetic concern. Eliminating or reducing major lighting impacts merits a high priority.

CULTURAL IMPACTS

Recreation

Wind-energy facilities create both positive and negative recreational impacts. On the positive side, many wind-energy projects are listed as tourist sights: some offer tours or provide information areas about the facility and wind energy in general; and several are considering incorporating visitor centers. Some developers allow open access to project sites that may provide additional opportunities for hunting, hiking, snowmobiling, and other activities.

There are two types of potential negative impacts on recreational opportunities: direct and indirect. Direct impacts can result when existing recreational activities are either precluded or require rerouting around a wind-energy facility. Indirect impacts include aesthetic impacts (addressed above) that may affect the recreational experience. These impacts can occur when scenic or natural values are critical to the recreational experience.

Most wind projects to date have been located on or proposed for private land. Policies vary regarding public use around wind turbines on both private and public lands. At project sites, access roads are often gated to prevent public access along roads, but projects are not usually fenced from public use, although signage may discourage use.

Evaluating Recreational Impacts

• In most cases, recreational uses will be identified in state and local documents and often on maps, although there may be times when recreational uses are only locally known. Some developers conduct recreation surveys to determine recreational uses in the study area and attitudes of users toward the development of wind-energy projects. Recreational concerns and interests are often identified in informal meetings and at public hearings. The USFS ranks recreational facilities as shown in Table 4-1. This provides an example that may need to be adapted by states or local communities in evaluating the impacts of wind-energy facilities.

Most aesthetic and recreational-assessment methods identify relative "sensitivity levels" of recreational uses related to factors such as the amount of use and the expectations of users. A high sensitivity level does not necessarily mean that a wind-energy facility should not be visible, but instead is an indication that further study is needed. The USFS defines the following levels for evaluating impacts on USFS recreational experiences:

• Sensitivity Level 1 areas (highly sensitive areas) include all areas seen from primary travel routes, use areas, and water bodies where a minimum of one-fourth of the forest visitors have a major concern for the scenic qualities. Areas specifically considered to be highly sensitive include roads providing access to highly sensitive recreation sites (i.e., sites where a natural environment, non-motorized use, and quiet are characteristic); National Scenic or Recreation Trails; heavily used seasonal trails through areas recognized as scenic attractions; significant recreational streams; water bodies with heavy fishing, boating, swimming, and other uses highly dependent on viewing scenery; wilderness and primitive areas; and observation sites along highly sensitive travelways.

TABLE 4-1 U.S. Forest Service Recreational Facilities Rankings

Primary Use Areas/Travel Routes	Secondary Use Areas/Travel Routes
National importance	Local importance
High use volume	Low use volume
Long use duration	Short use duration
Large size	Small size

SOURCE: Adapted from Visual Management System (USFS 1974) and the later Scenery Management System (USFS 1995).

• Sensitivity Level 2 areas ("moderately sensitive locations") include roads and trails on National Forest recreation maps that are not Level 1 or Level 3 and water bodies receiving low to moderate use.

• Sensitivity Level 3 areas (least sensitive areas) include travelways constructed primarily for non-recreation purposes such as timber access roads and utility line clearings, and areas where uses primarily depend little on scenic viewing (e.g., hunting or gathering fuel wood, Christmas trees, or berries).

Historic, Sacred, and Archeological Sites

In analyzing impacts on historic, sacred, and archeological sites, the primary concern is that no permanent harm should be done that would affect the integrity of the site. Archeological inventories are generally required in most states before construction can begin. Some Native American tribes have sacred sites that may not be known to outsiders. Direct impacts (actual removal or physical harm) to historic, sacred, or archeological sites can be easily avoided in most instances.

Some states and localities have designated certain landscapes as having particular historical significance. For example, a proposed wind project in Otsego County, New York, that would have been located within the Lindesay Patent Historic District was later withdrawn.[3] Designation of a historic district provides a reasonable indication of historic value, uniqueness, and public concern for the resource. Whether or not a wind-energy project would damage the resource may depend on the specific nature of the historic resources involved.

The indirect effects on the experience of a historic or sacred site or area resulting from either seeing or hearing a wind-energy project nearby are not as well documented. Most historic sites are assumed to be part of evolving landscape contexts. Concerns generally would arise only when specific aesthetic or landscape attributes of the surrounding area are identified in the documentation of the site's historic value. A setting where a multisensory experience has been re-created, such as at Plimoth Plantation in Massachusetts, might also warrant consideration. There, the visitor expects not just to see pre-revolutionary structures but to actually experience life at the time of the early settlers. A recent and currently unresolved case in Vermont concerned a historic Civilian Conservation Corps bath-house that was documented as having been sited to take advantage of scenic views down a lake where a proposed wind-energy facility would be visible. Unlike

---

[3]The proposed Global Harvest Wind Project was later withdrawn. Another currently proposed project would be visible from sensitive resources within a historic district, but a determination on that project has yet to be made.

housing developments, wind-energy projects cannot be screened from view, except behind intervening topography and vegetation. Such issues are likely to arise as wind projects are proposed in cultural landscapes, and guidance as to what constitutes an undue impact to historic or sacred sites and areas will be necessary.

Evaluating Impacts on Historic, Sacred, and Archeological Sites

Historic, sacred, and archeological sites and settings must be regarded as sensitive sites. In most states, key historic sites are well documented and rated regarding their local, state, or national significance. State Offices of Historic Preservation, along with local historical societies, provide detailed information on historic sites and properties, and usually are involved in the review of proposed wind-energy projects. State archaeologists generally recommend specific guidelines for archaeological surveys, depending on the site involved. Archeological and sacred sites may be less well known. Documentation of these sites is essential. Good descriptive documentation will identify the particular values involved and the extent to which the context or setting contributes to the structure or landscape and in what way. Generally, the documentation of historic sites offers useful guidance to the value of the surrounding landscape to the interpretation of the resource, although the final determination probably should be done by experts. Most states are only now beginning to develop methods for reviewing onsite and offsite impacts of wind-energy facilities on historic sites (e.g., Vermont Division for Historic Preservation 2007). Siting wind-energy projects in the vicinity of identified and documented historic or sacred landscapes as well as historic, sacred, and archeological sites is likely to "raise red flags." The impacts of viewing wind facilities from historic or sacred landscapes will require similar kinds of analyses to those noted in Appendix D for aesthetic impacts; however, additional guidance from relevant experts is needed in this area.

IMPACTS ON HUMAN HEALTH AND WELL-BEING

Wind-energy projects can have positive as well as negative impacts on human health and well-being. The positive impacts accrue mainly through improvements in air quality, as discussed previously in this report. These positive impacts (i.e., benefits) to health and well-being are diffuse; they are experienced by people living in areas where conventional methods of electricity generation are used less because wind energy can be substituted in the regional market.

In contrast, to the extent that wind-energy projects create negative impacts on human health and well-being, the impacts are experienced mainly by people living near wind turbines who are affected by noise and shadow flicker.

Noise

As with any machine involving moving parts, wind turbines generate noise during operation. Noise from wind turbines arises mainly from two sources: (1) mechanical noise caused by the gearbox and generator; and (2) aerodynamic noise caused by interaction of the turbine blades with the wind. As described below (see "Noise Levels"), noise of greatest concern can be generally classified as being of one of these three types: broadband, tonal, and low-frequency.

The perception of noise depends in part on the individual—on a person's hearing acuity and upon his or her subjective tolerance for or dislike of a particular type of noise. For example, a persistent "whoosh" might be a soothing sound to some people even as it annoys others. Nevertheless, it appears that subjective impressions of the noise from wind turbines are not totally idiosyncratic. A 1999 study (Kragh et al. 1999) included a laboratory technique for assessing the subjective unpleasantness of wind-turbine noise. Preliminary findings indicated that *noise tonality* and *noise-fluctuation strength* were the parameters best correlated with unpleasantness (Kragh et al. 1999).

Broadband, tonal, and low-frequency noise have all been addressed to some degree in modern upwind horizontal wind turbines, and turbine technologies continue to improve in this regard. With regard to the design of a wind-energy project, one is generally interested in assessing whether the additional noise generated by the wind turbines (relative to the ambient noise) might cause annoyance or a hazard to human health and well-being.

Noise impacts also can result from project construction and maintenance. These are generally of relatively short duration and occurrence but can include equipment operation, blasting, and noise associated with traffic into and out of the facility. These are not addressed in detail in this section. In the following, a brief review of wind-turbine noise and its impacts is presented along with suggested methods for assessing such impacts and mitigation measures.

Noise Levels

Noise from wind turbines, at the location of a receptor, is described in terms of sound pressure levels (relative to a reference value, typically 2×10^{-5} Pa) and is typically expressed in dB(A), decibels corrected or A-weighted for sensitivity of the human ear. Note that there is a difference between sound *power* used to describe the source of sound and sound *pres-*

sure used to describe the effect on a receptor. The sound power level from a single turbine is usually around 90-105 dB(A); such a turbine creates a sound pressure of 50-60 dB(A) at a distance of 40 meters (this is about the same level as conversational speech). Noise (sound-pressure) levels from an onshore wind project are typically in the 35-45 dB(A) range at a distance of about 300 meters (BWEA 2000; Burton et al. 2001). These are relatively low noise or sound-pressure levels compared with other common sources such as a busy office (~60 dB(A)), and with nighttime ambient noise levels in the countryside (~20-40 dB(A)). While turbine noise increases with wind speed, ambient noises—for example, due to the rustling of tree leaves— increase at a higher rate and can mask the turbine noise (BLM 2005a).

In addition to the amplitude of the noise emitted from turbines, its frequency content is also important, as human perception of sounds is different at different frequencies. Broadband noise from a wind turbine typically is a "swishing" or "whooshing" sound resulting from a continuous distribution of sound pressures with frequencies above 100 Hz. Tonal noise typically is a "hum" or "pitch" occurring at distinct frequencies. Low-frequency noise (with frequencies below 100 Hz) includes "infrasound," which is inaudible or barely audible sound at frequencies below 20 Hz.

Mechanical sounds from a turbine are emitted at "tonal" frequencies associated with the rotating machinery, while aerodynamic sounds are typically broadband in character. Mechanical noise is generated from rotating components in the nacelle, including the generator and gearbox, and to a lesser extent, cooling fans, pumps, compressors, and the yaw system. Aerodynamic noise, produced by the flow of air over blades, is created by blades interacting with eddies created by atmospheric inflow turbulence. This broadband aerodynamic noise is generally the dominant type of wind-turbine noise, and it generally increases with tip speed. Both mechanical and aerodynamic noise often are loud enough to be heard by people.

With older downwind turbines, some infrasound also is emitted each time a rotor blade interacts with the disturbed wind behind the tower, but it is believed that the energy at these low frequencies is insufficient to pose a health hazard (BWEA 2005). Nevertheless, a recent study by van den Berg (2004, 2006) suggests that, especially at night during stable atmospheric conditions, low-frequency modulation (at around 4 Hz) of higher frequency swishing sounds is possible. Note that this is not infrasound, but van den Berg (2006) states that it is not known to what degree this modulated fluctuating sound causes annoyance and deterioration in sleep quality to people living nearby.

Low-frequency vibration and its effects on humans are not well understood. Sensitivity to such vibration resulting from wind-turbine noise is highly variable among humans. Although there are opposing views on the subject, it has recently been stated (Pierpont 2006) that "some people feel disturbing amounts of vibration or pulsation from wind turbines, and can count in their bodies, especially their chests, the beats of the blades passing the towers, even when they can't hear or see them." More needs to be understood regarding the effects of low-frequency noise on humans.

Assessment

Guidelines for measuring noise produced by wind turbines are provided in the standard, IEC 61400-11: Acoustic Noise Measurement Techniques for Wind Turbines (IEC 2002), which specifies the instrumentation, methods, and locations for noise measurements. Wind-energy developers are required to meet local standards for acceptable sound levels; for example, in Germany, this level is 35 dB(A) for rural nighttime environments. Noise levels in the vicinity of wind-energy projects can be estimated during the design phase using available computational models (DWEA 2003a). Generally, noise levels are only computed at low wind speeds (7-8 m/s), because at higher speeds, noise produced by turbines can be (but is not always) masked by ambient noise.

Noise-emission measurements potentially are subject to problems, however. A 1999 study involving noise-measurement laboratories from seven European countries found, in measuring noise emission from the same 500 kW wind turbine on a flat terrain, that while apparent sound power levels and wind speed dependence could be measured reasonably reliably, tonality measurements were much more variable (Kragh et al. 1999). In addition, methods for assessing noise levels produced by wind turbines located in various terrains, such as mountainous regions, need further development.

Mitigation Measures and Standards

Noise produced by wind turbines generally is not a major concern for humans beyond a half-mile or so because various measures to reduce noise have been implemented in the design of modern turbines. The mechanical sound emanating from rotating machinery can be controlled by sound-isolating techniques. Furthermore, different types of wind turbines have different noise characteristics. As mentioned earlier, modern upwind turbines are less noisy than downwind turbines. Variable-speed turbines (where rotor speeds are lower at low wind speeds) create less noise at lower wind speeds when ambient noise is also low, compared with constant-speed turbines. Direct-drive machines, which have no gearbox or high-speed mechanical components, are much quieter.

Acceptability standards for noise vary by nation, state, and locality. They can also vary depending on time of day—nighttime standards are generally stricter. In the United States, the U.S. Environmental Protection Agency only provides noise guidelines. Many state governments issue their own regulations (e.g., Oregon Department of Environmental Quality

2006), and local governments often enact noise ordinances. Standards of acceptability need to be understood in the context of ambient (background) noise resulting from all other nearby and distant sources.

Shadow Flicker

As the blades of a wind turbine rotate in sunny conditions, they cast moving shadows on the ground resulting in alternating changes in light intensity. This phenomenon is termed shadow flicker. Shadow flicker is different from a related strobe-like phenomenon that is caused by intermittent chopping of the sunlight behind the rotating blades. Shadow flicker intensity is defined as the difference or variation in brightness at a given location in the presence and absence of a shadow. Shadow flicker can be a nuisance to nearby humans, and its effects need to be considered during the design of a wind-energy project.

In the United States, shadow flicker has not been identified as causing even a mild annoyance. In Northern Europe, on the other hand, because of the higher latitude and the lower angle of the sun, especially in winter, shadow flicker can be a problem of concern.

Assessment

Shadow flicker is a function of several factors, including the location of people relative to the turbine, the wind speed and direction, the diurnal variation of sunlight, the geographic latitude of the location, the local topography, and the presence of any obstructions (Nielsen 2003). Shadow flicker is not important at distant sites (for example, greater than 1,000 feet from a turbine) except during the morning and evening when shadows are long. However, sunlight intensity is also lower during the morning and evening; this tends to reduce the effects of shadows and shadow flicker. The speed of shadow flicker increases with wind-turbine rotor speed.

Shadow flicker may be analytically modeled, and several software packages are commercially available for this purpose (e.g., WindPro and GH WindFarmer). An online tool for simple shadow calculations for flat topography is also available (DWEA 2003b). These software packages generally provide conservative results as they typically ignore the numerous influencing factors listed above and only consider a worst-case scenario (i.e., no shadow or full shadow). Inputs to a shadow-flicker model in WindPro, for example, include a description of the turbine and site, the topography, the joint wind speed and wind direction distribution, and an average or distribution of sunshine hours. Typical output results include the number of shadow-hours per year; these are often represented by iso-lines or contours of equal annual shadow-hours on a topographical map. Daily and annual shadow variations may also be a part of the result (DWEA 2003b). A typical result might indicate, for example, that a house 300 meters from a 600 kW wind turbine with a rotor diameter of 40 meters will be exposed to moving shadows for approximately 17-18 hours annually, out of a total of 8,760 hours in a year (Andersen 1999).

Impacts

Shadow flicker can be a nuisance to people living near a wind-energy project. It is sometimes difficult to work in a dwelling if there is shadow flicker on a window. In addition to its intensity, the frequency of the shadow flicker is of importance. Flicker frequency due to a turbine is on the order of the rotor frequency (i.e., 0.6-1.0 Hz), which is harmless to humans. According to the Epilepsy Foundation, only frequencies above 10 Hz are likely to cause epileptic seizures. (For reference, frequencies of strobe lights used in discotheques are higher than 3 Hz but lower than 10 Hz.) If a turbine is close to a highway, the movement of the large rotor blades and possible resulting flicker can distract drivers. Irish guidelines, for example, recommend that turbines be set back from the road at least 300 meters (MSU 2004).

Mitigation Measures

Shadow flicker is not explicitly regulated. When a maximum number of hours of allowed shadow flicker per year is imposed for a neighbor's property (such as 30 hours/year for one wind-energy project in Germany), this number refers to those hours when the property is actually used by the people there and when they are awake. Denmark has no legislation regarding shadow flicker, but it is generally recommended that there be no more than 10 hours per year when flicker is experienced.

Even in the worst situations, shadow flicker only lasts for a short time each day—rarely more than half an hour. Moreover, flicker is observed only for a few weeks in the winter season. To avoid even limited periods of shadow flicker, a possible solution is to not run the turbines during this time. Obviously, another solution is to site the turbines such that their shadow paths avoid nearby residences.

Since tools for estimation of shadow flicker are readily available, such calculations are routinely done while planning a wind-energy project. One such study was performed for the Wild Horse project in the state of Washington (Nielsen 2003). Using results presented in the form of shadow flicker maps and distributions, one can determine suitable locations for wind turbines. Recently, tools have become available (GH WindFarmer) that not only compute shadow flicker in real time during turbine operation, but also convey information to the turbine control system to enable shutdown if the shadow flicker at a particular location becomes particularly problematic. However, the committee is unaware if such real-time systems have been implemented at any specific wind-energy project.

LOCAL ECONOMIC AND FISCAL IMPACTS

Wind-energy projects can have a range of economic and fiscal impacts, both positive and negative. Some of those impacts are experienced at the national or regional level, as discussed in Chapter 2. These involve, for example, tax credits and other monetary incentives to encourage wind-energy production, as well as effects of wind energy on regional energy pricing. In this section, the focus is on the local level: on private economic impacts, positive and negative, as well as on public revenues and costs.

Lease and Easement Arrangements

As discussed in Chapter 5, most of the onshore wind-energy projects in the United States have been sited on private land. Typically, the developer of a wind-energy project acquires rights to the use of land through negotiations with landowners. Rarely is land purchased in fee simple; instead, the developer purchases leases or easements for a specified duration. While a uniform offer may be made to landowners, contract prices may be negotiated individually and privately. The power of eminent domain is not available to non-government wind-energy developers.

Assessment

According to the American Wind Energy Association (AWEA 2006f), leasing arrangements can vary greatly, but a reasonable estimate for a lease payment to a landowner from a single utility-scale turbine is currently about $3,000 per year. Lease and easement arrangements can be a financial boon to landowners, providing a steady albeit modest income, but only if the financial and other contractual terms are fair.

A number of guides are now available for landowners who are considering either lease arrangements or granting easements for wind-energy projects. Some of these, such as the guides of the Wind Easement Work Group of Windustry, located in Minnesota, have been prepared by collaborations of wind-energy industry, government, and other partners (Nardi and Daniels 2005a). This work group has provided extensive guidance addressing such questions as:

- How much of my land will be tied up and for how long?
- What land rights am I giving up? What activities can I continue?

- How much will I be paid and how will I receive payments?
- Are the proposed payments adequate now and will they be adequate in the future?
- Does the proposed method of payment or the agreement itself present adverse tax consequences to me?
- Are there firm plans to develop my land, or is the developer just trying to tie it up?
- If payments are to be based on revenues generated by the wind turbines, how much information is the developer willing to disclose concerning how the owners' revenue will be determined?
- What rights is the developer able to later sell or transfer without my consent?
- Does the developer have adequate liability insurance?
- What are the developer's termination rights?
- What are my termination rights?
- If the agreement is terminated either voluntarily or involuntarily, what happens to the wind-energy structures and related facilities on my land?

Policies to Protect the Parties Involved

In a companion document, Windustry's Wind Easement Work Group issued a short set of best practices and policy recommendations regarding easements and leases (Nardi and Daniels 2005b). These included:

- *Public disclosure of energy production from wind turbines:* In order to facilitate transparency for production-based payments, increase public knowledge about the wind resource, and provide information to the state on the economic contribution of wind power.
- *Public filing of lease documents and public disclosure of terms (or include a "no gag" clause):* In order to reduce competition among neighbors, encourage developers to give fair deals, and lower the possibility of a single holdout among landowners.
- *Limiting easement periods to 30 years and option periods to 5 years:* To avoid tying up either the landowners or the developer for unduly long periods.

Property Values

It has been claimed that wind-energy projects do not adversely affect property values (Associated Press 2006). In contrast, it has been asserted that "adverse impacts on environmental, scenic and property values are often overlooked" (Schleede 2003, p. 1).

It is difficult to generalize about the effects of wind-energy projects on property values. A 2003 Renewable Energy Policy Project (REPP) study of the effect of wind development on property values found no statistical effects of changes in property values over time from wind-energy projects (Sterzinger et al. 2003). This study examined changes in property values within 5 miles of 10 wind-energy projects that came online between 1998 and 2001, looking at the three-year period before and after each project came online and using a simple linear-regression analysis. The study found no major pre-post differences, and it also found no major differences when property-value changes in the 5-mile areas around the wind-energy projects were compared with selected "comparable communities."

The REPP study, however, examined only *average* price changes. The authors noted that "it would be desirable in future studies to expand the variables incorporated into the analysis and to refine the view shed in order to look at the relationship between property values and the precise distance from development" (Sterzinger et al. 2003, p. 3). A 2006 study (Hoen 2006) more closely examined the effects on property values between 1996 and 2005 within 5 miles of a 20-turbine, 30-MW project in Madison County, New York. This study used a hedonic regression analysis method and found no measurable effects on property values, positive or negative, even on residences within a mile of the facility. In contrast, a 2005 analysis by the Power Plant Research Program of the Maryland Department of Natural Resources concerning a proposed wind energy facility—the Roth Rock facility in Garrett County, Maryland—concluded that the facility would have an uncertain impact on the property values of neighboring properties. It reached this conclusion after reviewing the 2003 REEP study as well as a 2004 study in the United Kingdom by the Royal Institution of Chartered Surveyors (RICS), which found negative impacts, especially on non-farm properties (RICS 2004), and after analyzing the property-value impacts of the Allegheny Heights (Clipper) wind-energy project located north of the Roth Rock project and permitted in 2003 (MDDNR 2006).

Property values are affected by many variables. Thus, empirically isolating the impacts of one variable (a wind-energy project) is extremely difficult unless one or more turbines are located close to a specific property, and even then, there may be confounding factors. Forecasts of property values in prospective host areas that are based on comparisons with existing host areas are of questionable validity, especially if there are significant differences between the areas.

Assessment

Despite the difficulty of reaching widely generalizable conclusions about the effects of wind-energy projects on property values, it is possible to theorize about important variables. The discussion of aesthetic impacts earlier in this chapter is relevant. On the one hand, to the extent that a property is valuable for a purpose incompatible with wind-energy projects, such as to experience life in a remote and relatively untouched area, a view that includes a wind-energy project—especially one with many turbines—may detract from property values. On the other hand, to the extent that the wind-energy project contributes to the prosperity of an area, it may help to bring in amenities and so may enhance property values.

Because wind installations in the United States are a relatively recent phenomenon and are only now beginning to burgeon, the long-term effects of wind-energy projects on property values also are difficult to assess. While property values may be initially affected by a wind-energy project, the effect may diminish as the project becomes an accepted part of the landscape. On the other hand, the effects on local and regional property values of a few projects with 20 to 50 turbines may be quite different from the effects of numerous projects with 100 to 200 turbines.

Mitigation Measures

When siting facilities that provide a public benefit but may be undesirable as neighbors, one mitigation measure that has been explored, for example, with waste facilities, is to provide property-value guarantees to property owners within a specified distance from the facility when they want to sell their properties (Zeiss and Atwater 1989; Smith and Kunreuther 2001). An issue in this arrangement is the fair level of the guaranteed selling price, as adjusted over time by an inflation factor.

Employment and Secondary Economic Effects

A wind-energy project is a source of jobs throughout its life cycle: for parts manufacturers and for researchers seeking to improve wind-turbine performance; for workers who transport and construct wind turbines and related infrastructure; for workers employed in the operation and maintenance of turbines, transmission lines, etc.; and for workers involved in project decommissioning. The number, skill and pay level, and location of the jobs will vary depending upon the scale, location, and stage of the project. Some of the jobs may be in the area that will host the wind turbines; some may be in a manufacturing plant several states away. At all locations, in addition to direct employment impacts, employment may be indirectly fostered through secondary economic effects, including indirect impacts (e.g., changes in inter-industry purchasing patterns) and induced impacts (e.g., changes in household spending patterns).

In addition, however, it is conceivable that a wind-energy project will have some adverse impacts on the economy of its host area. While it has been argued that wind-energy facilities can be a tourist attraction (AusWEA 2004), it also has been argued that wind projects are seen by people as undesirable in national forest areas (Grady 2004) and can damage tourism in areas of high scenic beauty (Schleede 2003). It is also possible that, while one or a few wind-energy facilities may be a tourist attraction, a proliferation may have the reverse effect.

Assessment

According to the AWEA's "Wind Energy and Economic Development: Building Sustainable Jobs and Communities," the European Wind Energy Association has estimated that in total, every MW of installed wind capacity directly and indirectly creates about 60 person-years of employment and 15 to 19 jobs. The fact sheet notes that the rate of job creation will decline as the industry grows and is able to take advantage of economies of scale (AWEA 2006f).

Of greatest interest at the local level, however, are not these totals but rather the jobs that become available to local or regional workers because of a wind-energy project in their vicinity. These jobs are likely to involve site preparation and facility construction during the project start-up period; skeleton crews for facility, grounds, and transmission line maintenance during facility operation typically about 20 years; and crews to perform decommissioning and site restoration work when the facility is closed.

The size of crews will vary depending upon the project scale, site characteristics, etc., but estimates of the number of employees, pay scales, skill requirements, and duration of employment can be made with reasonable accuracy. The secondary effects of wind-energy projects on the economy (both positive and negative) are much harder to estimate. On the one hand, a wind-energy project may increase the need for service sector businesses and jobs (gas stations, motels, restaurants, etc.). On the other hand, it may deter economic growth that would otherwise occur in the area (e.g., second homes, recreational facilities, and related amenities).

To estimate the secondary effects of a wind-energy project on a region's economy, the region first must be geographically defined. Changes in its economic activity generally are then measured in terms of changes in either (1) employment, including part-time and seasonal employment; (2) regional income, i.e., the sum of worker wages and salaries plus business income and profits; or (3) changes in sales or spending. A regional economic multiplier may be used to estimate the secondary economic effects of new money flowing into the region. In conducting the impact analysis, the aim is to estimate the changes that would occur if the project is built versus if it is not built (not just the before/after changes).

While regional economic models have been available for some time, they generally are not well suited to assessing the secondary economic impacts of a single project on a small region or area. Recently, however, an economic model was developed specifically to estimate the economic benefits from a new wind-energy facility. This model, which was developed for the National Renewable Energy Laboratory (NREL), is called JEDI (Jobs and Economic Development Impacts). JEDI is an input-output model that calculates the direct, indirect, and induced economic benefits from new wind-energy facilities. (A new JEDI model, JEDI II, estimates the local economic benefits from new coal and natural gas facilities as well.) JEDI II uses input data such as the size of the project, its plant-construction cost, the length of the construction period, and fixed and variable operation and maintenance costs to estimate impacts (direct, indirect, and induced) in terms of jobs, wage and salary income, and output (economic activity) both during the construction period and during the operating years (Goldberg et al. 2004).

Models such as JEDI can improve understanding of the economic impacts of new energy facilities, especially when those impacts are considered at the macro level. Similarly, assessments of the actual economic impacts of wind-energy facilities, in addition to forecasts of economic impacts, can improve our collective understanding of the economic benefits of wind-energy facilities and how those benefits are distributed. Surveying 13 studies of economic impacts (actual and forecast) of wind facilities on rural economies, one NREL report concluded that these facilities have a large direct impact on the economies of rural communities, especially those with few other supporting industries; however, such communities also see greater "leakage" of secondary economic effects to outside areas. In addition, the report concluded that the number of local construction and operations jobs created by the facility depends on the skills locally available (Pedden 2006).

More studies are needed of the economic impacts of wind facilities, both actual and estimated. The NWCC (2001) provides these guidelines for assessing the economic development impacts of wind energy:

- The audience for the study and the objectives to be pursued should receive primary consideration.
- The assumptions and scenarios used to analyze economic development impacts should be clearly stated.
- The model used to calculate impacts should use regional economic input data. The data should be representative of the study region (country, state, county, reservation, or multiple states and counties).
- Both the potential positive and negative (i.e., displacement) economic impacts of wind-energy development should be considered.

- The evaluation should consider the ownership, equity and sources of capital, and markets for the project for their relative impacts on the local community, reservation, state, region, or county.
- The evaluation should consider the timing and scale of the project in relation to other wind-energy development in the state, region, or country. Pioneering projects in new areas face economic considerations different from those of incremental projects in mature wind-resource areas.
- The evaluation should distinguish between short-term and long-term impacts.
- The evaluation should consider relative impacts on the economy at a level appropriate to the scope of the study.
- For both wind development and the displaced alternative, the evaluation should consider how new labor, material, and services would be supplied.

These guidelines are apt but demanding. From the perspective of the local affected area, it may be best to focus on the jobs that will be directly created by the project—what skills they require, what their pay levels are, what their duration will be, and what the company's hiring practices are—as well as on reasonably anticipated effects—positive and negative—on the local economy.

Employment Commitments

A developer seeking to be favorably received by a host area may explore with local officials the possibility of a commitment to give hiring preferences to local workers. As Pedden (2006) noted in a report on the economic impacts of wind facilities in rural communities, "some local governments offer incentives to developers in return for the developer agreeing to hire local labor."

Public Revenues and Costs

Like other industries, a wind-energy project generates tax dollars for the local government. According to the AWEA, "Wind Energy and Economic Development: Building Sustainable Jobs and Communities" (AWEA 2006f):

- Alameda County, in California, collected $725,000 in property taxes in 1998 from wind-turbine installations valued at $66 million.
- 240 MW of wind capacity installed in Iowa in 1998 and 1999 produced $2 million annually in tax payments to counties and school districts.

• The director of economic development in Lake Benton, Minnesota, said that each 100 MW of wind development generates about $1 million annually in property-tax revenue.

In addition, as with the private economy, the wind-energy project may indirectly generate taxes for the local government. However, as discussed above with regard to the private economy, an assessment of fiscal benefits in the form of tax revenue should be based on changes that would occur if the project was built versus if it was not built. The project may encourage some forms of economic development that generate taxes, but it may deter others.

A wind-energy facility also may entail public costs. Some of these, such as improvements of local public roads accessing the facility, will be obvious. Others, such as improved community services that may be expected in the wake of the development, will be indirect and less obvious. Taken together, the costs to a small, rural government have the potential to be significant.

Fiscal Commitments

The developer and the local government should have a clear mutual understanding of both the basis for tax revenues and what public expenditures are expected to make the project possible.

ELECTROMAGNETIC INTERFERENCE

Through electromagnetic interference (EMI), wind-energy projects conceivably can have negative impacts on various types of signals important to human activities: television, radio, microwave/radio fixed links, cellular phones, and radar.

EMI is electromagnetic (EM) disturbance that interrupts, obstructs, or otherwise degrades or limits the effective performance of electronics or electrical equipment. It can be induced intentionally, as in some forms of electronic warfare, or unintentionally, as a result of spurious emissions and responses and intermodulation products. In relation to wind turbines, two issues are relevant: (1) possible passive interference of the wind turbines with existing radio or TV stations, and (2) possible electromagnetic emissions produced by the turbines.

There are several ways in which electromagnetic waves can deviate from their intended straight-line communication paths. These include:

• Blocking the path with an obstacle, thus creating a "shadow" or area where the intended EM wave will not occur. To a large extent, the "blocking" depends on the size of the obstacle as a function of the wavelength of the electromagnetic wave.

   • Refraction of the EM wave. Refraction is the turning or bending of any wave, such as a light or sound wave, when it passes from one medium into another with different refractive properties.

In the context of wind-energy projects, EMI often is discussed in relation to the following telecommunications facilities:

   • Television broadcast transmissions (approx 50 MHz-1 GHz)
   • Radio broadcast transmissions (approx 1.5 MHz [AM] and 100 MHz [FM])
   • Microwave/radio "fixed links" (approx 3-60 GHz)
   • Mobile phones (approx 1 or 2 GHz)
   • Radar

Television

The main form of interference to TV transmission caused by wind-energy projects is the scattering and reflection of signals by the turbines, mainly the blades. In relation to the components that make up a wind turbine, the tower and nacelle have very little effect on reception (i.e., only a small amount of blocking, reflection, and diffraction occurs). This is backed up by laboratory measurements that show that the tower introduces only a small, localized (up to approximately 100 m) attenuation of the signal (Buckley and Knight Merz 2005).

The British Broadcasting Corporation has issued recommendations based on a simple concept for calculating the geometry associated with reflected signals from wind turbines and how directional receiving aerials can provide rejection of the unwanted signals (BBC 2006).

Typical mitigation requirements include:

   • Re-orientation of existing aerials to an alternative transmitter
   • Supply of directional aerials to mildly affected properties
   • Switch to supply of cable or satellite television (subject to parallel broadcast of terrestrial channels)
   • Installation of a new repeater station in a location where interference can be avoided (this is more complex for digital but also less likely to be required for digital television)

Radio

Available literature indicates that effects of wind projects on both Amplitude Modulated (AM) and Frequency Modulated (FM) radio transmission systems are considered to be negligible and only apply at very small distances from the wind turbine (i.e., within tens of meters). For AM transmissions, this is due to low broadcast frequencies and long (100+ meter) signal wavelengths, which makes distortion difficult even for very large wind turbines. For FM transmissions, this is due to the fact that ordinary FM receivers are susceptible to noise interference only while operating in the threshold regions relative to signal-to-noise ratios. Thus, a distorted audio signal may be superimposed on the desired sound close to a wind turbine, potentially causing interference, only if the primary FM signal is weak.

Fixed Radio Links

Fixed radio links, also known as point-to-point links, are by definition a focused radio transmission directed at a specific receiver. Fixed links are not intended to be picked up by any receivers other than those at which they are directed. They typically rely on the use of a parabolic reflector antenna (like satellite dishes) to transmit a direct narrow beam of radio waves to a receiving antenna. A direct line of sight is required between the transmitter and receiver, and any obstructions within the line of sight may degrade the performance or result in the loss of the link.

A wind turbine may degrade the performance of a fixed link, not only if it is within the line of sight of the link but also if it is within a certain lateral distance of the link, known as the "Fresnel Zone."

Cellular Phones

Mobile-phone reception depends greatly on the position of the mobile receiver. Therefore, the movement of the receiver and the topography—including both natural and unnatural obstacles—have a major impact on the quality of the signal. The mere movement of the receiver can ensure that wind turbines will have a very minimal effect, if any, on communication quality.

Radar

The potential for interference of wind turbines with radar is only partially understood. If there is such interference, it would primarily affect military and civilian air-traffic control. In addition, National Weather Service weather radars might be affected.

Two recent reports treated the problems in some detail. The first is a report by the U.S. Department of Defense to the U.S. Congressional Defense Committees (DOD 2006). The second is a British report on the impacts of wind-energy projects on aviation radar (Poupart 2003).

The DOD report concludes that "[w]ind farms located within radar line of sight of air defense radar have the potential to degrade the ability of the radar to perform its intended function. The magnitude of the impact will depend upon the number and locations of the turbines. Should the impact prove sufficient to degrade the ability of the radar to unambiguously detect objects of interest by primary radar alone this will negatively influence the ability of U.S. military forces to defend the nation." It concludes further that "[t]he Department has initiated research and development efforts to develop additional mitigation approaches that in the future could enable wind turbines to be within radar line of sight of air defense radars without impacting their performance."

The U.K. report focused on the development and validation of a computer model that can be used to predict the radar reflection characteristics, which are a function of the complex interaction between radar energy and turbines. These effects are described by the Radar Cross Section (RCS). The report concludes that the model enables a much more detailed quantification of the complex interaction between wind turbines and radar systems than was previously available. Among the findings are:

- Wind-turbine towers and nacelles can be designed to have a small RCS.
- Blade RCS returns can be effectively controlled only through the use of absorbing materials (stealth technology).
- The key factors influencing the effect of wind-energy facilities on radar are spacing of wind turbines within a facility, which needs to be considered in the context of the radar cross-range and down-range resolutions.
- No optimal layout or format can be prescribed, because each wind-energy facility will have its own specific requirements that depend on many factors.

The report concludes that the model has a large potential for further use, such as the following:

- It can generate the detailed data required for sophisticated initial screening of potential facility sites.
- It can support the development of mitigation and solutions, in-

cluding siting optimization, control of wind-turbine RCS, and the development of enhanced radar filters that are able to remove returns from wind turbines.

• It is clear that as of late 2006, the interference of wind turbines with radars is a problem as yet unsolved. Research and larger-scale investigations are currently under way in several countries; they may eventually lead to standardization and certification procedures.

CONCLUSIONS AND RECOMMENDATIONS

Aesthetic Impacts

Wind-energy facilities often are highly visible. Responses to proposed wind projects based on aesthetics are among the most common reasons for strong reactions to them. Reactions to the alteration of places that contribute to the beauty of our surroundings are natural and should be acknowledged. Excellent methods exist for identifying the scenic resource values of a site and its surroundings, and they should be the basis for visual impact assessments of proposed projects. Tools are available for understanding project visibility and appearance as well as the landscape characteristics that contribute to scenic quality. Lists of potential mitigation measures are also readily available. Nevertheless, the difficult step of determining under what circumstances and why a project may be found to have undue visual impacts is still poorly handled by many reviewing boards. The reasons include a lack of understanding of visual methods for landscape analysis and a lack of clear guidelines for decision making.

Current Best Practices

Information concerning best practices in the United States is found through the NWCC and its sponsored proceedings and links. Europe and Canada generally have done a more thorough job in providing definitive best-practice guidelines. The integration of local, regional, and national planning and review efforts in those countries contributes to the success of their review processes. Funding in those nations for planning and more extensive surveys of public perceptions of wind energy is also far ahead of that in the United States. Here, standards for best practices are evolving as communities and states recognize the need for a more systematic approach to evaluating visual impacts. There is considerable variability in the review of proposed projects.

Information Needs

Processes for evaluating the aesthetic impacts of wind-energy projects should be developed with a better understanding of the aesthetic principles that influence people's experience of scenery. Comparative studies are needed of wind-energy projects that have relatively widespread acceptance of their aesthetic impacts and those that do not. These studies could provide useful information about a range of factors that contribute to acceptability within different landscape types. These studies should take into account that sites and projects vary dramatically in the types of scenic resources involved; the proximity and sensitivity of views; and the particular project characteristics, including scale.

The tradeoffs between placing wind-energy projects close to population centers where they are closer to electricity users but visible to more people, and placing them in remote areas where they are less visible but where the wilderness, remote, and undeveloped qualities of the landscape may hold value need discussion as well as a clear understanding of the tradeoffs involved. These issues need to be addressed broadly, not only singling out aesthetic concerns.

Impacts on Recreational, Historic, Sacred, and Archeological Sites

Wind-energy projects can be compatible with many recreational activities, but concerns may arise when they are close to recreational activities for which the enjoyment of natural scenery is an important part of the experience. Historic, sacred, and archeological resources can be harmed by direct impacts that affect the integrity of the resource or future opportunities for research and appreciation. The experience of certain historic or sacred sites or landscapes can also be indirectly affected by wind-energy projects, especially if particular qualities of the surrounding landscape have been documented as important to the experience, interpretation, and significance of the proximate historic or sacred site. Greater clarity is needed about how such situations should be evaluated. For example, the importance and special qualities of the experience must be assessed within the context of the relative visibility and prominence of the proposed wind-energy project.

Current Best Practices

Useful methods exist for evaluating both the relative sensitivity of recreational areas and recreational users, and for determining valuable scenic resources. Siting to avoid impacts on highly sensitive recreational uses, and project design to mitigate both direct and indirect impacts can be important. Mitigation techniques can include relocation of project design elements, relocation of recreational activities (such as a trail), and enhancement of existing recreational activities.

State Historic Preservation Offices (SHPOs) generally identify all known historic sites of state and national significance. Local historical societies or comprehensive plans may identify additional sites of local significance. The SHPO typically requires a Class II survey to determine the existence of unknown resources in areas where such surveys are lacking. Guidelines for evaluating direct impacts on historic sites and structures often are available, and many states require archeological surveys for certain sites. Few guidelines currently exist, however, for evaluating indirect impacts of wind-energy projects on historic or sacred sites and landscapes.

Information Needs

• Research examining the perceptions of recreational users toward wind-energy projects that are located near dispersed and concentrated recreational activities would provide useful data for decision makers. However, aesthetic impacts are very site-specific, so the results of such research likely will be able to guide site-specific assessments but not substitute for them.

• Guidelines are needed concerning distances at which recreational activities can occur safely around wind turbines.

• Policy makers and decision makers need better guidance from historic-preservation experts and others concerning the methods for evaluating the effects of wind-energy projects on historic, sacred, and archeological resources.

Noise and Shadow Flicker

Noise can be monitored by various measurement techniques. However, an important issue to consider, especially when studying noise, is that its perception and the degree to which it is considered objectionable depend on individuals exposed to it.

Shadow flicker caused by wind turbines can be an annoyance, and its effects need to be considered during the design of a wind-energy project. In the United States, shadow flicker has not been identified as even a mild annoyance. In Northern Europe, because of the higher latitude and the lower angle of the sun, especially in winter, shadow flicker has, in some cases, been noted as a cause for concern.

Best (or Good) Practices

Good practices for dealing with the potential impacts of noise from a wind-energy project could include the following:

• Analysis of the noise should be made based on the operating characteristics of the specific wind turbines, the terrain in which the project will be located, and the distance to nearby residences.

• Pre-construction noise surveys should be conducted to determine pre-project background noise levels and to determine later on what, if any, changes the wind project brought about.

• If regulatory threshold levels of noise are in place, a minimum distance between any of the wind turbines in the project and the nearest residence should be maintained so as to reduce the sound to the prescribed threshold.

• To have a process for resolving potential noise complaints, a telephone number should be provided through which a permitting agency can be notified of any noise concern by any member of the public. Then, agency staff can work with the project owner and concerned citizens to resolve the issue. This process can also include a technical assessment of the noise complaint to ensure its legitimacy.

Shadow flicker is reasonably well understood. With a little careful planning and the use of available software, the potential for shadow flicker can be assessed at any site, and appropriate strategies can be adopted to minimize the time when it might be an annoyance to residents nearby.

Information Needs

Recent research studies regarding noise from wind-energy projects suggest that the industry standards (such as the IEC 61400-11 guidelines) for assessing and documenting noise levels emitted may not be adequate for nighttime conditions and projects in mountainous terrain. This work on understanding the effect of atmospheric stability conditions and on site-specific terrain conditions and their effects on noise needs to be accounted for in noise standards. In addition, studies on human sensitivity to very low frequencies are recommended.

Computational tools have become available that not only compute shadow flicker in real time during turbine operation, but also convey information to the turbine-control system to allow shutdown if the shadow flicker at a particular location becomes particularly problematic. Hence, the development and implementation of a real-time system at a wind-energy project to take such actions when shadow flicker is indicated might be useful.

Local Economic and Fiscal Impacts

When assessing the economic and fiscal impacts of a wind-energy project, the main issues that arise include (1) fair treatment of both landowners who lease land for the project and other affected but uncompensated owners and occupants; (2) a fine-grained understanding of how wind-energy facilities may affect property values; (3) a realistic appraisal of the net economic effects of the wind-energy facility, during its construction and over its lifetime; and (4) a similarly realistic assessment of the revenues the local government can expect and the costs it will have to assume.

Current Best Practices

The guidelines referred to in the text—of Windustry, regarding leasing and easement arrangements; and of the NWCC, regarding assessments of economic development impacts of wind power—contain good advice and are examples of current standards for best practices. In addition, best practices include:

- Gathering as much "hard" information as possible: the terms of the lease and easement arrangements; the type, pay scale, and duration of jobs that are likely to be generated for local workers; the taxes that the project will directly generate; and the known public costs that it will entail.
- Qualitatively taking into account other, less tangible economic factors: opportunity costs that may arise from the project; the duration of benefits from the project; and the likelihood of an uneven distribution of benefits (e.g., one landowner may realize income by leasing land for a turbine while another may be within close range of the turbine but receive no income).
- Adopting guarantees and mitigation measures that are tailored to the facility, the surrounding community members, and the local government and are fair to all involved.

Information Needs

Large wind-energy facilities are fairly new in the United States. Many current analyses of their economic impacts are fueled by enthusiasm or skepticism. There is a need for systematic collection and analysis of economic data on a facility-by-facility and region-by-region basis. These data should take into account the type of facility, including the number of turbines at the facility and elsewhere in the region. The data should cover the following types of information:

- Leasing arrangements
- Jobs directly created (including skill and pay levels, duration, hiring policies)
- Local government revenue and costs
- Economic mitigation and enhancement measures

More studies also are needed of public attitudes toward specific wind-energy facilities and how they affect economic behavior (e.g., property values, tourism, new residential development). To allow for cross-facility and longitudinal comparisons, the methods of data collection and analysis used in these studies should be replicable.

Electromagnetic Interference

With the exception of radar, the main EMI effects of wind-energy projects are well understood. Wind turbines have the potential to cause interference to television broadcasts, while the audio parts of TV broadcasts are less susceptible to interference. The data available are adequate to predict interference effects and areas and to minimize interference at the planning stage or propose suitable mitigation requirements.

Information Needs

Regarding radar, more research is needed to understand the conditions under which wind turbines can interfere with radar systems and to develop appropriate mitigation measures.

In addition, while EMI is not an issue in all countries (e.g., it is not an issue in Denmark), EMI issues should be given sufficient coverage in environmental impact statements and assessments to provide adequate evaluation of wind-energy project applications.

GENERAL CONCLUSIONS AND RECOMMENDATIONS

Well-established methods are available for assessing the positive and negative impacts of wind-energy projects on humans; these methods enable better-informed and more-enlightened decision making by regulators, developers, and the public. They include systematic methods for assessing aesthetic impacts, which often are among the most-vocalized concerns expressed about wind-energy projects.

Because relatively little research has been done on the human impacts of wind-energy projects, when wind-energy projects are undertaken, routine documentation should be made of processes for local interactions and impacts that arise during the lifetime of the project, from proposal through decommissioning. Such documentation will facilitate future research and therefore help future siting decisions to be made.

The impacts discussed in this chapter should be taken within the context of both the environmental impacts discussed in Chapter 3 and the broader contextual analysis of wind energy—including its electricity production benefits and limitations—presented in Chapter 2. Moreover, the conclusions and recommendations presented by topic here should not be taken in isolation; instead, they should be treated as part of a process. Chapter 5 elaborates on processes for planning and evaluating wind-energy projects and for public involvement in these processes.

Finally, the text of this chapter describes many specific questions to be asked and issues to be considered in assessing various aspects of the effects of wind-energy projects on humans, especially concerning aesthetic impacts, and those questions and issues should be covered in assessments and regulatory reviews of wind-energy projects.

5

Planning for and Regulating
Wind-Energy Development

The purpose of this chapter is to describe the current status of planning and regulation for wind energy in the United States, with an emphasis on the Mid-Atlantic Highlands, and then critique current efforts with an eye to where they might be improved. To accomplish this purpose, we reviewed guidelines intended to direct planning and regulation of wind-energy development as well as regulatory frameworks in use at varying geographic scales. To enhance our interpretation of wind-energy planning and regulation in the United States, we drew on the experiences of other countries with longer histories of wind-energy development and different traditions of land-use planning and development regulation. We focused on onshore wind energy, although many elements of planning and regulation that influence onshore wind-energy developments apply to offshore installations as well.

As with other human endeavors that engage both private and public resources, wind-energy development is influenced by an interconnected, but not necessarily well-integrated, suite of policy, planning, and regulatory tools. "Policy" can be broadly defined to encompass a variety of goals, tools, and practices—some codified through laws; some less formally specified. (For a discussion of traditional and new policy tools, see, e.g., NRC 2002.) Policies encompass, but are not limited to, planning and regulation. Policy tools related to wind energy, including national and regional goals, tax incentives, and subsidies, have been discussed in Chapter 2, so we concentrate on planning and regulation here. "Planning"—whether legally mandated or not—is a process that typically involves establishing goals; assessing resources and constraints, as well as likely future conditions; and then developing and refining options. "Regulation," as understood within a legal framework, typically consists of methods and standards to implement laws. Regulation is created and carried out by public agencies charged with this responsibility by law. The scope of agency discretion in establishing and administering regulations depends largely upon whether the law is highly detailed or more general.

The chapter begins with a review of guidelines that have been developed to direct wind-energy planning and/or regulation. Some of these have been promulgated by governmental or non-governmental organizations concerned with limited aspects of wind energy, such as the guidelines for reducing wildlife impacts developed by the U.S. Fish and Wildlife Service (USFWS 2003). Some are more comprehensive in scope, such as those developed by the National Wind Coordinating Committee (NWCC 2002). We also consider guidelines developed by states to direct wind-energy development toward areas judged most suitable and to assist local governments in carrying out their regulatory responsibilities with respect to wind energy. Then we review regulation of wind-energy development via federal laws, including development on federal lands, in situations where there is a federal nexus by reason of federal funding or permitting, and where there is no such nexus. Next we review regulation of wind-energy development at the state and local levels by concentrating on recurring themes: the locus of regulatory authority (state, local, or a combination thereof); the locus of review for environmental effects; information required for review; public participation in the review; and balancing positive and negative effects of wind-energy development. In these sections we also report on the interaction between planning and regulation, although that interaction is generally less well developed in the United States than in some of the other countries we examined. Then we critique what we have learned about regulation of wind energy by examining some of the tensions in regulation, for example between local and broader-level interests and between flexibility and predictability of regulatory processes. Finally, we present a set of recommendations for improving wind-energy planning and regulation in the United States.

GUIDELINES FOR WIND-ENERGY PLANNING AND REGULATION

In the United States and, notably, in other nations with considerable wind-energy experience, governmental and non-governmental organizations working at various geographic scales have adopted guidelines to help those developing wind-energy projects and those regulating wind-energy development to meet a mix of public and private interests in a complex, and often controversial, technical environment. Here we review U.S. guidelines for different jurisdictional levels (e.g., state, local), for different environmental components (e.g., wildlife), and for the different purposes of guiding planning versus guiding regulatory review. We also draw on the experience of other countries where guidelines for wind-energy development have a longer history.

Some guidelines are for proactive planning of wind-energy development. "Planning" is an ambiguous term, however. Within the context of wind-energy development, it can refer to highly structured processes that carry considerable legal weight and result in identifying certain areas as suitable for wind turbines (as in the Denmark example below). Alternatively, it can refer to loosely structured processes that are largely advisory and result in criteria for evaluating the favorable and unfavorable attributes of prospective sites (as in the Berkshire example below). In addition, planning for wind-energy development may take a broad view of the incremental impacts of multiple wind-energy projects in a region, or it may take a narrow view and focus primarily on a single project. And, geographic scales for planning range from the national to the local level.

Other guidelines focus on regulation. They prescribe for regulatory authorities reviewing wind-energy developments what procedures should be followed, what kinds of information should be examined, and what criteria should be used to make permitting decisions. Many guidelines mingle the two functions of planning and project-specific regulatory review. In practice, planning guidelines that suggest where and how wind-energy development should be done may become criteria for regulatory permitting decisions if projects inconsistent with planning guidelines are rejected.

The United States is in the early stages of learning how to plan for and regulate wind energy. The experiences of other countries, where debates over wind energy have been going on for much longer, can be instructive for bringing U.S. frameworks to maturity. For example, Britain and Australia have dealt with controversies about wind-energy development by working with stakeholder groups, including opponents of wind energy, to develop "Best Practice Guidelines" (BWEA 1994; AusWEA 2002). BWEA (British Wind Energy Association) and AusWEA (Australian Wind Energy Association) were convinced that "they needed to become more transparent and more engaged with the public than any other industry" (Gipe 2003). In Ireland, the Minister of the Environment, Heritage and Local Government released an extensive "Planning Guidelines" document on wind energy in June 2006 (DEHLG 2006). This document advises local authorities on planning for wind energy in order to ensure consistency throughout the country in identifying suitable locations and in reviewing applications for wind-energy projects. Not only are these guidelines prescriptive—that is, they express procedures and approaches that *should* be taken—but they also are linked to other government policies.

A prescriptive national approach is less likely in the United States, where each state is governed by different laws and policies regarding the regulation of wind-energy projects, but the comprehensive approaches used in other countries could be adopted at the state level. The highly structured approach of Denmark, for example (see Box 5-1), could be informative for states wishing to develop integrated frameworks to plan for and regulate wind-energy development. We note, however, that Denmark is smaller and more homogeneous than many U.S. states, has a much stronger tradition of central planning of land use, and has many wind projects owned by local cooperatives, rather than private developers.

National Wind Coordinating Committee Guidelines

The NWCC was established in 1994 as a collaboration among representatives of wind equipment suppliers and developers, green power marketers, electric utilities, state utility commissions, federal agencies, and environmental organizations. Its permitting handbooks propose guidelines for how wind-energy developments should be reviewed. In 2002, the Siting Subcommittee of the NWCC revised its *Permitting of Wind Energy Facilities: A Handbook*, originally issued in 1998 (NWCC 2002). Intended for those involved in evaluating wind-energy projects, the handbook describes the five typical phases of permitting processes for energy facilities, including wind turbines and transmission facilities:

(1) *Pre-Application:* This phase occurs before the permit application is filed, during which the developer meets with the permitting agency(ies) and others immediately affected, such as nearby landowners, and local agencies. This phase may be mandatory or voluntary, and it may involve public notice and/or public meetings.

(2) *Application review:* This phase begins when the permit application is filed. Its activities, required documents, and public involvement requirements depend upon the application review process, as does its duration. In some cases, agencies may be required to reach a decision within a specified period.

(3) *Decision making:* In this phase, the lead agency determines not only whether to allow the project but also whether to impose measures constraining the project's construction, operation, monitoring, and decommissioning. During this phase, public hearings are likely to be required.

(4) *Administrative appeals and judicial review:* Appeals of permit decisions may be made to the decision maker, to the administrative review board, or to the courts.

(5) *Permit compliance:* This phase extends through the project's life-

BOX 5-1
Planning for Wind-Energy Development in Denmark

Until the beginning of the 1990s, the approach in Denmark (like the U.S. approach today) was: "Find yourself a site, and then apply for permission to erect your wind turbine(s)" (DWEA 2004). This laissez-faire approach changed in the early 1990s, when Denmark's third energy plan—Energy 2000—was put forward. It included the goal of 1,500 MW of installed wind energy by 2005. In 1994, an executive order, the "Wind Turbine Circular," made cities responsible for planning for wind turbines, including looking for appropriate sites. In 1999, with a new Wind Turbine Circular, the planning responsibility was redirected to county (amt) authorities. These county-level efforts, and corresponding local efforts, target areas considered suitable for wind farms. The original goal of 1,500 MW nationally was met several years before the 2005 deadline. In 2002, the Danish government indicated that further onshore wind-energy development would not be encouraged but that offshore wind-energy development would be allowed. Denmark's success in installing so much wind-energy capacity has been attributed to numerous factors including (1) the relatively small size of wind projects (1-30 turbines), (2) cooperative ownership of many wind projects with direct benefits to local citizens, and (3) comprehensive planning and review in which localities directly participate (J. Lemming, RISØ, personal communication 2006; Nielsen 2002).

Regulatory Review Within a Planning Context

In Denmark, planning for land-based wind development is linked to the regulatory review process. The centerpiece of the review process is a mandatory environmental impact assessment (EIA) (if a project involves more than three wind turbines or wind turbines over 80 meters in height). The regional planning authority—typically, the county—is responsible for initiating the EIA and ensuring its quality. The EIA is a joint effort of the developer, the developer's consultants, and the county. The EIA must describe the project and establish that the site is appropriate from a wind resource standpoint. The site is further described, including working areas and roads to be used during construction. Alternative sites must be investigated, as well as the no-build alternative, and the developer must substantiate why the proposed site is preferred.

time. It may include monitoring, inspection, addressing local complaints during the project's operation, as well as following up on requirements for closure and decommissioning.

These five phases are included in the handbook as descriptions of what typically happens (given a great deal of variation among states), not as recommendations of what should happen. However, the authors of the handbook suggest eight principles that should be followed when structuring a permit review process:

The EIA must describe the landscape surrounding the site, with emphasis on anything that may be affected during the construction or operation phases of the project. Protected species of flora and fauna require special consideration, as do birds protected under international agreements. Any adverse effects on water reserves must be noted. If the project is to be located outside areas previously designated for wind farms, conflicts over the use of land (e.g., because of protected species, arable land, scenic resources) must be given special attention. The EIA also should assess the project's positive environmental impacts (e.g., reduced CO_2, NO_x, and SO_2.

Standards have been set for evaluating impacts on the human environment. No turbine may be sited closer to the nearest residence than a distance of four times the height of the turbine. The EIA must address adverse noise and visual impacts, including cumulative impacts from multiple turbines within a radius of 1 to 2 km. The noise level from the wind turbine(s) must be estimated, using a protocol described in the Noise Declaration (a national-level regulation). The EIA must describe how far shadows from the turbines will reach at all times of year, and the layout of the turbines in relation to major landscape features must be described. Possible adverse effects on property values, tourism, and other commercial activities in the vicinity should be described.

Prior to preparation of the EIA and the planning documents, the project must be publicized for at least four weeks, with opportunities for private citizens and organizations to submit suggestions and comments. These submissions must be included when preparing the EIA. Following completion of the EIA and the planning documents (or their amendments), this material must be publicized for at least eight weeks, after which a public hearing is held where suggestions or objections are again gathered. Following the final decision on the project, anyone who has submitted objections to the project must receive a written answer to the objections.

When the EIA is completed, authorities at the county and local levels formulate amendments to their wind-energy plans, using the EIA as a common point of reference. During the subsequent public comment period, the state can veto the project (this is a national-level decision) but must substantiate why it is exercising its veto power. After public hearings, the plans are presented to county and local political bodies (the county council and the city council). The county council or city council may approve or reject the project. Construction begins only if the project has been approved (Ringkøbing Amt, Møller og Grønborg, and Carl Bro 2002).

- *Significant public involvement:* Including early and meaningful information and opportunities for involvement.
- *An issue-oriented process:* One which focuses the decision on issues that can be dealt with "in a factual and logical manner" (NWCC 2002, p. 16).
- *Clear decision criteria:* As well as clear specification of factors that must be considered and minimum requirements that must be met.
- *Coordinated permitting process:* Including both horizontal coordination among various agencies and vertical coordination between state and local decision makers.

- *Reasonable time frames:* In part to provide the developer with known points for providing information, making changes, and receiving a decision.
- *Advance planning:* In particular, early communication on the part of the developers and the permitting agencies.
- *Timely administrative and judicial review:* Including addressing issues such as who has standing to initiate a review and time limits within which reviews must be initiated.
- *Active compliance monitoring:* Including specifying reports that must be submitted and establishing site inspection timetables, non-compliance penalties, a complaint resolution process, etc.

Federal Government Guidelines

Concerns about the effects of wind-energy projects on bird and bat mortality, in combination with federal laws protecting some wildlife species, led the U.S. Fish and Wildlife Service (USFWS) to provide interim guidelines for evaluating wildlife impacts (technical aspects of which are reviewed in Chapter 3). We know of no other federal-level guidelines addressing the review of wind-energy projects on private land. However, the Federal Aviation Administration (FAA) reviews all structures 200 feet or taller for compliance with aviation-safety guidelines. There have not been uniform standards until fairly recently (see Box 5-2). Both the Bureau of Land Management (BLM) and the U.S. Forest Service (USFS) provide guidance for the review of wind-energy projects on lands under their jurisdictions. These are described below under federal regulatory approaches to wind energy.

U.S. Fish and Wildlife Service Interim Guidelines

On May 13, 2003, the USFWS released "Interim Guidance on Avoiding and Minimizing Wildlife Impacts from Wind Turbines" (USFWS 2003). Adherence to the guidelines is voluntary, as the guidelines note:

> . . . the wind industry is rapidly expanding into habitats and regions that have not been well studied. The Service therefore *suggests* a precautionary approach to site selection and development and will employ this approach in making recommendations and assessing impacts of wind-energy developments. We *encourage* the wind-energy industry to follow these guidelines and, in cooperation with the Service, to conduct scientific research to provide additional information on the impacts of wind-energy development on wildlife. We further *encourage* the industry to look for opportunities to promote bird and other wildlife conservation when planning wind-energy facilities (e.g., voluntary habitat acquisition or conservation easements) (USFWS 2003, emphasis added).

BOX 5-2
Federal Aviation Agency (FAA) Obstruction Lighting Guidelines

To determine lighting requirements, each site and obstruction is reviewed by the FAA for particular safety concerns, such as distance from nearby airports. Negative effects of required lighting on night-flying birds and bats, and sometimes also on people near wind-energy projects, have prompted revisions of initial lighting standards for wind turbines. A recent study conducted by the FAA Office of Aviation Research resulted in recommendations for obstruction lighting that considerably reduced earlier lighting guidelines (Patterson 2005). The following chart summarizes the former and current guidelines, though individual site requirements may vary. Source of current guidelines: FAA (2007).

Former FAA Guidelines	Current FAA Guidelines
• Lights mounted on the nacelle of every turbine • Two flashing or pulsed light fixtures for night lighting • Two flashing white light fixtures during daytime • Flashing can be synchronous or random	• Lights needed only to mark periphery (ends or edges) of project or cluster; with maximum lighting distance of a half-mile; highest turbines must also be lit. • Single red flashing or pulsing light fixture at night. • White strobe lights may be used but not in conjunction with red lights (one or the other) • Flashing must be synchronous (all at the same time) • No daytime lighting required as long as turbines are a white color (not gray). • Preferred light is a red flashing (L-864) with minimum light intensity of 2000 candelas. • Lights should be mounted above the nacelle height for visibility (hub may obscure) • Turbine locations should be noted on aviation maps.

The guidelines include recommendations regarding:

• A two-step site evaluation protocol (first, identify and evaluate reference sites—i.e., high-quality wildlife areas; second, evaluate potential development sites to determine risk to wildlife and rank sites against each other using the highest-ranking reference site as a standard).

• Site development (e.g., placement and configuration of turbines, development of infrastructure, planning for habitat restoration); and turbine design and operation (USFWS 2003).

The guidelines direct wind-energy development away from concentrations of birds and bats and toward fragmented or degraded habitat (rather than areas of intact and healthy wildlife habitat). The guidelines also address some desirable features of regulatory review processes, such as recommending multiyear, multiseason pre-construction studies of wildlife use at proposed project sites; multiyear, multiseason post-construction studies to monitor wind-project impacts; and involvement of independent wildlife agency specialists in development and implementation of pre- and post-construction studies. The guidelines were circulated to the public with a request for review and the Service recently announced the development of a Federal Advisory Committee Act-compliant collaborative effort to revise the guidelines based on public comment.

State and Regional Guidelines

Several states with wind resources have developed guidelines for siting and/or permitting wind-energy projects. This has been particularly true for states where review occurs at the local level, since the projects may be complex and very different from the kinds of projects most local governing bodies are used to addressing. The NWCC Guidelines (NWCC 2002) appear to have provided a useful template for states, with basic information that can be adapted to the particular needs and conditions of states that have wind potential. Below we describe a few of these to illustrate the types of provisions that state guidelines may include.

Kansas's wind-energy guidelines were adapted from the NWCC Guidelines and are intended to assist local communities in regulating land use for wind-energy projects. The guidelines recognize landscape features that are important in Kansas. Under "Land Use Guidelines," (KSREWG 2003, p. 3) native tallgrass prairie landscapes are singled out as having particular value, especially where they remain unfragmented. Cumulative impacts are noted because there is intense interest in wind-energy development in certain areas of the state. Kansas includes guidelines on "Socioeconomic, Public Service and Infrastructure," as well as on public interaction (KSREWG 2003, p. 6).

South Dakota also adapted NWCC permitting guidelines (SDGFP 2005). In sections concerning "natural and biological resources," South Dakota's guidelines call attention to areas of the state that have been identified as potential sites for wind-energy development, but are considered "unique/rare in South Dakota" (SDGFP 2005, p. 1). Developers are urged to use environmental experts to make an early evaluation of the biological setting and to communicate with agency, university, and environmental organizations. They are warned that "if a proposed turbine site has a large potential for biological conflicts and an alternative site is eventually deemed appropriate, the time and expense of detailed wind resource evaluation work may be lost" (SDGFP 2005, p. 3). In sections on "visual resources," developers are told to inform stakeholders about what to expect from a wind-energy project, target areas already modified by human activities, and be prepared to make tradeoffs and coordinate planning across juris-dictions and with all stakeholders. Under "socioeconomic, public services, and infrastructure," developers are admonished not to take advantage of municipalities that lack zoning or permitting processes for wind-energy development.

Wisconsin's guidelines (WIDNR 2004) focus on natural resource issues with minimal guidance in other areas. The guidelines direct wind-energy development away from wildlife areas, migration corridors, current or proposed major state ecosystem acquisition and restoration projects, state and local parks and recreation areas, active landfills (because they attract birds), wetlands, wooded corridors, major tourist/scenic areas, and airports and landing strips clear zones. USFWS guidelines are cited as models for pre-construction studies, with two years of post-construction monitoring recommended for the first wind-energy projects in a particular area.

In contrast with guidelines focused exclusively on wildlife issues, some guidelines reflect a much more comprehensive approach. As illustrated in the accompanying box (Box 5-3), the wind-energy siting guidelines devel-oped by the Berkshire Regional Planning Commission in Massachusetts are multifaceted and proactive, as is an assessment methodology prepared by the Appalachian Mountain Club for wind energy in the Berkshires (BRPC 2004; Publicover 2004).

Regulatory review processes could possibly use such a method to evalu-ate proposed wind-energy projects to see if they met a threshold for suit-ability. Similar procedures have been proposed for other states, such as Virginia, where Boone et al. (2005) proposed a land-classification database for use in screening out sites that are likely to be deemed unsuitable for wind-energy development, such as designated wilderness areas or concen-trations of birds or bats.

Guidelines Directed Toward Local Regulation

In some cases, the guidelines that states have developed are intended to serve as models for local ordinances and local-level review processes. The "Michigan Siting Guidelines for Wind Energy Systems" (MIDLEG 2005) is a model zoning ordinance for local governments, although it notes that "the Energy Office, DLEG (Department of Labor and Economic Growth) has no authority to issue regulations related to siting wind energy systems"

BOX 5-3
Guidelines for Planning and Regulatory Review of Wind Energy in the Berkshires, Massachusetts

Perhaps as a result of interest in wind-energy project development in the Massachusetts Berkshire region, the Berkshire Regional Planning Commission developed "Wind Power Siting Guidelines" (BRPC 2004). Most of these guidelines refer to desired features of application and regulatory review procedures. For example, the guidelines direct that viewshed analyses should be done "with the most accurate elevation data available from the state using a GIS such as ArcGIS Spatial Analyst or 3D Analyst" (BRPC 2004, p. 1) and that "important cultural locations, Shakespeare & Company, and Hancock Shaker Village should be located on the map to determine if they will be impacted by the visibility of the turbine development" (pp. 1-2). There are also safety guidelines, such as "Existing homes are not within potential safety impact areas from ice or blade throw or tower failure" (p. 2).

The commission also asked the state of Massachusetts to become involved in wind-energy development to provide "state-wide siting guidelines for the development of wind power facilities" (p. 3) and "financial assistance to municipalities with areas conducive to wind-energy development to develop adequate local land use regulations for wind energy facilities" (p. 4). The commission suggested that communities hosting wind-energy projects should require that applicants pay for consultants to assist the municipality in evaluating the possible and negative impacts of a proposed project and in establishing beneficial agreements for municipal revenue generation.

Also working in the Berkshires, the Appalachian Mountain Club (AMC) developed "A Methodology for Assessing Conflicts Between Windpower Development and Other Land Uses" (Publicover 2004). This document considers various ecological and sociocultural characteristics that make sites appropriate or inappropriate for wind-energy projects, beyond engineering or economic considerations. Beginning with Geographic Information System (GIS) layers identifying wind sites of Class 4 and above, the AMC methodology overlaid known ecological, recreational, and scenic resources onto the wind map. Resource data were assigned "conflict ratings" that included importance of a resource (local, state, federal significance), proximity to the site, and size of the area. These data can be examined with different subjective weightings on ecological and social factors to see how they might affect an overall site suitability rating. A trial application of this method of analysis suggested that certain sites had far fewer conflicts than others, but the authors cautioned that many variables that could be important to siting decisions were not included in the study.

(MIDLEG 2005, p. 1). Pennsylvania also has produced a model zoning ordinance for local communities (Lycoming County 2005), discussed below in the analysis of state and local regulatory review. Both the Michigan and Pennsylvania models are very basic in their requirements, with little detail about information required or how it will be judged.

REGULATION OF WIND-ENERGY DEVELOPMENT

In this section we move from guidelines for planning and regulating wind-energy development to a review of regulatory frameworks that have been put in place at different jurisdictional levels and for different land ownerships. First, we review federal regulation of wind energy: most narrowly, federal regulation of wind-energy development on federal lands; then federal regulation of wind-energy development that has a federal "nexus" via federal funding or permitting; then, most broadly, federal regulation of wind-energy development regardless of land ownership.

To better understand regulation of wind-energy development, we review regulatory frameworks for a number of states. Because the focus of this document is the Mid-Atlantic Highlands, we include all four states in this region (Pennsylvania, Maryland, Virginia, and West Virginia). These four states vary in the intensity of their review processes, thus giving a picture of the range of regulatory oversight in the United States today. We also review wind-energy regulation for states outside the Mid-Atlantic Highlands, choosing some from northeastern states that share many landscape, social, and wind-energy characteristics with the Mid-Atlantic Highlands, and some from contrasting landscapes.

In reviewing regulatory frameworks at all levels, we emphasize regulations that are likely to be particularly salient for wind-energy projects, and especially regulations that are likely to affect wind development in the Mid-Atlantic Highlands region. We give rather little attention to regulations that apply equally to any type of construction or industrial operation, wind energy or other.

Federal Regulation of Wind Energy

Federal Regulation of Wind-Energy Development on Federal Lands

As of mid-2005, all of the wind-energy facilities erected on federal lands were in the western United States on land managed by the BLM; they included about 500 MW of installed wind-energy capacity under right-of-way authorizations (GAO 2005). At that time, the BLM developed its Final Programmatic Environmental Impact Statement on Wind Energy Development (BLM 2005a) in order to expedite wind-energy development in response to National Energy Policy recommendations. The Wind Energy Development Program, which is to be implemented on BLM land in 11 western states, establishes policies and best-management practices addressing impacts to natural and cultural resources (BLM 2005b).

As of mid-2006, other federal land management agencies such as the USFS had not developed general policies regarding wind-energy development and were reviewing proposals on a case-by-case basis. No wind-energy projects currently exist on USFS lands but two were under review as of mid-2006,[1] one in southern Vermont in the Green Mountain National Forest and another in Michigan in the Huron Manistee National Forest. National Forests operate under the guidance of Land and Resource Management Plans, which form the basis for review of all proposed actions. Recent updates of Forest Land and Resource Management Plans address wind-energy projects. In most cases a project would require a "special use authorization" (Patton-Mallory 2006). If an application is accepted, the project undergoes National Environmental Policy Act (1969) (NEPA) review (see next section), but the process will vary depending on the agencies and states involved.

Section 388 of the Energy Policy Act of 2005 gave the U.S. Department of the Interior's Minerals Management Service (MMS) responsibility for reviewing offshore wind-energy development proposals that occur on the outer continental shelf. As of the fall of 2006, the MMS was drafting a programmatic environmental impact statement (EIS) for renewable energies on the outer continental shelf. No offshore wind-energy project was operational or even under construction in the United States at the end of 2006.

Federal Regulation of Wind-Energy Projects with a Federal "Nexus"

Wind-energy developments are subject to the NEPA if they are considered "federal actions" because a federal agency is conducting an activity, permitting it, or providing funds for it. (Another potential federal "nexus" for wind energy—the federal production-tax credit for renewable energy facilities [see Chapter 2]—does not trigger review under the NEPA.) The Council on Environmental Quality has promulgated regulations that include provisions for establishing categorical exclusions from NEPA requirements (NEPA Task Force 2003). Otherwise, the NEPA requires that federal agencies prepare an environmental assessment or, if significant impacts are anticipated, the much more extensive EIS. An EIS must describe the proposed action and provide an analysis of its impacts as well as alternatives to that action, and it must include public involvement in the EIS process. If an EIS is undertaken, socioeconomic impacts must be analyzed as part of the EIS. Otherwise, socioeconomic/cultural impacts of wind-energy projects are given little explicit attention at the federal level. Wind-energy projects on BLM land are under a programmatic EIS, as described above (BLM 2005a).

---

[1]Based on interview with Robert Bair of the Green Mountain National Forest Service, 2006.

Federal Regulation of Wind-Energy Development in General

Federal regulation of wind-energy facilities is minimal if the facility does not receive federal funding or require a federal permit; this is the situation for most energy development in the United States. The Federal Energy Regulatory Commission (FERC) regulates the interstate transmission of electricity, oil, and natural gas, but it does not regulate the construction of individual electricity-generation (except for non-federal hydropower), transmission, or distribution facilities (FERC 2005).

Apart from the FAA guidelines, the threat of enforcement of environmental laws protecting birds and bats is the main federal constraint on wind-energy facilities not on federal lands, because—as discussed in Chapter 3—bird and bat fatalities have been observed at a number of existing facilities. The Migratory Bird Treaty Act applies to all migratory birds native to the United States, Canada, and Mexico; this includes many species that use the Mid-Atlantic Highlands, including for migration. The Bald and Golden Eagle Protection Act (16 U.S.C. §§ 668a-d, last amended in 1978) protects two raptor species. Bald eagles nest in isolated parts of the Mid-Atlantic Highlands whereas golden eagles are mainly migrants or winter residents, although a few may nest in the region (Hall 1983). Permits to "take" species protected under the Migratory Bird Treaty Act (16 U.S.C. §§ 703-712) and to take golden eagles under the Bald and Golden Eagle Act can be issued by the USFWS in very limited circumstances. The Endangered Species Act (ESA) (7 U.S.C. 136; 16 U.S.C. 460 et seq. [1973]), protects species that have been listed as being in imminent danger of extinction throughout all or a significant portion of their range (endangered) or those that are likely to become endangered without appropriate human intervention (threatened). There are federally listed species from many taxa in the Mid-Atlantic Highlands, some of which may be affected by wind-energy projects (Chapter 3). The ESA also protects habitat designated as "critical" to the survival of listed species. Non-critical habitat is protected indirectly in that if habitat destruction would lead to the direct take of an individual of the protected species, destruction of the habitat would be a violation of the ESA. In this situation the species is receiving the protection, not the habitat. Thus, the ESA provisions could affect wind-energy development not only via mortality of birds and bats due to collisions with wind turbines, but also via mortality or habitat loss for endangered or threatened species due to construction and operation of wind-energy facilities. The ESA does allow incidental taking of a protected species (i.e., taking that is incidental to an otherwise legal activity) if a permit has been granted by the USFWS. This provision has been applied to a wind-energy development via incidental take permits that have been approved as part of the Habitat Conservation Plan submitted during the permitting process for the Kaheawa

Pastures Wind Energy Generation Facility on Maui (HIDLNR 2006). The USFWS is responsible for implementation and enforcement of these three laws. Violations are identified in several ways, including receiving citizen complaints and self-reporting by individuals or industry. Although the USFWS investigates the "take" of protected species, the government, as of mid-2005, had not prosecuted industry, including wind-energy companies, for most violations of wildlife laws (GAO 2005).

Like other construction and operation activities, wind-energy projects are subject to federal regulations protecting surface waters and wetlands, such as the Clean Water Act. If a project disturbs one acre or more, or is part of a larger project disturbing one acre or more, the project developer must comply with National Pollutant Discharge Elimination System (NP-DES) requirements. Compliance involves preparing a Storm Water Pollution Prevention Plan in order to obtain an NPDES permit, which is issued by the state's environmental regulatory agency. Section 404 of the Clean Water Act may also apply if the waters of the United States are potentially affected. Before construction begins, the developer also must ensure that the requirements of various federal laws and regulations protecting historic and archeological resources are met. Provisions such as these apply to all types of construction, not just wind energy, and we will not consider them in any detail here.

State and Local Regulation of Wind-Energy Development

Most regulatory review of wind-energy development takes place at the state or local level, or some combination of them, and most energy development has been on private land, although a few states have anticipated that wind-energy projects could be proposed for state-owned land. In reviewing state and local regulatory frameworks, the committee found it difficult to be sure that we understood these frameworks and their implementation accurately. There are several reasons for our uncertainty:

- The written regulations themselves are often complex and sometimes apparently contradictory.
- Aspects of their implementation are often discretionary, making it hard to summarize the true effects of the written regulations.
- Regulation of wind-energy development is new for most jurisdictions, so both the regulations themselves and the procedures for implementing them are evolving and precedents are being set gradually through experience.

Because of the rapidly changing nature of regulation of wind-energy development, the committee examined records from several recent wind-energy proposals to see how the regulatory process is working in practice, as well as reviewing the regulations themselves.

State-Owned Lands

Some states have developed policies with regard to the use of state-owned lands for wind-energy development. The Pennsylvania Department of Conservation and Natural Resources has completed draft criteria for siting wind-energy projects and a GIS screening tool to guide consideration of the appropriateness of commercial-scale wind-energy development on a small portion of state forestlands (PADCNR 2006).

The state of Vermont has decided that commercial wind-energy development is not an appropriate use for state-owned lands, but that small-scale individual turbines would be appropriate for powering state facilities (VTANR 2004).

Privately Owned Lands

All of the federal regulations described in the previous section as applying to wind-energy developments or other construction activities, regardless of ownership or funding, apply in addition to the state and local regulations discussed here. In some cases, there are state and local regulations that parallel federal requirements. Many states have their own regulations for endangered species, water quality, and so forth. In the Mid-Atlantic Highlands, Pennsylvania, Maryland, and Virginia have state laws protecting various animals and plants (Musgrave and Stein 1993); West Virginia does not (WVDNR 2003). States assemble their own lists of species protected under these laws and may include species not listed at the federal level.

Also, most wind-energy projects undergo some type of local review through local zoning and related ordinances. These local ordinances will not be discussed in detail, unless they are the only level of review or when the local provisions are particularly salient for wind-energy projects (e.g., noise or height ordinances). State and local regulations that govern construction and development projects typically apply to wind-energy projects as well.

Rather than summarize the regulatory process for particular state or local jurisdictions, we concentrate on several recurring themes, some of which came to our attention during public presentations to our committee and some of which we identified as we examined the regulations for numerous states and municipalities. These themes are: (1) the locus of regulatory review (state, local, or mixed); (2) separation or integration of utility and environmental issues in the review process; (3) the information required for review; (4) the procedures for public participation in the review process;

and (5) balancing the positive and negative effects of wind-energy development. In the following sections, we describe these themes using examples from the Mid-Atlantic Highlands states and elsewhere. Then we critique and interpret some of the same themes, along with some others, in order to identify potential improvements to regulatory processes.

Locus of Regulatory Authority: State, Local, Mixed

Regulatory review of wind-energy development varies considerably. It tends to follow one of three patterns: (1) all projects are handled entirely at the state level, (2) larger projects are handled at the state level and smaller projects at the local level (with the size cutoff varying among states), or (3) all projects are handled primarily at the local level. Many states have some state-level permitting of electrical generation facilities, especially transmission lines. Three of the four states in the Mid-Atlantic Highlands have state utility commissions that oversee proposals for electricity generation and transmission. In Virginia, siting (or expanding) a wind-energy facility falls under State Corporation Commission regulation of electric generation facilities (VASCC 2006a). In Maryland, the Public Service Commission must approve construction of electricity-generating facilities and all overhead electric transmission lines of more than 69 kV (MDPSC 1997). In May 2005, West Virginia finalized specific provisions pertaining to wind-energy facilities in its Public Service Commission procedures (WVPSC 2005). Other states are in the process of incorporating specific language concerning wind-energy projects into regulatory rules and guidelines.

In addition to systems for permitting construction and operation of electricity-generating facilities and transmission lines, approvals are often required to connect wind-generated electricity to regional transmission grids, such as the PJM Interconnect in the Mid-Atlantic Highlands. (The PJM Interconnect covers central and eastern Pennsylvania, virtually all of New Jersey, Delaware, western Maryland, and Washington, D.C. A new control area called PJM West is now covered by PJM Interconnect and covers the northern two-thirds of West Virginia, portions of western and central Pennsylvania, western Maryland, and small areas of southeastern Ohio [Bartholomew et al. 2006]).

In some cases, the developer must obtain a variety of state permits before final review by a local planning or governing body. Sometimes the state regulatory authority coordinates or consolidates these permits. The Oregon Office of Energy encourages developers of smaller wind-energy facilities to obtain permits through the Energy Facility Siting Council rather than dealing separately with the variety of state and local permits otherwise required. They argue that at the state-level siting process there is "a defined set of objective standards," while "local-level siting is subject to local procedures and ordinances that vary from county to county and city to city" (White 2002, p. 4). In addition, the Oregon Energy Office states, "Most local land use ordinances address energy facility siting in a superficial way, if they address it at all. It may not be clear what standards the local jurisdiction will apply in deciding whether or to issue a conditional use permit" (p. 4). It notes that "most planning departments around the state have no experience siting large electric generating facilities" (White 2002, p. 5).

Local governments (counties and towns or cities) regulate wind-energy development via local ordinances that apply to any construction proposal. Local regulations, such as zoning of land uses, rights-of-way, building permits, and height restrictions, may constrain wind-energy development. In Virginia, for example, the following local permits were required for the proposed Highlands New Wind Development: (a) a conditional-use permit from the County Board of Supervisors (conditions on height, setback, lighting, color, fencing, screening, signs, operations, erosion control, decommissioning, bonding); (b) a building permit from Highland County; and (c) a site plan approved by the Highland County Technical Review Committee (VASCC 2005).

In Pennsylvania, local regulations constitute the only review, and county governments that issue zoning recommendations and permits for land development and subdivision plans are the regulatory authorities. The Pennsylvania Wind Working Group (which included representatives of the Pennsylvania Department of Environmental Protection, the Clean Air Council, municipal governments, environmental advocacy organizations, and wind-energy companies) has developed a model ordinance to help local governments carry out this responsibility. The Pennsylvania model ordinance contains no environmental provisions except during decommissioning, when re-seeding after grading is required. It does provide guidance on visual appearance of wind turbines and related infrastructure, sound levels, shadow flicker, minimum property setbacks, interference with communications devices, protection of public roads, liability insurance, decommissioning, and dispute resolution. The model ordinance contains language about waivers of the provisions of the ordinance (PAWWG 2006).

As another example, Manitowoc County, Wisconsin, has developed an ordinance regulating large wind-energy projects, defined as projects with more than 100 kW capacity or a total height of more than 170 feet (Kirby Mountain 2006). This ordinance puts limits on noise (less than or equal to 5 dB(A) above the ambient level at any point on neighboring property). It restricts wind-energy development to areas zoned "agricultural" and puts a one-quarter-mile buffer around any area that is zoned C1-Conservancy or NA-Natural Area or within one-quarter mile of any state or county forest, hunting area, lake access, natural area, or park. It requires setbacks of towers from neighboring properties and from public roads and power lines. Other requirements include minimum lighting needed to satisfy FAA guidelines, uniform design for towers within one mile, and steps to reduce shadow flicker at occupied structures on neighboring property.

Locus of Review of Environmental Impacts

Another source of variation in wind-energy regulation among different states is how the review of environmental impacts takes place (here, we are treating "environmental" broadly, to include sociocultural effects, as well as effects on the non-human environment). In some states, environmental permitting of wind-energy projects—including their biological, aesthetic, historic, air quality, and water quality considerations—is under the aegis of the public-utility regulatory authority. In other states, this function is performed by another state agency or by a regional or local body. Most wind-energy projects do not have a federal nexus that triggers NEPA review (see above), but some states have their own environmental-review processes that may come into play when wind-energy developments are proposed. New York and California both have State Environmental Quality Review processes (e.g., NYSERDA 2005a) that trigger required EISs in certain circumstances. In New York, for example, the Department of Environmental Conservation classifies actions as Type I (likely to have significant impact, EIS required), or Type II (only local permits required), or Unlisted (may fall into either category). Projects that are 100 feet or taller in an area without zoning regulations that alter an area 10 acres or larger trigger an EIS process; most commercial wind-energy projects would fall in this category.

In many states, the state utilities commission is charged with the authority to weigh environmental impacts, along with other factors, in deciding whether to permit a wind-energy facility to be built and operated, but with provisions for input from other state and federal agencies more knowledgeable about the environment. In Virginia, the Department of Environmental Quality coordinates the environmental review of electricity-generation facilities and may be responsible for issuing certain permits, such as an Erosion and Sedimentation Control Plan or a section 401 permit from the State Water Control Board. It coordinates input from the Departments of Game and Inland Fisheries, Conservation and Recreation, Historic Resources, Transportation and Mines, Minerals and Energy, and the Virginia Marine Resources Commission. However, the Virginia State Corporation Commission (SCC) has the ultimate responsibility for reviewing and issuing construction permits for wind-energy facilities and other electricity-generating units (VASCC 2006a). In West Virginia, the Division of Natural Resources may become involved if permits related to impacts on endangered and threatened species are required (WVPSC 2005). In Maryland, the Department of Natural Resources (DNR) Power Plant Research

Program is responsible for coordinating the review of proposed energy facilities and transmission lines with other units within the DNR as well as with other state agencies (MDDNR 2006).

The manner in which environmental information is presented to state regulatory authorities varies as well. In West Virginia, input from the Division of Natural Resources and from the USFWS is presented during the public comment period, which would seem to give it less "weight" than if it were presented in a separate stage of the review process. However, the West Virginia process also requires the applicant to file an affidavit listing any permits required by federal or state wildlife authorities due to anticipated impacts on wildlife (WVPSC 2005). In Vermont, where regulatory review of energy facilities is a quasi-judicial process, the Vermont Agency Natural Resources is automatically a party in the case and makes recommendations during hearings on wildlife studies and other natural resource issues (VTANR 2006).

It is not always clear what roles the environmental agencies will play in permitting decisions. In Virginia, the Department of Game and Inland Fisheries (DGIF) coordinates evaluation of effects of proposed projects on wildlife. Although generally supportive of alternative energy sources, including wind, the DGIF voiced substantive concerns about possible effects on birds and bats from the proposed Highland New Wind Development in Highland County to the Virginia SCC. The DGIF asked that the developer provide additional wildlife information and visual analysis, referring to the USFWS guidelines as a standard for wildlife studies that should be provided. The DGIF later wrote to the SCC that the proposed project presents unacceptable risks to wildlife, given that it lacks pre- and post-construction studies of birds, bats, and some other species groups requested by the DGIF, and that it lacks binding requirements for mitigation of adverse effects on wildlife populations. This is strong language from the DGIF, but authority to decide what requirements or conditions to impose on the developer remains with the Virginia SCC (Virginia State Corporation Commission, Case No. PUE-2005-00101, Hearing Examiner's Ruling, July 11, 2006).

Information Required for Review

Regulatory authorities are charged with weighing a complex mix of environmental, socioeconomic, and cultural factors in deciding whether to permit wind-energy development. Even states that have only local review of wind-energy projects, such as Pennsylvania, prescribe a long list of factors for which the applicant should provide information to the review process (e.g., Lycoming County 2005). Generally, little direction is provided about what and how much information to provide, which leads to a wide variance in the amount and quality of information provided by industry for different projects. Some states are developing clearer standards through accumulated practice. West Virginia's recent additions to its utilities review process to address wind-energy development are unusual in prescribing the duration and time of year for studies on birds and bats near proposed wind-energy projects (GAO 2005, p. 31). In some states, pre-hearing conferences are used to identify the types and extent of information that should be provided by the applicant. As illustrated by the Highland New Wind Development case in Virginia (VASCC 2006b), regulatory authorities, other agencies, and parties can request additional information if there appear to be gaps or insufficient information on which to make a decision.

The burden of proof for compliance with regulatory criteria rests almost entirely with the applicant, who usually delegates responsibility for demonstrating compliance to contractors with specialized knowledge. In some cases, regulatory authorities have staff that can provide additional information and review the application for accuracy. In some instances, regulatory authorities may hire independent experts, sometimes at the expense of the developer. In general, it is up to the applicant to provide sufficient information that a decision can be reached, but up to the opposing parties to demonstrate why the standards for acceptance have not been reached.

Public Participation in the Review

It is a well-accepted democratic principle that those whose well-being may be affected by decisions should have a chance to provide input to regulatory processes (see discussion above regarding eight principles for wind-energy regulation, also NWCC 2002). Participation (other than by the applicant and the decision-making authority) is important for securing additional technical expertise, giving a voice to those who might be affected, and conveying information about public values that the decision makers need to carry out the balancing act that the decision procedures require. However, the manner in which input is received varies greatly at all phases of participation. In cases where a proposal for wind-energy development triggers an environmental impact process, whether a NEPA or a state process, as in New York, requirements for public participation may be spelled out as part of the environmental review procedure, although this participation is often late in the process. Elsewhere, requirements for public participation are part of the utilities-review procedure.

The first prerequisite for public participation is that the relevant "publics" should be informed of proposed wind-energy developments. Some state regulations spell out in great detail who should receive notice (and who should give notice) via what media and at what point in the application process. Sometimes the requirements differ according to the size of the proposed project. Some regulatory processes require notification to selected state, federal, and local government agencies, in addition to adjoining property owners and the general public.

There may be different categories of participation, depending partly on the type of decision process followed in a particular jurisdiction. The most common include an information meeting, in which the developer provides information to the public and answers questions about the project; a site visit, to which both regulators and the public are invited and which may include visits to points from which the project would be visible; a public hearing, during which members of the public can provide comments (usually written comments also are accepted over a designated period of time); and participation in the hearing process. On the less formal end of the spectrum, developers and local and regional governments often organize forums for discussing either specific projects or issues of wind-energy development generally. More formally, in contested cases affected parties can apply for "intervener" status. In Vermont, participants become interveners by demonstrating that they will be materially affected by the project. Interveners often include abutting property owners, town or county governments (e.g., planning commissions), as well as public interest groups, environmental organizations, and business groups that can demonstrate that they have a substantive interest in the outcome, are not adequately represented by another party in the case, and would not unduly delay the proceedings. In states like Vermont where quasi-judicial rules apply to the hearing process, interveners receive all mailings concerning written testimony, design changes, etc. They are entitled to present their own witnesses and to cross-examine witnesses (VTPSB 2006).

In all the processes the committee reviewed, input from participants is advisory to the decision authorities. When agencies or other governing bodies that hold permitting responsibilities could refuse to issue a permit required for construction or operation to begin (e.g., local construction permit), they also function as decision authorities. In other instances, their input to the overall decision authority is advisory and is weighed along with other inputs.

Some jurisdictions, both state and local, have formal processes to receive protests from those who disagree with decisions to permit wind-energy development. Decisions may be appealed to a higher board or the state supreme court. Those who can demonstrate that they have been harmed by wind-energy development may be able to seek damages. Those who are concerned about effects on public resources, such as wildlife or cultural resources, may be able to request modifications of the wind-energy installation or of operating procedures to mitigate harm to these resources, especially if they are in violation of specific provisions of a permit. There may also be processes by which the public can provide notice to public officials if a permit violation has been observed.

Although all the required participation processes we reviewed fall into the more passive, one-way communication end of the participatory spectrum (e.g., officials informing the public or officials receiving input from the public), it appears that both applicants and decision authorities are sometimes taking the initiative to convene more active participatory processes with multiway communication among applicant, decision authorities, other government entities, and affected individuals and organizations. Indeed, the permitting guidelines developed by the NWCC urge proponents of wind energy to begin working with affected communities well before submitting formal applications in order to reduce the likelihood of crippling public opposition later in the process (NWCC 2002). In other countries, such as Britain, Australia, and Denmark, early negotiation with affected communities and likely opponents of wind-energy developments has been identified as essential to eventual success in siting wind-generation facilities (BWEA 1994; AusWEA 2002; Ringkøbing Amt, Møller og Grønborg, and Carl Bro 2002). In Germany a government program designed to provide incentives for public acceptance of wind projects gave residents the right to become investors in local wind-energy projects with direct benefits to their own electric bills (Hoppe-Kilpper and Steinhauser 2002).

In addition to participation as an element of regulatory review, participation in proactive planning for wind-energy development is another part of the public-participation spectrum. At least in theory, comprehensive plans form the basis for zoning ordinances and may inform regulatory processes at the state or local level, especially when there is clear language concerning particular resources and land uses. Some states, such as Oregon, require towns to develop comprehensive plans (White 2002). Wind-energy plans have been critical for siting wind-energy projects in Denmark, as described earlier (Ringkøbing Amt, Møller og Grønborg, and Carl Bro 2002). Public participation at the planning stage helps ensure that the values important to stakeholders and general citizens are reflected in the comprehensive plans that seek to guide wind-energy development.

Balancing Pluses and Minuses

Once regulatory authorities receive information on environmental effects, costs, and technical specifications for proposed wind-energy developments, they are charged to decide whether to allow the development to go forward, and with what, if any, conditions to ameliorate negative effects. Directions for this complex weighing of pluses and minuses of using wind energy are scant and generally limited to general statements about "balancing" interests and acting "in the public good," resulting in a holistic balancing of positive and negative impacts of the proposed development, rather than a decision based on clearly stated decision criteria. Often, the direction to regulators appears to presume approval unless serious difficulties with the proposed development become evident. In Virginia, an applicant must show the effects of the facility on the reliability of electric service and the effects on the environment and on economic development, and why the construction and operation of the facility would not be contrary to the public interest (VASCC 2006a). In Vermont, the Public Services Board weighs overall public benefits (need, reliability, economic benefit) against impacts to the natural and cultural environment (VTPSB 2003). In West Virginia, the utilities commission is directed to balance the public interest, the general interest of the state and local economy, and the interests of the applicant (WVPSC 2006a). In some cases the general public good inherent in providing electricity may be judged to outweigh some level of other impacts. This weighing of public good against impacts can be informed by review criteria; by evidence presented; by state energy plans or policy, if they exist; or by precedent. The state may apply conditions to minimize adverse impacts on the environment, including scenic and cultural resources. The applicant can be required to mitigate adverse impacts involving views, noise, traffic, etc.

Some states have articulated standards for making wind-energy regulatory decisions (e.g., State of Oregon 2006). However, specific criteria for different elements of the regulatory review, such as assessment of environmental effects, often are lacking. In Maryland, applicants are required to comply with environmental regulations, and conditions may be imposed to mitigate adverse impacts on environmental and cultural resources, but what constitutes compliance and what may be required for mitigation are open to interpretation in particular cases. Maryland's Wind Power Technical Advisory Group, a non-regulatory body from the Power Plant Research Program, has recommended standards for siting, operating, and monitoring wind-energy projects to minimize negative effects on birds and bats (MD Windpower TAG 2006). Sometimes more specific criteria are found in case law rather than in statutes. Vermont, for example, developed a much more detailed process known as the "Quechee Analysis" for analyzing visual impacts as part of case history, which has become an integral part of the regulatory criteria (Vissering 2001). Maine has developed guidance for review of development within the Unincorporated Territories (Maine Land Use Regulatory Commission 1997), but it has not been updated to address some of the specific attributes of wind-energy projects. The best processes provide a detailed framework that asks critical questions, along with a framework for determining how the outcome should be judged.

The same lack of definite criteria applies to post-construction operation, although some jurisdictions are working on specific monitoring criteria. In Virginia, the DGIF supports setting a threshold for implementation of mitigation measures if more than 1.8 bats or 3.5 birds are killed per turbine per year. Research is currently being conducted on new technologies for deterrents or mechanisms that reduce mortality of bats and birds. As these mitigation measures become available, the DGIF recommends their pre- and post-construction implementation in consultation with natural-resources agencies (VADEQ 2006).

A Critique of Planning and Regulatory Review

Wind energy is a recent addition to the energy mix in most areas, and regulation of wind-energy development is evolving rapidly. Our review of current regulatory practices captures only a snapshot of a changing landscape. Regulatory authorities, wind-energy developers, affected citizens, and non-governmental organizations promoting and opposing wind-energy projects are learning as they go. In this section we move beyond simply describing the current status of planning and regulation of wind-energy development to evaluating the merits and deficiencies of current processes and suggesting where and how they might be improved. We call attention to some cross-cutting themes affecting regulatory review of wind-energy development: (1) the interactions among choosing the locus of review, balancing competing goals, and facilitating public participation; (2) the merits of flexible versus more rigidly specified review processes; (3) cumulative effects of wind-energy development; (4) long-term accountability for both positive and negative effects of wind-energy development; and (5) assistance to improve the quality of decisions about wind-energy development.

Interaction of Locus of Review, Balancing of Interests, and Public Participation

In analyzing different types of regulatory processes, the committee found variation ranging from reviews conducted almost entirely at the state level to those conducted almost entirely at the local level. Choosing a level for reviewing wind-energy development is likely to imply some corresponding consequences for the balance of competing interests and for the structure and content of public participation in decisions. These corresponding consequences may not be intentional and the connection to level of review not explicit. Several states seem to be moving toward state-level review, perhaps because of concerns about potentially inequitable decisions in different locations and the inexperience inherent in local review. Oregon, for example, encourages developers to select state rather than local review by offering a more streamlined process at the state level (White 2002). Review at a scale larger than local allows implementation of a rational power-generation network with oversight of potential cumulative impacts.

Putting utility regulation at the state rather than local level implies that there is a public interest that is broader in scale, and greater in importance, than strictly local interests. If the preceding is true, then environmental and societal costs of wind-energy development, evaluated at site-specific, local, and regional scales, must be weighed against public benefits that might be realized at the state level or beyond. Some states have examined this tension between local and broader interests quite explicitly. For example, Vermont created a Commission on Wind Energy Regulatory Policy in 2004 to recommend changes to the current regulatory process (Vermont Commission on Wind Energy Regulatory Policy 2004). One issue of concern was whether wind-energy projects should be reviewed under the State's Public Service Board, which reviews all public-utility projects, or whether review should be made under the more localized District Environmental Commissions, which focus on land use. That report represents a thoughtful and deliberate consideration of the implications of level of review for how local versus broader-scale interests are to be weighed in decisions about wind-energy development. The Vermont analysis confirmed the choice of a state-level review process, where public interest on a broad scale is weighed against possibly adverse effects at the local level, but it also recommended increased protection for local interests during the process through aggressive public notification and public participation.

One of those increased protections concerns the manner of public participation in the review process, another arena where choosing the level of review may implicitly determine who has standing as a participant in the review process and how they can participate. Where review is strictly local, broader interests may have less opportunity to be heard. These broader interests may include people beyond the wind-energy development site who would like to receive the benefits of wind energy, and regional or national organizations advocating the protection of wildlife and humans from possibly harmful effects of wind-energy development. Some more-formally constituted participatory processes, such as quasi-judicial hearings, specify how individuals or organizations may petition for an enhanced status. For example, they can be designated "interveners," which entitles them to privileges such as cross-examining experts and receiving copies of all filings in a contested case. The Vermont Commission on Wind Energy Regulatory Policy made numerous recommendations concerning public participation in the regulatory process, addressing issues such as advance notice to communities and affected individuals prior to filing, the number and timing of public hearings, the definition of "affected communities," and information and assistance to increase public understanding of and participation in the regulatory process.

Another matter that may be affected by level of review is equity with respect to socioeconomic class, race, or ethnicity of citizens living near wind-energy facilities who are most susceptible to local adverse effects. Environmental-justice issues most often are raised where locally controversial facilities are sited disproportionately in low-income or otherwise politically weak neighborhoods, where citizens may lack educational and political resources to represent their own interests effectively. Here, level of review may cut both ways: developers might take advantage of strictly local review to site facilities where oversight is weak, or state-level review might consistently place the interests of the larger public ahead of the interests of a politically weak local population. These concerns may be less likely to arise for wind-energy facilities than for other types of locally controversial facilities, because the technical requirements for successful wind-energy development constrain the location of facilities so tightly (at least on land).

Both developers and regulatory authorities can take the initiative to foster public participation in wind-energy development, rather than stopping at the minimum needed to satisfy regulatory requirements. Local and state governments can invite public participation in proactive planning for wind-energy development to learn how stakeholder groups and the general citizenry view opportunities and obstacles. Developers could meet with adjoining landowners, community groups, and environmental organizations during the pre-application phase to hear concerns about a proposed project, giving them the opportunity to make changes that decrease the likelihood of public opposition. To prepare for this involvement, developers may benefit from providing descriptions of the proposed project and rationale for selecting the proposed site rather than an alternative for the public to review. Regulatory authorities can solicit public participation beyond required public notices and public hearings to bring local knowledge about environmental and cultural resources into the decision-making process and to satisfy procedural justice concerns for representation of those affected by regulatory decisions.

Optimizing Flexibility, Rigor, and Predictability of Regulatory Review

Processes for reviewing wind-energy proposals vary in the formality of the process and in the degree to which timelines and decision criteria are specified in advance. There are tradeoffs between the predictability and rigor that may be achieved with processes that are more formal and more clearly specified, and the flexibility and adaptability that may be achieved with processes that are less formal and less clearly specified. For example, many review processes specify a timeline for various stages of the review (e.g., submission of technical information, notification to affected publics) or specify a deadline for the regulatory authority to respond to the request for permission to construct a facility. Having specific timelines and deadlines protects developers, regulators, and the public from the extended uncertainty that might accompany a drawn-out review process. However, one notable characteristic of wind-energy proposals is that they vary enormously in the complexity of potential effects. This complexity suggests that a more-flexible timeline would allow both complex and simple projects to meet common standards for quality of information submitted and quality of evaluation of that material by regulators and the public. In Vermont, rather than specifying the same deadline for all utility proposals, state statutes require the utilities board to set timelines for each proposal based on its complexity; once set, all the parties to the review are held to the timelines (Vermont Commission on Wind Energy Regulatory Policy 2004).

In evaluating current regulatory-review processes, the committee was struck by the minimal guidance offered about the kind and amount of information that should be provided for review; the degree of adverse or beneficial effects of proposed developments that should be considered critical for approving or disapproving a proposed project; and how competing costs and benefits of a proposed project should be weighed, either with regard to that single proposal or in comparison with likely alternatives if that project is not built. This lack of guidance leaves a lot to the discretion of regulatory authorities and the other agencies that review elements of the proposed project, making both developers and the public vulnerable to inconsistent requirements among proposed projects and among potential locations. It also has limited our knowledge of the impacts of wind-energy development on human and natural resources. As regulatory authorities accumulate experience with wind-energy proposals, conventions are developing for how much pre-project study of bird and bat activity should be done or what level of bird or bat mortality at operating wind-energy projects will be considered cause for remedial action, as Virginia DGIF has done in recommending limits for bird and bat mortality in comments on the proposed New Highland Wind Development (VADEQ 2006). Nevertheless, there is still something to be said for letting the context of a particular wind-energy proposal set the requirements for information and the thresholds for regulatory decisions, as the Vermont process does for setting the timeline for review. Such flexibility could optimize the expenditure of both private and public resources on information collection and review by focusing on the particular elements most likely to be troublesome for a particular project. However, this degree of flexibility requires a great deal of trust in the judgment of the regulatory authority by developers and the public.

Proactive planning for wind-energy development at state and local levels could give valuable direction to regulatory review by articulating public values that might be affected by projects (e.g., local aesthetic values or socioeconomic concerns, such as effects on tourism). These values, as translated into planning guidelines and local zoning ordinances, help set standards for regulatory review.

There are advantages and disadvantages to giving regulators more direction on how to weigh competing costs and benefits of proposed wind-energy projects to make decisions that advance "the public good," as is required of many regulatory authorities. Having thresholds of positive or negative effects may make regulatory decisions easier to defend from criticism, but specification of such thresholds can inhibit regulators from weighing a complex suite of factors to make a combined index of how much a particular project advances the public good. Tools from multicriteria decision making (e.g., Hammond et al. 2002) can help structure this process by representing preferences for possible outcomes and weighting various decision criteria in numerical form. However, the assessment of those weights and preferences are expressions of value, raising the critical question of whose values should inform decisions about the public good. Some argue that citizens have authorized regulatory bodies, such as utilities commissions, to represent public values taken as a whole. Others argue that only through participatory processes, including negotiation of regulatory rules, or through overtly political processes, such as public forums, can the diverse values of different constituencies be expressed. Public involvement in areas affected by wind-energy proposals is one mechanism for eliciting that diversity of values, but the complex task of combining them into a single decision remains with the regulatory authority.

There are, similarly, pros and cons to more versus less formal review processes. On the formal end of the spectrum, quasi-judicial processes have such merits as producing written records of deliberations, prescribing who can speak in what capacities during hearings, providing opportunities to cross-examine expert witnesses and challenge evidence, and requiring authorities to respond to public comments to indicate how an issue has been addressed. These merits are, to some extent, offset by constraints on who may qualify to participate in hearings and what roles they can play. In addition, more formal processes, although providing a basis for appeal when parties question a decision, may solidify conflicting views and inhibit the more creative give-and-take that can sometimes help resolve contentious issues.

Assuring Long-Term Project-Permit Compliance

Post-construction monitoring for compliance with permit conditions is a critical part of the regulatory process. It is needed to ensure that projects are built according to approved plans and that required post-construction studies and mitigation measures are being carried out properly. Full access to project sites is needed for those charged with conducting studies or monitoring activities. Access has been problematic in the past. For example, access to the Mountaineer Project in West Virginia to conduct studies of bird and bat fatalities was discontinued by the project owner (E. Arnett, Bat Conservation International, personal communication 2005).

The application for the proposed Jack Mountain/Liberty Gap project was dismissed without prejudice (i.e., the application could be resubmitted) by the West Virginia Public Service Commission because the applicant refused to allow access to the property for hydrological studies (WVPSC 2006b). Well-defined processes for addressing post-construction monitoring and potential permit violations are needed at both local and state levels. Public confidence in facility compliance would be enhanced if site operators designated an accessible contact person who could respond to inquiries or complaints. In addition to monitoring for adverse environmental effects, including adverse socioeconomic effects, documenting the energy benefits of wind-energy facilities over the lifespan of the installation also is important. For this purpose, data on electricity generated, which must be reported monthly to the Department of Energy's Energy Information Agency for electricity-generating plants of 1 MW or greater, should be more easily accessible by the public than they currently are on the agency's web site. To ensure long-term compliance with monitoring, mitigation, and reporting requirements, commitments made by the initial site developer should be passed to subsequent operators of the site, including those responsible for maintaining, refurbishing, or re-powering during the project's lifetime, and decommissioning after its lifetime. To ensure transparency, state public-service commissions, with the corresponding state environmental or natural-resources offices, could evaluate pre- and post-construction monitoring as part of the permitting process.

Proactive Planning and Evaluation of Cumulative Effects

The positive and negative cumulative effects of wind-energy development across space and over time generally receive little attention in current regulatory-review processes, although developers have sometimes been asked to provide information about cumulative effects (e.g., Highland New Wind Development in Virginia [VADEQ 2006]). As the Vermont Commission on Wind Energy Regulatory Policy (2004) noted, broader review may facilitate better consideration of cumulative effects than strictly local review. In addition, wind turbines can be large in relation to natural landscape features, extending their effects (e.g., visual impact) beyond the boundaries of the municipality where the turbine itself is located. Broader review would capture effects that extend beyond local jurisdictions.

Consideration of cumulative effects would be facilitated by more proactive planning for wind-energy development at scales ranging from national to regional between or within states. Resistance to centralized planning and devotion to private-property rights and individual autonomy in the United States may rule out the type of integrated planning and regulation that northern European countries and Australia have pursued. Nevertheless, there is room in the United States for better integration of these functions to the benefit of wind-energy developers and for protection of the public good. It is a waste of private and public resources when developers invest in projects that cannot be sited successfully. Planning at state and local levels works with regulatory review to direct wind-energy development to locations and site designs that minimize adverse effects. Clear planning documents set the stage for predictable and defensible review actions.

There often are thresholds for project or turbine size, below which regulatory scrutiny either is not required or is much reduced. If several small projects are installed in a small area, their effects could accumulate without the benefit of regulatory review. For example, several individual businesses or farms may install small turbines, on the order of 40 kW. Although a single turbine meeting relevant construction and zoning requirements might have little effect on local wildlife, aesthetics, and cultural resources, several of them might have significant effects, but they would not be regulated. This is a gap in current regulatory policy.

Improving the Quality of Review

Evaluating the merits and drawbacks of wind-energy proposals strains the resources of regulatory authorities in state utilities commissions and even more in local governments. Although experience is accumulating, wind energy still is new and unfamiliar. Local decision authorities are unlikely to learn by experience very rapidly because they see relatively few wind-energy proposals. Regulatory guidelines, both from nationwide efforts (e.g., NWCC 2002) and state-level efforts (e.g., KSREWG 2003; KSEC 2004), are one form of assistance to state and local decision makers. Many states, including California, Colorado, Maryland, Pennsylvania, New York, and states in the Great Lakes Region, have sponsored or established wind-energy working groups, bringing together stakeholders such as environmental groups, industry, academia, and state agencies to set goals and guidelines for wind-energy development. In some states, efforts such as Maryland's Wind Power Technical Advisory Group help fill technical gaps at the local level (MD Windpower TAG 2006). In Vermont, the state utilities board can hire independent experts at the expense of the developer to assist the state in its review (Vermont Commission on Wind Energy Regulatory Policy 2004). Similar assistance would be even more beneficial to local decision makers.

FRAMEWORK FOR REVIEWING WIND-ENERGY PROPOSALS

Part of the committee's charge was to develop an analytical framework for reviewing environmental and socioeconomic effects of wind-energy proposals. Ideally, this framework would (1) detail not only the types of effects to be considered, but also how those effects are to be evaluated as desirable or undesirable and how positive and negative effects are to be weighed in an overall assessment of a particular proposal; (2) address wind-energy development across a range of spatial and temporal scales; (3) integrate technical information on wind-energy effects with expressions of relevant public values; and (4) enable comparisons of wind-energy projects with other forms of electricity production. We have stopped short of this ideal for several reasons.

Although in theory it seems sensible to weigh the comparative environmental performance of different electricity sources, in practice the generally piecemeal nature of U.S. policy making and regulation offers few opportunities for such comparisons. Energy policies (expressed through such means as tax credits and other financial incentives) usually are the result of considering particular energy sources by themselves rather than the result of weighing the advantages and disadvantages of different energy sources. Regulatory review of energy facilities almost always is a yes/no judgment on a single proposal (perhaps with modifications or conditions imposed), not a comparative judgment of the merits of different energy sources, sites, or facility designs. There is little planning that addresses particular mixes of energy sources, particular sites for wind-energy development, or particular designs for wind-energy facilities. Even if such planning were done, it would have limited impact on proposed wind-energy facilities and their approval, because proposals usually arise one at a time. The review of individual proposals usually is quite limited in scope, both temporally and spatially, with little opportunity for a full life-cycle analysis or for consideration of effects that accumulate across space and time.

In addition, the U.S. system, with its private ownership of most energy facilities and with its prevailing emphasis on markets as the best arbiters of balancing the costs and benefits of energy projects, offers few opportunities for thorough public deliberation on the full spectrum of positive and negative effects of a particular energy facility. At present, if a proposed project meets regulatory requirements (which generally do not include a comprehensive balancing of positive and negative effects), it usually must be approved. Setting regulatory thresholds (e.g., for noise, number of birds killed, visibility) implies that some tradeoffs among costs and benefits are addressed, but even if the tradeoffs are addressed, it usually is not in a transparent and comprehensive way. Instead, these implicit tradeoffs evolve more or less invisibly as projects are proposed, reviewed, modified, and implemented. Eventually, this evolution may result in changes to regulatory processes and standards, but even then, the weighing of tradeoffs does not necessarily become transparent.

There is, moreover, currently no social consensus on how the advantages and disadvantages of wind-energy projects should be traded off or whose value systems should prevail in making such judgments. Instead, these decisions usually take place through a combination of citizen participation, political advocacy, and regulatory decision making. As discussed earlier in this chapter, both predictable but also more rigid regulatory-review procedures, and less predictable but also more flexible procedures, have their advantages. In addition, maintaining the flexibility to tailor the intensity of regulatory review to the complexity and controversy associated with particular wind-energy proposals makes more efficient use of society's resources than a "one size fits all" process that does not provide opportunities for exceptions.

For all of these reasons, we focus our efforts on incrementally improving the way wind-energy decisions are made today. We offer an evaluation guide that aids vertical coordination of regulatory review by various levels of government and helps to ensure that regulatory reviews are well grounded procedurally and evaluate the many facets of the human and non-human environment that may be affected by wind-energy development.

Coordinating Levels of Governmental Responsibility

To assist those responsible for planning and regulating wind-energy development and to facilitate the coordination of their work, we suggest using a two-dimensional matrix of jurisdictional levels and areas of responsibility. Jurisdictional levels range in scale from international (occasionally) and national to regional, state, and local. Areas of responsibility include formulation and execution of policy, planning, and public relations; legal and regulatory activities; and impact evaluation. In Figure 5-1, these two dimensions are displayed as a matrix.

The details of how this matrix is filled out will vary from state to state, and to a lesser extent, from project to project. Nonetheless, using the matrix and considering each of its cells will help to ensure that important elements of governmental responsibilities have not been overlooked and that review efforts are well coordinated across geographic areas and jurisdictional levels. Once the respective responsibilities of the various jurisdictions are clearly identified and articulated, a checklist of questions like those in Box 5-4 below can serve as a template for evaluation.

Evaluation Guide

The evaluation guide presented here represents a step toward a realistic, workable framework for reviewing proposed and evaluating existing wind-energy projects. If this guide is followed and adequately documented, the results will provide a basis not only for evaluating an individual wind-

	Federal	Regional/State	Local
Policy, Planning, and Public Relations			
Legal and Regulatory			
Evaluation of Impacts			
Environmental			
Human Health and Well-Being			
Aesthetic			
Cultural			
Economic and Fiscal			
Electromagnetic Interference			

FIGURE 5-1 Matrix for organizing review of wind-energy projects

energy project, but also for comparing two or more proposed projects and for undertaking an assessment of cumulative effects of existing and proposed facilities. In addition, following this guide may facilitate rational documentation of the most important areas for research.

The guide first addresses procedural considerations—policy, planning, and public relations—and relevant laws and regulations. It then addresses the main potential effects of wind-energy facilities, organizing them into six categories drawn from Chapters 3 and 4: (1) impacts on the environment, (2) impacts on human health and well-being, (3) aesthetic impacts, (4) cultural impacts, (5) economic and fiscal impacts, and (6) electromagnetic interference. A seventh cross-cutting category concerning cumulative impacts is added. All these potential effects should be considered also in light of the benefits of any proposed project, including environmental benefits. The guide (Box 5-4) is presented as sets of questions to aid evaluation at various jurisdictional levels.

CONCLUSIONS AND RECOMMENDATIONS

The committee concludes that a country as large and as geographically diverse as the United States, and as wedded to political plurality and private enterprise, is unlikely to plan for wind energy at a national scale in the same way as some other countries are doing. Nevertheless, national-level energy policies (implemented through mechanisms such as incentives, subsidies, research agendas, and federal regulations and guidelines) to enhance the benefits of wind energy while minimizing negative impacts would help in planning and regulating wind-energy development at smaller scales. Uncertainty about what policy tools will be in force hampers proactive planning for wind development. More specific conclusions and recommendations follow.

BOX 5-4
Guide for Evaluating Wind-Energy Projects

Policy, Planning, and Public Relations

1. Are the relevant energy policies and planning processes clearly defined at all jurisdictional levels, and are they coordinated and aligned among federal, state, and local levels? Are national-level energy policies available and being used? Are well-reasoned planning documents available to make regulatory-review actions predictable and defensible?

2. Have mechanisms been established to provide necessary information to interested and affected parties, and to seek meaningful input from them as wind-energy projects are planned and implemented? Are developers required to provide early notification of their intent to develop wind energy?

3. Are procedures—including policies and regulations—in place for evaluating the impacts of wind-energy projects that cross jurisdictional boundaries, especially for those that involve more than one state?

4. Is guidance available to developers, regulators, and the public about what kinds of information are needed for review, what degrees of adverse and beneficial effects of proposed wind-energy developments should be considered critical in evaluating a proposed project, and how competing costs and benefits of a proposed project should be weighed with regard to that proposal only, or by comparison with likely alternatives? Are there mechanisms in place through which interested parties can obtain the pertinent available information?

5. Are regional planning documents available that provide guidance on the quality of wind resources, capacity of transmission options, potential markets, major areas of concern, and tradeoffs that should be considered?

Legal and Regulatory Considerations

1. Are wind-energy guidelines and regulations issued by different federal agencies compatible, are those guidelines and regulations aligned with other federal regulating rules and regulations, and do the guidelines and regulations follow acceptable scientific principles when establishing data requirements?

2. Does the review process include steps that explicitly address the cumulative impacts of wind-energy projects over space and time; that is, by reviewing each new project in the context of other existing and planned projects in the region?

Evaluation of Impacts

General

1. Are the biological, aesthetic, cultural, and socioeconomic attributes of the region sufficiently well known to allow an accurate assessment of the environmental impacts of the wind-energy project, and to distinguish among the potential sites considered during the site-selection process? Are there species, habitats, recreational areas, or cultural sites of special interest or concern that will be affected by the project? How will this descriptive information be collected, who will judge its quality and reliability, and how will the information be shared with stakeholders? Are there key gaps in the needed information that should be addressed with further research before a project is approved or to guide the operation of an approved project?

Environmental Impacts

1. What environmental mitigation measures will be taken and how will their effectiveness be measured? Are there any legal requirements for such measures (e.g., habitat conservation plans)? Are any listed species at risk from the proposed facility?
2. How and by whom will the environmental impacts be evaluated once the project is in operation? If these evaluations indicate needed changes in the operation of the facility, how will such a decision be made and how will their implementation be assured?
3. What pre-siting studies for site selection and pre-construction studies for impact assessment and mitigation planning are required?
4. What post-construction studies, with appropriate controls, are required to evaluate impacts, modify mitigation if needed, and improve future planning?

Impacts on Human Health and Well-Being

1. Have pre-construction noise surveys been conducted to determine the background noise levels? Will technical assessments of the operational noise levels be conducted? Is there an established process to resolve complaints from the operation of the turbines?
2. Is there a process in place to address complaints of shadow flicker and does the operator use the best software programs to minimize any flicker?

Aesthetic Impacts

1. Has the project planning involved professional assessment of potential visual impacts, using established techniques such as those recommended by the U.S. Forest Service or U.S. Bureau of Land Management?
2. How have the public and the locally affected inhabitants been involved in evaluating the potential aesthetic and visual impacts?

Cultural Impacts

1. Has there been expert consideration of the possible impacts of the project on recreational opportunities and on historical, sacred, and archeological sites?

Economic and Fiscal Impacts

1. Have the direct economic impacts of the project been accurately evaluated, including the types and pay scales of the jobs produced during the construction and operational phases, the taxes that will be produced, and costs to the public?
2. Has there been a careful explication of the indirect economic costs and benefits, including opportunity costs and the distribution of monetary and non-monetary benefits and costs?
3. Are the guarantees and mitigation measures designed to fit the project and address the interests of the community members and the local jurisdictions?

Electromagnetic Interference

1. Has the developer assessed the possibility of radio, television, and radar interference?

Cumulative Effects

1. How will cumulative effects be assessed, and what will be included in that assessment (i.e., the effects only of other wind-energy installations, or of all other electricity generators, or of all other anthropogenic impacts on the area)? Have the spatial and temporal scales of the cumulative-effects assessment been specified?

Conclusion

Because wind energy is new to many state and local governments, the quality of decisions to permit wind-energy developments is uneven in many respects.

Recommendation

Guidance on planning for wind energy and on data requirements and on procedures for reviewing wind-energy proposals should be developed. In addition, technical assistance with gathering and interpreting information needed for decision making should be provided. This guidance and technical assistance, conducted at appropriate jurisdictional levels, could be developed by working groups composed of wind-energy developers, non-governmental organizations with diverse views of wind-energy development, and local, state, and federal government agencies.

Conclusion

There is little anticipatory planning for wind-energy projects, and it is not clear whether mechanisms currently exist that could incorporate such planning in regulatory decisions even if such planning occurred.

Recommendation

Regulatory reviews of individual wind-energy projects should be preceded by coordinated, anticipatory planning whenever possible. Such planning for wind-energy development coordinated with regulatory review of wind-energy proposals would benefit developers, regulators, and the public because it would prompt developers to focus proposals on locations and site designs most likely to be successful. This planning could be implemented at scales ranging from state and regional levels to local levels. Anticipatory planning for wind-energy development also would help researchers target their efforts where they will be most informative for future wind-development decisions.

Conclusion

Choosing the level of regulatory authority for reviewing wind-energy proposals carries corresponding implications for how the following issues are addressed:

- Cumulative effects of wind-energy development.

- Balancing negative and positive environmental and socioeconomic impacts of wind energy.
- Incorporating public opinions into the review process.

Recommendation

In choosing the levels of regulatory review of wind-energy projects, agencies should review the implication of those choices to all three issues listed above. Decisions about the level of regulatory review should include procedures for ameliorating the disadvantages of a particular choice (e.g., enhancing opportunities for local participation in state-level reviews).

Conclusion

Well-specified, formal procedures for regulatory review enhance predictability, consistency, and accountability for all parties to wind-energy development. However, flexibility and informality also have advantages, such as matching the time and effort expended on review to the complexity and controversy associated with a particular proposal; tailoring decision criteria to the ecological and social contexts of a particular proposal; and fostering creative interactions among developers, regulators, and the public to find solutions to wind-energy dilemmas.

Recommendation

When consideration is given to formalizing review procedures and specifying thresholds for decision criteria, this consideration should include attention to ways of retaining the advantages of more flexible procedures.

Conclusion

Using an evaluation guide to organize regulatory review processes—such as the guide we have provided here—can help achieve comprehensive and consistent regulation, coordinated across jurisdictional levels and across types of effects.

Recommendation

Regulatory agencies should adopt and routinely use an evaluation guide in their reviews of wind-energy projects. The guide should be available to developers and the public.

Conclusion

The environmental benefits of wind energy, mainly reductions in atmospheric pollutants, are enjoyed at wide spatial scales, while the environmental costs, mainly aesthetic impacts and ecological impacts such as increased mortality of birds and bats, occur at much smaller spatial scales. There are similar, if less dramatic, disparities in the scales of occurrence of economic and other societal benefits and costs. The disparities in scale, while not unique to wind energy, complicate the evaluation of tradeoffs.

Recommendation

Representatives of federal, state, and local governments should work with wind-energy developers, non-governmental organizations, and other interest groups and experts to develop guidelines for addressing tradeoffs between benefits and costs of wind-energy generation of electricity that occur at widely different scales, including life-cycle effects.

References

Able, K.P. 1977. The flight behavior of individual passerine nocturnal migrants: A tracking radar study. Anim. Behav. 25(4):924-935.

Able, K.P., and S.A. Gauthreaux, Jr. 1975. Quantification of nocturnal passerine migration with a portable ceilometer. Condor 77(1):92-96.

Adams, D.C., J. Gurevitch, and M.S. Rosenberg. 1997. Resampling tests for meta-analysis of ecological data. Ecology 78(5):1277-1283.

Ahlén, I. 2002. Bats and birds killed by wind power turbines [in Swedish]. Fauna och Flora 97(3):14-21.

Ahlén, I. 2003. Wind Turbines and Bats: A Pilot Study [in Swedish]. Final Report Dnr 5210P-2002-00473, P-nr P20272-1. Swedish National Energy Commission, Eskilstuna, Sweden (English translation by I. Ahlén, March 5, 2004).

Ahlén, I. 2004. Heterodyne and time-expansion methods for identification of bats in the field and through sound analysis. Pp. 72-79 in Bat Echolocation Research: Tools, Techniques, and Analysis, R.M. Brigham, E.K.V. Kalko, G. Jones, S. Parsons, and H.J.G.A. Limpens, eds. Austin, TX: Bat Conservation International.

Aldridge, H.D.J.N., and I.L. Rautenbach. 1987. Morphology, echolocation and resource partitioning in insectivorous bats. J. Anim. Ecol. 56(3):763-778.

Alliant Energy. 2007. Wind Power. Alliant Energy [online]. Available: http://www.powerhousekids.com/stellent2/ groups/public/documents/pub/phk_ee_re_001502.hcsp [accessed April 19, 2007].

Amstrup, S.C., T.L. McDonald, and B.F.J. Manly. 2005. Handbook of Capture-Recapture Analysis. Princeton, NJ: Princeton University Press.

Andersen, D.P. 1999. Review of Historical and Modern Utilization of Wind Power. RISØ National Laboratory, Roskilde, Denmark [online]. Available: http://www.risoe.dk/rispubl/ VEA/dannemand.htm [accessed Aug. 24, 2006].

Anderson, D., D. Curry, E. DeMeo, S. Enfield, T. Gray, L. Hartman, K. Sinclair, R. Therkelsen, and S. Ugoretz. 2002. Permitting of Wind Energy Facilities: A Handbook Revised 2002. Prepared by the National Wind Coordinating Committee Siting Subcommittee, c/o RESOLVE, Washington, DC. August 2002 [online]. Available: http://www.nationalwind.org/publications/siting/permitting2002.pdf [accessed June 1, 2006].

Anderson, R.L., and J.A. Estep. 1988. Wind Energy Development in California: Impact, Mitigation, Monitoring, and Planning. California Energy Commission, Sacramento, CA.
Anderson, R.L., J. Tom, N. Neumann, J. Noone, and D. Maul. 1996a. Avian risk assessment
methodology. Pp. 74-87 in Proceedings of the National Avian-Wind Power Planning Meeting II, September 20-22, 1995, Palm Springs, CA. King City, Ontario: LGL Ltd Environmental Research Associates. October 1996 [online]. Available: http://www.nationalwind.
org/publications/wildlife/avian95/avian95-09.htm [accessed May 19, 2006].
Anderson, R.L., J. Tom, N. Neumann, and J.A. Cleckler. 1996b. Avian Monitoring and Risk
Assessment at Tehachapi Pass Wind Resource Area, California. California Energy Commission, Sacramento, CA. November 1996. 40 pp.
Anderson, R.L., H. Davis, W. Kendall, L.S. Mayer, M.L. Morrison, K. Sinclair, D. Strickland,
and S. Ugoretz. 1997. Standard metrics and methods for conducting avian/wind energy
interaction studies. Pp. 265-272 in Proceedings of the 1997 American Wind Energy As-
sociation Annual Meeting, November 5, 1997, Boston, MA. Washington, DC: American
Wind Energy Association.
Anderson, R.L., M. Morrison, K. Sinclair, and M.D. Strickland. 1999. Studying Wind Energy/Bird
Interactions: A Guidance Document. Prepared for Avian Subcommittee and National Wind
Coordinating Committee. December 1999 [online]. Available: http://www.nationalwind
.org/publications/wildlife/avian99/Avian_booklet.pdf [accessed May 19, 2006].
Anderson, R.L., D. Strickland, J. Tom, N. Neumann, W. Erickson, J. Cleckler, G. Mayorga, G.
Nuhn, A. Leuders, J. Schneider, L. Backus, P. Becker, and N. Flagg. 2000. Avian monitor-
ing and risk assessment at Tehachapi Pass and San Gorgonio Pass wind resource areas,
California: Phase I preliminary results. Pp. 31-46 in Proceedings of the National Avian-
Wind Power Planning Meeting III, May 1998, San Diego, CA. National Wind Coordinat-
ing Committee/RESOLVE, Washington, DC [online]. Available: http://www.nationalwind.
org/publications/wildlife/avian98/06-Anderson_etal-Tehachapi_San_Gorgonio.pdf [ac-
cessed May 24, 2006].
Anderson, R.L., N. Neumann, J. Tom, W.P. Erickson, M.D. Strickland, M. Bourassa, K.J.
Bay, and K.J. Sernka. 2004. Avian Monitoring and Risk Assessment at the Tehachapi
Pass Wind Resource Area: Period of Performance: October 2, 1996–May 27, 1998.
NREL/SR-500-36416. National Renewable Energy Laboratory, Golden, CO. September
2004 [online]. Available: http://www.nrel.gov/docs/fy04osti/36416.pdf [accessed May
19, 2006].
Anderson, R.L., J. Tom, N. Neumann, W.P. Erickson, M.D. Strickland, M. Bourassa, K.J.
Bay, and K.J. Sernka. 2005. Avian Monitoring and Risk Assessment at the San Gorgonio
Wind Resource Area: Phase I Field Work: March 3, 1997-May 29, 1998; Phase II Field
Work: August 18, 1999-August 11, 2000. NREL/SR-500-38054. National Renewable
Energy Laboratory, Golden, CO. August 2005 [online]. Available: http://www.nrel.gov/
docs/fy05osti/38054.pdf [accessed May 19, 2006].
Appleton, J. 1975. The Experience of Landscape. New York: Wiley.
Arbogast, B.S., R.A. Browne, P.D. Weigl, and G.J. Kenagy. 2005. Conservation genetics of
endangered flying squirrels (Glaucomys) from the Appalachian mountains of eastern
North America. Anim. Conserv. 8(2):123-133.
Arnett, E.B. 2005. Relationships Between Bats and Wind Turbines in Pennsylvania and West
Virginia: An Assessment of Fatality Search Protocols, Patterns of Fatality, and Behavioral
Interactions with Wind Turbines. Final Report. Prepared for the Bats and Wind Energy
Cooperative, by Bat Conservation International, Austin, TX. June 2005 [online]. Avail-
able: http://www.batcon.org/wind/BWEC2004finalreport.pdf [accessed May 25, 2006].
Arnett, E.B. 2006. A preliminary evaluation on the use of dogs to recover bat fatalities at wind
energy facilities. Wildlife Soc. Bull. 34:1440-1445.
Arnett, E.B., and J.P. Hayes. 2000. Bat use of roosting boxes installed under flat-bottom
bridges in western Oregon. Wildlife Soc. Bull. 28(4):890-894.

Arnett, E.B., K. Brown, W.P. Erickson, J. Fiedler, T.H. Henry, G.D. Johnson, J. Kerns, R.R. Koford, C.P. Nicholson, T. O'Connell, M. Piorkowski, and R. Tankersley, Jr. In press. Ecological impacts of wind power development on bats: Case studies on the patterns of bat
fatalities at wind power facilities in North America. Journal of Wildlife Management.

Ash, A.N. 1997. Disappearance and return of plethodontid salamanders to clearcut plots in
the southern Blue Ridge Mountains. Conserv. Biol. 11(4):983-989.

Associated Press. 2006. Developer: Wind farm won't hurt property values. Associated Press,
May 17, 2006 [online]. Available: http://www.windaction.org/news/3068 [accessed Sept.
8, 2006].

Aubrey, K.B., J.P. Hayes, B.L. Biswell, and B.G. Marcot. 2003. Ecological role of tree-dwelling
mammals in western coniferous forests. Pp. 405-443 in Mammal Community Dynamics:
Management and Conservation in the Coniferous Forests of Western North America, C.J.
Zabel and R.G. Anthony, eds. Cambridge: Cambridge University Press.

AusWEA (Australian Wind Energy Association). 2002. Best Practice Guidelines for Implementation of Wind Energy Projects in Australia. March 2002 [online]. Available: http://www.
auswea.com.au/auswea/downloads/AusWEAGuidelines.pdf [accessed Sept. 15, 2006].

AusWEA (Australian Wind Energy Assocation). 2004. Wind Farming and Tourism. Fact Sheet
4. Wind Industry Development Project. AusWEA, Melbourne, Australia [online]. Available: http://www.auswea.com.au/WIDP/assets/4Tourism.pdf [accessed Aug. 24, 2006].

Avery, M., P.F. Springer, and J.F. Cassel. 1976. The effects of a tall tower on nocturnal bird
migration—a portable ceilometer study. Auk 93(2):281-291.

Avise, J.C. 1992. Molecular population structure and the biogeographic history of a regional
fauna: A case history with lessons for conservation biology. Oikos 63(1):62-76.

Avise, J.C. 2004. Molecular Markers, Natural History, and Evolution, 2nd Ed. Sunderland,
MA: Sinauer.

AWEA (American Wind Energy Association). 1995. Avian Interactions with Wind Energy
Facilities: A Summary. Prepared by Colson & Associates for AWEA, Washington, DC.
January 1995.

AWEA (American Wind Energy Association). 2006a. U.S. Wind Industry Ends Most Pro-
ductive Year, Sustained Growth Expected for at Least Next Two Years. AWEA News
Releases: January 24, 2006 [online]. Available: http://www.awea.org/news/US_Wind
_Industry_Ends_Most_Productive_Year_012406.html [accessed May 18, 2006].

AWEA (American Wind Energy Association). 2006b. State-Level Renewable Energy Portfolio
Standards (RPS). American Wind Energy Association [online]. Available: http://www.
awea.org/legislative/pdf/RPS_Fact_Sheet.pdf [accessed Sept. 10, 2006].

AWEA (American Wind Energy Association). 2006c. National Renewable Energy Portfolio
Standards (RPS). American Wind Energy Association [online]. Available: http://www.
awea.org/legislative/pdf/Federal_RPS_Factsheet.pdf [accessed Sept. 10, 2006].

AWEA (American Wind Energy Association). 2006d. Energy Department, Wind Industry Join
to Create Action Plan to Realize National Vision of 20% Electricity from Wind. American Wind Energy Association News Room: June 5, 2006 [online]. Available: http://www.
awea.org/newsroom/releases/Energy_Dept_Wind_Industry_Action_Plan_060506.html
[accessed Sept. 19, 2006].

AWEA (American Wind Energy Association). 2006e. Comparative Air Emissions of Wind and
Other Fuels. Wind Energy Fact Sheet. American Wind Energy Association [online]. Available: http://www.awea.org/pubs/factsheets/EmissionKB.PDF [accessed Sept. 19, 2006].

AWEA (American Wind Energy Association). 2006f. U.S. Wind Energy Installations Reach
New Milestone. American Wind Energy Association News Room: August 14, 2006 [on-
line]. Available: http://www.awea.org/newsroom/releases/US_Wind_Energy_Installations
_Milestone_081006.html [accessed Sept. 28, 2006].

AWEA (American Wind Energy Association). 2007. Wind Energy Projects throughout the United States of America. American Wind Energy Association [online]. Available: http://www.awea.org/projects/index.html [accessed March 13, 2007].

Bäckman, J., and T. Alerstam. 2003. Orientation scatter of free-flying nocturnal passerine migrants: Components and causes. Anim. Behav. 65(5):987-996.

Balcom, B.J., and R.H. Yahner. 1996. Microhabitat and landscape characteristics associated with the threatened Allegheny woodrat. Conserv. Biol. 10(2):515-523.

Ball, S.C. 1952. Fall Bird Migration on the Gaspé Peninsula. Bulletin Peabody Museum of Natural History 7. New Haven, CT: Peabody Museum of Natural History, Yale University.

Barbour, R.W., and W.H. Davis. 1969. Bats of America. Lexington: University Press of Kentucky.

Barclay, R.M.R., and R.M. Brigham. 2004. Geographic variation in the echolocation calls of bats: A complication for identifying species by their calls. Pp. 144-149 in Bat Echolocation Research: Tools, Techniques, and Analysis, R.M. Brigham, E.K.V. Kalko, G. Jones, S. Parsons, and H.J.G.A. Limpens, eds. Austin, TX: Bat Conservation International.

Barclay, R.M.R., and L.M. Harder. 2003. Life histories of bats: Life in the slow lane. Pp. 209-253 in Bat Ecology, T.H. Kunz and M.B. Fenton, eds. Chicago, IL: University of Chicago Press.

Barclay, R.M.R., and A. Kurta. 2007. Ecology and behavior of bats roosting in tree cavities and under bark. Pp. 17-57 in Bats in Forests: Conservation and Management, M.J. Lacki, J.P. Hayes, and A. Kurta, eds. Baltimore, MD: Johns Hopkins University Press.

Bartholomew, E., C. Bolduc, K. Coughlin, B. Hill, A. Meier, and R. Van Buskirk. 2006. Current Energy: Supply of and Demand for Electricity for PJM Interconnect. Environmental Energy Technologies Division, Lawrence Berkeley National Lab [online]. Available: http://currentenergy.lbl.gov/pjm/index.php [accessed Sept. 14, 2006].

Batcalls.org. 2006. Bat Sounds [online]. Available: http://www.batcalls.org/ [accessed June 28, 2006].

Battin, J. 2004. When good animals love bad habitats: Ecological traps and the conservation of animal populations. Conserv. Biol. 18(6):1482-1491.

Baur, A., and B. Baur. 1990. Are roads barriers to dispersal in the land snail *Arianta arbustorum*? Can. J. Zool. 68(3):613-617.

BBC (British Broadcasting Company). 2006. The Impact of Large Buildings and Structures (including Wind Farms) on Terrestrial Television Reception [online]. Available: http://www.bbc.co.uk/reception/factsheets/index.shtml [accessed March 28, 2007].

Beavers, S.C., and F.L. Ramsey. 1998. Detectability analysis in transect surveys. J. Wildl. Manage. 62(3):948-957.

Bednarz, J.C., D. Klem, L.J. Goodrich, and S.E. Senner. 1990. Migration counts of raptors at Hawk Mountain, Pennsylvania, as indicators of population trends, 1934-1986. Auk 107(1):96-109.

Bell, J.L., and R.C. Whitmore. 1997. Eastern towhee numbers increase following defoliation by gypsy moths. Auk 114(4):708-716.

Bell, J.L., and R.C. Whitmore. 2000. Bird nesting ecology in a forest defoliated by gypsy moths. Wilson Bull. 112(4):524-531.

Berthold, P. 1991. Patterns of avian migration in light of current global "greenhouse" effects: A central European perspective. Pp. 780-786 in Acta: XX Congressus Internationalis Ornithologici: Christchurch, December 2-9, 1990, B.D. Bell, R.O. Cossee, J.E.C. Flux, B.D. Heather, R.A. Hitchmough, C.J.R. Robertson, and M.J. Williams, eds. Wellington: New Zealand Ornithological Trust Board.

Berthold, P. 2001. Bird Migration: A General Survey, 2nd Ed. New York: Oxford University Press. 253 pp.

Betke, M., D. Hirsh, S. Crampton, J. Horn, N. Hristov, C.J. Cleveland, and T.H. Kunz. In press. Thermal imaging reveals significantly smaller Brazilian free-tailed bat colonies than previously estimated. J. Mammal.

Betts, B.J. 1996. Roosting behaviour of silver-haired bats (*Lasionycteris noctivagans*) and big brown bats (*Eptesicus fuscus*) in northeast Oregon. Pp. 55-61 in Bats and Forests Symposium, October 19-21, 1995, Victoria, British Columbia, Canada, R.M.R. Barclay and R.M. Brigham, eds. Working Paper 23. Victoria, BC: British Columbia, Ministry of Forests Research Branch.

Bevanger, K. 1994. Three questions on energy transmission and avian mortality. Fauna Norv. Ser. C 17(2):107-114.

Bibby, C.J., D.A. Hill, N.D. Burgess, and S.H. Mustoe. 2000. Bird Census Techniques, 2nd Ed. London: Academic Press. 302 pp.

Biewald, B. 2005. Using Electric System Operating Margins and Build Margins in Quantification of Carbon Emission Reductions Attributable to Grid Connected CDM Projects. Prepared for United Nations Framework Convention on Climate Change, by Synapse Energy Economics, Cambridge, MA. September 19, 2005 [online]. Available: http://www.synapse-energy.com/Downloads/SynapseReport.2005-09.UNFCCC.Using-Electric-System-Operating-Margins-and-Build-Margins-.05-031.pdf [accessed Sept. 19, 2006].

Bingman, V.P., K.P. Able, and P. Kerlinger. 1982. Wind drift, compensation, and the use of landmarks by nocturnal bird migrants. Anim. Behav. 30(1):49-53.

Bisbee, D.W. 2004. NEPA review of offshore wind farms: Ensuring emission reduction benefits outweigh visual impacts. Boston Coll. Environ. Aff. Law Rev. 31(2):349-384.

Black, J.E., and N.R. Donaldson. 1999. Comments on "Display of bird movements on the WSR-88D: Patterns and quantification." Weather Forecast. 14(6):1039-1040.

BLM (Bureau of Land Management). 1995. Draft Environmental Impact Statement Kenetech/PacificCorp Windpower Project, Carbon County, Wyoming. Prepared by Mariah Associates, Inc., for U.S. Department of the Interior, Bureau of Land Management, Rawlins District, Great Divide Resource Area, Cheyenne, WY.

BLM (Bureau of Land Management). 2005a. Final Programmatic Environmental Impact Statement (DPEIS) on Wind Energy Development on BLM-Administered Lands in the Western United States. FES 05-11. U.S. Department of Interior, Bureau of Land Management. June 2005 [online]. Available: http://windeis.anl.gov/documents/fpeis/index.cfm [accessed May 18, 2006].

BLM (Bureau of Land Management). 2005b. Record of Decision: Implementation of a Wind Energy Development Program and Associated Land Use Plan Amendments. U.S. Department of Interior, Bureau of Land Management. December 2005 [online]. Available: http://windeis.anl.gov/documents/docs/WindPEISROD.pdf [accessed Sept. 21, 2006].

Bluestein, J., E. Salerno, L. Bird, and L. Vimmerstedt. 2006. Incorporating Wind Generation in Cap and Trade Programs. Technical Report NREL/TP-500-4006. National Renewable Energy Laboratory, Golden, CO [online]. Available: http://www.nrel.gov/docs/fy06osti/40006.pdf [accessed Sept. 19, 2006].

Boone, D.D. 2006. Wind Energy Application Filed for Grid Interconnection Study Within Mid-Atlantic Highland Region of PJM (PA, WV, MD, DC & VA). PJM Wind Energy Project Summary Covering PA, MD, WV and VA. August 20, 2006 [online]. Available: http://www.vawind.org/Assets/Docs/PJM_windplant_queue_summary_073106.pdf [accessed Sept. 22, 2006].

Boone, D.D., J.K. Dunscomb, R. Webb, and C. Wulf. 2005. A Landscape Classification System: Addressing Environmental Issues Associated with Utility-Scale Wind Energy Development in Virginia. Prepared by the Virginia Wind Energy Collaborative Environmental Working Group. April 21, 2005 [online]. Available: http://www.vawind.org/#javascript [accessed May 26, 2006].

Bowen, G.J., and B. Wilkinson. 2002. Spatial distribution of $\delta^{18}O$ in meteoric precipitation. Geology 30(4):315-318.

Braun, C.E., ed. 2005. Techniques for Wildlife Investigations and Management, 6th Ed. Bethesda, MD: Wildlife Society.

Brittingham, M.C., and L.M. Williams. 2000. Bat boxes as alternative roosts for displaced bat maternity colonies. Wildlife Soc. Bull. 28(1):197-207.

Britzke, E.R. 2004. Designing monitoring programs sing frequency-division bat detectors: Active versus passive sampling. Pp. 79-83 in Bat Echolocation Research: Tools, Techniques, and Analysis, R.M. Brigham, E.K.V. Kalko, G. Jones, S. Parsons, and H.J.G.A. Limpens, eds. Austin, TX: Bat Conservation International.

Britzke, E.R., M.J. Harvey, and S.C. Loeb. 2003. Indiana bat, *Myotis sodalis*, maternity roosts in the southern United States. Southeast. Nat. 2(2):235-242.

Brody, A.J., and M.R. Pelton. 1989. Effects of roads on black bear movement in western North Carolina. Wildlife Soc. Bull. 17(1):5-10.

Brooks, M. 1965. The Appalachians. Boston: Houghton Mifflin. 346 pp.

Brown, W.S. 1993. Biology, Status and Management of the Timber Rattlesnake (*Crotalus horridus*): A Guide for Conservation. Herpetological Circular 22. Lawrence, KA: Society for the Study of Amphibians and Reptiles.

BRPC (Berkshire Regional Planning Commission). 2004. Wind Power Policy Siting Guidelines. September 16, 2004 [online]. Available: http://www.mtpc.org/Project%20Deliverables/GP_CP_BRPC_guide.pdf [accessed Sept. 12, 2006].

Bruderer, B. 1978. Effects of alpine topography and winds on migrating birds. Pp. 252-265 in Animal Migration, Navigation, and Homing: Symposium held at the University of Tübingen, August 17-20, 1977, K. Schmidt-Koenig and W.T. Keeton, eds. Berlin: Springer-Verlag.

Bruderer, B. 1997a. The study of bird migration by radar. Part 1: The technical basis. Naturwissenschaften 84(1):1-8.

Bruderer, B. 1997b. The study of bird migration by radar. Part 2: Major achievements. Naturwissenschaften 84(2):45-54.

Bruderer, B. 1999. Three decades of tracking radar studies on bird migration in Europe and the Middle East. Pp. 107-141 in Migrating Birds Know No Boundaries: Proceedings International Seminar on Birds and Flight Safety in the Middle East, Israel, April 25-29, 1999, Y. Leshem, Y. Mandelik, and J. Shamoun-Baranes, eds. Tel-Aviv, Israel: University of Tel-Aviv.

Bruderer, B., and L. Jenni. 1988. Strategies of bird migration in the area of the Alps. Pp. 2150-2161 in Acta: XIX Congressus Internationalis Ornithologici, Ottawa, Canada, June 22-29, 1986, H. Ouellet, ed. Ottawa, Ontario: University of Ottawa Press.

Bruderer, B., and A.G. Popa-Lisseanu. 2005. Radar data on wing-beat frequencies and flight speeds of two bat species. Acta Chiropterol. 7(1):73-82.

Bruderer, B., and P. Steidinger. 1972. Methods of quantitative and qualitative analysis of bird migration with tracking radar. Pp. 151-167 in Animal Orientation and Navigation, S.R. Galler, K. Schmidt-Koenig, G.J. Jacobs, and R.E. Belleville, eds. NASA SP-262. Washington, DC: National Aeronautics and Space Administration.

Bruderer, B., T. Steuri, and M. Baumgartner. 1995. Short-range high-precision surveillance of nocturnal migration and tracking of single targets. Israel J. Zool. 41(3):207-220.

Brundage, K.J., S.G. Benjamin, and M.N. Schwartz. 2001. Wind Energy Forecasts and Ensemble Uncertainty from the RUC. Preprints, 9th Conference on Mesoscale Processes, July 29-August 2, 2001, Fort Lauderdale. American Meteorological Society [online]. Available: http://ams.confex.com/ams/pdfpapers/23353.pdf [accessed May 16, 2007].

Brunet-Rossinni, A.K., and S.N. Austad. 2004. Ageing studies on bats: A review. Biogerontology 5(4):211-222.

Brush, R.O. 1979. The attractiveness of woodlands: Perceptions of forest landowners in Massachusetts. For. Sci. 25(3):495-506.

Buchler, E.R. 1976. A chemiluminescent tag for tracking bats and other small nocturnal animals. J. Mammal. 57(1):173-176.

Buchler, E.R., and S.B. Childs. 1981. Orientation to distant sounds by foraging big brown bats (*Eptesicus fuscus*). Anim. Behav. 29(2):428-432.

Buchler, E.R., and P.J. Wasilewski. 1985. Magnetic remanence in bats. Pp. 483-487 in Magnetite Biomineralization and Magnetoreception in Organisms: A New Biomagnetism, J.L. Kirschvink, D.S. Jones, and B.J. MacFadden, eds. New York: Plenum Press.

Buckelew, A.R., and G.A. Hall. 1994. The West Virginia Breeding Bird Atlas. Pittsburgh, PA: University of Pittsburgh Press. 232 pp.

Buckland, S.T., D.R. Anderson, K.P. Burnham, and J.L. Laake. 1993. Distance Sampling: Estimating Abundance of Biological Populations, 1st Ed. London: Chapman and Hall. 446 pp.

Buckland, S.T., D.R. Anderson, K.P. Burnham, J.L. Laake, D.L. Borchers, and L.J. Thomas. 2001. An Introduction to Distance Sampling: Estimating Abundance of Biological Populations. Oxford, UK: Oxford University Press.

Buckland, S.T., D.R. Anderson, K.P. Burnham, J.L. Laake, D.L. Borchers, and L. Thomas, eds. 2004. Advanced Distance Sampling. Oxford, UK: Oxford University Press.

Buckley, S., and S. Knight Merz. 2005. Wind Farms and Electromagnetic Interference-Dispelling the Myths. SKM Consulting [online]. Available: http://www.skmconsulting.com/Markets/energy/Wind_farms_Electromagnetic_Interference.htm [accessed Aug. 24, 2006].

Burke, D.M., and E. Nol. 1998. Influence of food abundance, nest-site habitat, and forest fragmentation on breeding ovenbirds. Auk 115(1):96-104.

Burke, H.S., Jr. 1999. Maternity colony formation in *Myotis septentrionalis* using artificial roosts: The rocket box, a habitat enhancement for woodland bats? Bat Research News 40(3):77-78.

Burnham, K.P., D.R. Anderson, and J. Laake. 1980. Estimation of Density from Line Transect Sampling of Biological Populations. Wildlife Monographs 72. Washington, DC: Wildlife Society.

Burton, T.M., and G.E. Likens. 1975a. Energy flow and nutrient cycling in salamander populations in the Hubbard Brook Experimental Forest, New Hampshire. Ecology 56(5):1068-1080.

Burton, T.M., and G.E. Likens. 1975b. Salamander populations and biomass in the Hubbard Brook Experimental Forest, New Hampshire. Copeia 1975(3):541-546.

Burton, T., D. Sharpe, N. Jenkins, and E. Bossanyi. 2001. Wind Energy Handbook. West Sussex, England: John Wiley & Sons. 617 pp.

Buss, I.O. 1946. Bird detection by radar. Auk 63(3):315-318.

Butchkoski, C.M., and J.M. Hassinger. 2002. Ecology of a maternity colony roosting in a building. Pp. 130-142 in The Indiana Bat: Biology and Management of an Endangered Species, A. Kurta and J. Kennedy, eds. Austin, TX: Bat Conservation International.

BWEA (British Wind Energy Association). 1994. Best Practice Guidelines for Wind Energy Development. London: Wind Energy Association. November 1994 [online]. Available: http://www.bwea.com/pdf/bpg.pdf [accessed Sept. 15, 2006].

BWEA (British Wind Energy Association). 2000. Noise from Wind Turbines—The Facts. British Wind Energy Association, London [online]. Available: http://www.bwea.com/pdf/noise.pdf [accessed Aug. 24, 2006].

BWEA (British Wind Energy Association). 2005. Low Frequency Noise and Wind Turbines. BWEA Briefing Sheet. British Wind Energy Association, London [online]. Available: http://www.bwea.com/pdf/briefings/lfn_summary.pdf [accessed Aug. 24, 2006].

Callahan, E.V., R.D. Drobney, and R.L. Clawson. 1997. Selection of summer roosting sites by Indiana bats (Myotis sodalis) in Missouri. J. Mammal. 78(3):818-825.

Carter, T.C., M.A. Menzel, and D.A. Saugey. 2003. Population trends of solitary foliage-roosting bats. Pp. 41-47 in Monitoring Trends in Bat Populations of the United States and Territories: Problems and Prospects, T.J. O'Shea and M.A. Bogan, eds. Biological Resources Discipline, Information and Technology Report USGS/BRD/ITR-2003-003. U.S. Geological Survey [online]. Available: http://www.fort.usgs.gov/products/publications /21329/21329.pdf [accessed May 25, 2006].

Castleberry, S.B. 2000. Conservation and Management of the Allegheny Woodrat in the Central Appalachians. Ph.D. Dissertation, West Virginia University, Morgantown, WV. 166 pp.

Castleberry, S.B., W.M. Ford, P.B. Wood, N.L. Castleberry, and M.T. Mengak. 2001. Movements of Allegheny woodrats in relation to timber harvesting. J. Wildl. Manage. 65(1):148-156.

Castleberry, S.B., W.M. Ford, P.B. Wood, N.L. Castleberry, and M.T. Mengak. 2002. Summer microhabitat selection by foraging Allegheny woodrats (Neotoma magister) in a managed forest. Am. Midl. Nat. 147(1):93-101.

CBD (Center for Biological Diversity, Inc.). 2004. Complaint for Violations of California Business and Professions Code Section 17200 et seq. Case No. RG04183113. Superior Court of the State of California, County of Alameda. November 1, 2004 [online]. Available: http://www.abanet.org/environ/committees/renewableenergy/teleconarchives/011905/ altamountcomplaint.pdf [accessed Sept. 11, 2006].

Chamberlain, C.P., J.D. Blum, R.T. Holmes, X. Feng, T.W. Sherry, and G.R. Graves. 1997. The use of isotope tracers for identifying populations of migratory birds. Oecologia 109(1):132-141.

Chamberlain, D.E., M.R. Rehfisch, A.D. Fox, M. Desholm, and S.J. Anthony. 2006. The effect of avoidance rates on bird mortality predictions made by wind turbine collision risk models. Ibis 148(S1):198-202.

Chambers, C.L., V. Alm, M.S. Siders, and M.J. Rabe. 2002. Use of artificial roost by forest-dwelling bats in northern Arizona. Wildlife Soc. Bull. 30(4):1085-1091.

Chautauqua Windpower, LLC, Ecology and Environment, Inc., Pandion Systems, Inc., and LeBoef, Lam, Green and Macrae. 2004. Draft Avian Risk Assessment for the Chautauqua Wind Project: Towns of Westfield and Ripley, Chautauqua County, New York. Prepared for Co-lead Agencies, Towns of Ripley and Westfield, Chautauqua County, New York. 26 pp.

Clark, A.M., P.E. Moler, E.E. Possardt, A.H. Savitzky, W.S. Brown, and B.W. Bowen. 2003. Phylogeography of the timber rattlesnake (Crotalus horridus) based on mtDNA sequences. J. Herpetol. 37(1):145-154.

Clark, B.S., D.M. Leslie, Jr., and T.S. Carter. 1993. Foraging activity of adult female Ozark big-eared bats (Plecotus townsendii ingens) in summer. J. Mammal. 74(2):422-427.

Clegg, S.M., J.F. Kelly, M. Kimura, and T.B. Smith. 2003. Combining genetic markers and stable isotopes to reveal population connectivity and migration patterns in a neotropical migrant, Wilson's warbler (Wilsonia pusilla). Mol. Ecol. 12(4):819-830.

Clemmer, S., D. Donovan, A. Nogee, and J. Deyette. 2001. Clean Energy Blueprint: A Smarter National Energy Policy for Today and the Future. Union of Concerned Scientists with American Council for an Energy-Efficient Economy, Tellus Institute. October 2001 [online]. Available: http://www.ucsusa.org/clean_energy/clean_energy_policies/clean-energy-blueprint.html [accessed Sept. 19, 2006].

Cleveland, C.J., M. Betke, P. Federico, J.D. Frank, T.G. Hallam, J. Horn, J.D. Lopez, Jr., G.F. McCracken, R.A. Medellin, A. Moreno-Valdez, S.G. Sansone, J.K. Westbrook, and T.H. Kunz. 2006. The economic value of the pest control services provided by the Brazilian free-tailed bats in south-central Texas. Front. Ecol. Environ. 4(5):238-243.

Cochran, W.G. 1977. Sampling Techniques, 3rd Ed. New York: Wiley.

Cochran, W.G., and G.M. Cox. 1957. Experimental Designs, 2nd Ed. New York: Wiley.

Conroy, M.J., J.R. Goldsberry, J.E. Hines, and D.B. Stotts. 1988. Evaluation of aerial transect surveys for wintering American black ducks. J. Wildlife Manage. 52(4):694-703.

Constantine, D. 1966. Ecological observation of lasiurine bats in Iowa. J. Mammal. 47(1):34-41.

Cooper, B.A. 1996. Use of radar for wind power-related avian research. Pp. 58-72 in Proceedings of National Avian-Wind Power Planning Meeting II, Palm Springs, California September 20-22, 1995, Palm Springs, CA. Prepared by LGL Ltd., Environmental Research Associates, King City, Ontario [online]. Available: http://www.nationalwind.org/publications/wildlife/avian95/avian95-08.htm [accessed June 12, 2006].

Cooper, B.A., and T.J. Mabee. 2000. Bird Migration near Proposed Wind Turbine Sites at Wethersfield and Harrisburg, New York. Prepared for Niagara. Mohawk Power Corporation, Syracuse, NY, by ABR, Inc., Forest Grove, OR. 46 pp.

Cooper, B.A., R.H. Day, R.J. Ritchie, and C.L. Cranor. 1991. An improved marine radar system for studies of bird migration. J. Field Ornithol. 62(3):367-377.

Cooper, B.A., C.B. Johnson, and E.F. Neuhauser. 1995a. The impact of wind turbines in upstate New York. Pp. 607-611 in Proceedings of Windpower '95 Annual Conference and Exhibition of the American Wind Energy Association, March 26-30, 1995. Washington, DC: American Wind Energy Association.

Cooper, B.A., C.B. Johnson, and R.J. Ritchie. 1995b. Bird Migration near Existing and Proposed Wind Turbine Sites in the Eastern Lake Ontario Region. Prepared for Niagara Mohawk Power Corporation, Syracuse, NY, by ABR, Inc., Forest Grove, OR. 71 pp.

Cooper, B.A., T.J. Mabee, A.A. Stickney, and J.E. Shook. 2004a. A Visual and Radar Study of Spring Bird Migration at the Proposed Chautauqua Wind Energy Facility, New York. Prepared for Chautauqua Windpower LLC, Lancaster, NY, by ABR, Inc., Forest Grove, OR. April 27, 2004 [online]. Available: http://www.abrinc.com/news/Publications_Newsletters/Visual%20and%20Radar%20Study%20of%20Bird%20Migration,%20Chautauqua%20%20%20Wind%20Energy%20Facility,%20NY,%20Spring%202003.pdf [accessed June 13, 2006].

Cooper, B.A., A.A. Stickney, and T.J. Mabee. 2004b. A Radar Study of Nocturnal Bird Migration at the Proposed Chautauqua Wind Energy Facility, New York, Fall 2003. Prepared for Chautauqua Windpower LLC, Lancaster, NY, by ABR, Inc., Forest Grove, OR. April 27, 2004 [online]. Available: http://www.abrinc.com/news/Publications_Newsletters/Radar%20Study%20of%20Nocturnal%20Bird%20Migration,%20Chautauqua%20Wind%20Energy%20Facility,%20NY,%20Fall%202003.pdf [accessed June 13, 2006].

Cooper, B.A., T.J. Mabee, and J.H. Plissner. 2004c. Case Studies: Application of Radar and Visual Study Techniques at Wind Energy Sites. Slide Presentation at National Avian-Wind Power Planning Meeting V, November 3-4, 2004, Lansdowne, VA [online]. Available: http://www.nationalwind.org/events/wildlife/2004-2/presentations/Cooper_Technology.pdf [accessed June 13, 2006].

Cooper, R.J., J.A. DeCecco, M.R. Marshall, A.B. Williams, G.A. Gale, and S.B. Cederbaum. 2005a. Bird studies. Chapter 6 in Long-Term Evaluation of the Effects of *Bacillus thuringiensis kurstaki*, Gypsy Moth Nucleopolyhedrosis Virus Product Gypchek, and *Entomophaga maimaiga* on Nontarget Organisms in Mixed Broadleaf-Pine Forests in the Central Appalachians, J.S. Strazanac and L. Butler, eds. FHTET 2004-14. U.S. Department of Agriculture, Forest Service, Forest Health Technology Enterprise Team, Morgantown, WV.

Cooper, B.A., T.J. Mabee, and J.H. Plissner. 2005b. Application of radar and other tools on avian species. Pp. 75-76 in Proceedings of the Onshore Wildlife Interactions with Wind Developments: Research Meeting V, November 3-4, 2004, Lansdowne, VA, S.S. Schwartz, ed. Prepared for the Wildlife Subcommittee of the National Wind Coordinating Committee by RESOLVE, Inc., Washington, DC [online]. Available: http://www.nationalwind.org/events/wildlife/2004-2/proceedings.pdf [accessed June 13, 2006].

Corben, C. 2004. Zero-crossing analysis for bat identification: An overview. Pp. 95-107 in Bat Echolocation Research: Tools, Techniques, and Analysis, R.M. Brigham, E.K.V. Kalko, G. Jones, S. Parsons, and H.J.G.A. Limpens, eds. Austin, TX: Bat Conservation International.

Cornell Laboratory of Ornithology. 2005. The Birds of North America Online. Cornell Lab of Ornithology and the American Ornithologists Union [online]. Available: http://bna.birds.cornell.edu/BNA/ [accessed Dec. 27, 2006].

Corten, G.P., and H.F. Veldkamp. 2001. Aerodynamics: Insects can halve wind-turbine power. Nature 12(6842):42-43.

Cox, D.R. 1958. Planning of Experiments. New York: Wiley.

CPC (Climate Prediction Center). 2004. Regional Climate Maps: USA, Twelve Month Total Precipitation: Jan.-Dec. 2005. Climate Prediction Center, National Weather Service, National Oceanic and Atmospheric Administration [online]. Available: http://www.cpc.ncep.noaa.gov/products/analysis_monitoring/regional_monitoring/us_12-month_precip.shtml [accessed Oct. 5, 2006].

Crockford, N.J. 1992. A Review of the Possible Impacts of Wind Farms on Birds and Other Wildlife. JNCC Report 27. Joint Nature Conservation Committee, Peterborough, UK.

Crum, T.D., and R.L. Alberty. 1993. The WSR-88D and the WSR-88D operational support facility. Bull. Am. Meteorol. Soc. 74(9):1669-1687.

Crum, T.D., R.L. Alberty, and D.W. Burgess. 1993. Recording, archiving, and using WSR-88D data. Bull. Am. Meteorol. Soc. 74(4):645-653.

Cryan, P.M. 2003. Seasonal distribution of migratory tree bats (*Lasiurus* and *Lasionycteris*) in North America. J. Mammal. 84(2):579-593.

Cryan, P.M., and A.C. Brown. In press. Does migration of hoary bats past a remote island offer clues toward the problem of bat fatalities at wind turbines. Biol. Conserv.

Curry & Kerlinger LLC, and ESS Group, Inc. 2004. Spring/Fall 2002 Avian Radar Studies for the Cape Wind Energy Project Nantucket Sound. Appendix 5.7-E in Cape Wind Energy Project Draft Environmental Impact Statement. ESS Project No. E159. Prepared for Cape Wind Associates, Boston, MA, by Curry & Kerlinger, LLC, Cape May Point, NJ, and ESS Group, Inc., Sandwich, MA. May 12, 2004 [online]. Available: http://www.nae.usace.army.mil/projects/ma/ccwf/app57e.pdf [accessed June 13, 2006].

Cyalume Light Technologies. 2006. Cyalume®. Cyalume Light Technologies, West Springfield, MA [online]. Available: http://www.cyalume.com/ [accessed Oct. 16, 2006].

Daily, G.C., S. Alexander, P.R. Ehrlich, L. Goulder, J. Lubchenco, P.A. Matson, H.A. Mooney, S. Postel, S.H. Schneider, D. Tilman, and G.M. Woodwell. 1997. Ecosystem services: Benefits supplied to human societies by natural ecosystems. Issues in Ecology 2. Washington, DC: Ecological Society of America.

Dalquest, W.W. 1943. Seasonal distribution of the hoary bat along the Pacific Coast. Murrelet 24:21-24.

Damborg, S. 2002. Public Attitudes Toward Wind Power. Danish Wind Industry Association [online]. Available: http://www.windpower.org/media(485,1033)/public_attitudes_towards_wind_power.pdf [accessed April 24, 2007].

Day, R.H., and L.C. Byrne. 1989. Avian Research Program for the Over-the-Horizon Backscatter Central Radar System, Spring 1989: Radar Studies of Bird Migration. Prepared for Metcalf & Eddy/Holmes and Narver, Inc., Wakefield, MA, by ABR, Inc., Forest Grove, OR. 102 pp.

Day, R.H., and L.C. Byrne. 1990. Avian Research Program for the Over-the-Horizon Back-scatter Central Radar System: Radar Studies of Bird Migration Fall 1989. Prepared for Metcalf & Eddy/Holmes and Narver, Inc., Wakefield, MA, by ABR, Inc., Forest Grove, OR. 111 pp.

DeCarolis, J.F., and D.W. Keith. 2006. The economics of large-scale wind power in a carbon constrained world. Energy Policy 34(4):395-410.

DEHLG (Department of the Environment, Heritage and Local Government). 2006. Wind Energy Development Guidelines for Planning Authorities. Department of the Environment, Heritage and Local Government, Dublin, Ireland [online]. Available: http://www.environ.ie/DOEI/DOEIPub.nsf/enSearchView/5559589989DC56328025719C004EB778?OpenDocument&Lang=en [accessed Sept. 12, 2006].

Delibes, M., P. Gaona, and P. Ferreras. 2001. Effects of an attractive sink leading into maladaptive habitat selection. Am. Nat. 158(3):277-285.

deMaynadier, P.G., and M.L. Hunter, Jr. 2000. Road effects on amphibian movements in a forested landscape. Nat. Area. J. 20:56-65.

Demong, N.J., and S.T. Emlen. 1978. Radar tracking of experimentally released migrant birds. Bird Banding 49(4):342-359.

Denholm, P., G.L. Kulcinski, and T. Holloway. 2005. Emissions and energy efficiency assessment of baseload wind energy systems. Environ. Sci. Technol. 39(6):1903-1911.

DeSante, D.F., K.M. Burton, P. Velez, and D. Froehlich. 2001. MAPS Manual: 2001 Protocol. The Institute for Bird Populations, Point Reyes Station, CA. 67 pp.

Desholm, M. 2003. Thermal Animal Detection System (TADS). Development of a Method for Estimating Collision Frequency of Migrating Birds at Offshore Wind Turbines. NERI Technical Report No. 440. National Environmental Research Institute, Denmark [online]. Available: http://www2.dmu.dk/1_viden/2_Publikationer/3_fagrapporter/rapporter/FR440.pdf [accessed June 7, 2006].

Desholm, M., and J. Kahlert. 2005. Avian collision risk at an offshore wind farm. Biol. Lett. 1(3):296-298.

Desholm, M., A.D. Fox, and P.D. Beasley. 2004. Best Practice Guidance for the Use of Remote Techniques for Observing Bird Behavior in Relation to Offshore Wind Farms. Report produced for Collaborative Offshore Wind Research into the Environment (COWRIE) Consortium, by National Environmental Research Institute, Rønde, Denmark, and QinetiQ, Malvern Technology Centre, Worcestershire, UK [online]. Available: http://www.offshorewindfarms.co.uk/Downloads/REMOTETECHNIQUES-FINALREPORT.pdf [accessed June 13, 2006].

Didham, R.K. 1997. The influence of edge effects and forest fragmentation on leaf-litter invertebrates in central Amazonia. Pp. 55-70 in Tropical Forest Remnants: Ecology, Management, and Conservation of Fragmented Communities, W.F. Laurance and R.O. Bierregaard, eds. Chicago, IL: University of Chicago Press.

Diefenbach, D.R., C. Butchkoski, and J.D. Hassinger. 2005. The Effects of Forest Fragmentation and Population Size on the Occupancy of Woodrat Habitat Sites and Habitat Management System Priorities, a First Cut. State Wildlife Grant Report. Pennsylvania Game Commission, Harrisburg, PA.

Diehl, R.H., and R.P. Larkin. 2005. Introduction to the WSR-88D (NEXRAD) for ornithological research. Pp. 876-888 in Bird Conservation Implementation and Integration in the Americas: Proceedings of the Third International Partners in Flight Conference, March 20-24, 2002, Asilomar, CA, Vol. 2, C.J. Ralph and T.D. Rich, eds. Gen. Tech. Rep. PSW-GTR-191. U.S. Department of Agriculture, Forest Service, Pacific Southwest Research Station, Albany, CA. June 2005 [online]. Available: http://www.fs.fed.us/psw/publications/documents/psw_gtr191/Asilomar/pdfs/876-888.pdf [accessed June 13, 2006].

Diehl, R.H., R.P. Larkin, and J.E. Black. 2003. Radar observation of bird migration over the Great Lakes. Auk 120(2):278-290.

DiMauro, D., and M.I. Hunter, Jr. 2002. Reproduction of amphibians in natural and anthropogenic temporary pools in managed forests. For. Sci. 48(2):397-406.

Dinsmore, S.J., and D.H. Johnson. 2005. Population analysis in wildlife biology. Pp. 154-169 in Techniques for Wildlife Investigations and Management, 6th Ed., C.E. Braun, ed. Bethesda, MD: Wildlife Society.

DOD (U.S. Department of Defense). 2006. Interim Policy on Proposed Windmill Farm Locations. Memorandum for DOD-JNDPG Participants and DOD/DHS LRR JPO Participants from Headquarters Air Combat Command/A3A-Department of Defense/Department of Homeland Security Long Range Radar Joint Program Office, Langley Air Force Base, VA. March 21, 2006 [online]. Available: http://www.af.mil/shared/media/document/AFD-060801-032.pdf [accessed April 23. 2007].

Duguay, J.P. 1997. Influence of Two-Age and Clearcut Timber Management Practices on Songbird Abundance, Nest Success, and Invertebrate Biomass in West Virginia. Ph.D. Dissertation, West Virginia University, Morgantown, WV.

Duguay, J.P., P.B. Wood, and G.W. Miller. 2000. Effects of timber harvests on invertebrate biomass and avian nest success. Wildlife Soc. Bull. 28(4):1123-1131.

Duguay, J.P., P.B. Wood, and J.V. Nichols. 2001. Songbird abundance and avian nest survival rates in forests fragmented by different silvicultural treatments. Conserv. Biol. 15(5):1405-1415.

Dürr, T., and L. Bach. 2004. Bat deaths and wind turbines: A review of current knowledge, and of information available in the database for Germany [in German]. Bremer Beiträge für Naturkunde und Naturschutz 7:253-264.

DWEA (Danish Wind Energy Association). 2003a. Wind Turbines and the Environment: Landscape. Danish Wind Energy Association, Copenhagen, Denmark [online]. Available: http://www.windpower.org/en/tour/env/index.htm [accessed Sept. 5, 2006].

DWEA (Danish Wind Energy Association). 2003b. Shadow Casting from Wind Turbines. Danish Wind Energy Association, Copenhagen, Denmark [online]. Available: http://www.windpower.org/en/tour/env/shadow/index.htm [accessed Sept. 5, 2006].

DWEA (Danish Wind Energy Association). 2004. Wind Turbine Planning—Heading for Structural Reorganization. Danish Wind Industry Association (Vindmølleindustrien). September 2004.

Eastwood, E. 1967. Radar Ornithology. London: Methuen.

Eby, P. 1991. Seasonal movements of grey-headed flying-foxes, *Pteropus poliocephalus* (*Chiroptera: Pteropodidae*), from two maternity camps in northern New South Wales. Wildlife Res. 18(5):547-559.

EC (European Commission). 1995. Externalities of Energy: ExternE Project, Volume 6: Wind & Hydro. EUR 16525. Prepared for the European Commission Directorate General XII Science, Research and Development, by Metroeconomica, CEPN, IER, Eyre Energy-Environment, ETSU, Ecole des Mines. Luxembourg: Office for Official Publications of the European Communities.

EERE (Energy Efficiency and Renewable Energy). 2006. Wind and Hydropower Technologies Program. Energy Efficiency and Renewable Energy, U.S. Department of Energy [online]. Available: http://www1.eere.energy.gov/windandhydro/about.html [accessed Oct. 24, 2006].

Efroymson, R.A., and G.W. Suter II. 2001. Ecological risk assessment framework for low-altitude aircraft overflights: II. Estimating effects on wildlife. Risk Anal. 21(2):263-274.

eGRID 2000. 2006. Emissions and Generation Resource Integrated Database. U.S Environmental Protection Agency [online]. Available: http://www.epa.gov/cleanenergy/egrid/index.htm [accessed Aug. 17, 2006].

EIA (Energy Information Administration). 2004a. Form EIA-860 Database: 2004 Annual Electric Generator Report. Energy Information Administration, U.S. Department of Energy [online]. Available: http://www.eia.doe.gov/cneaf/electricity/page/eia860.html [accessed Aug. 17, 2006].

EIA (Energy Information Administration). 2004b. Form EIA-906 and EIA-920 Databases: 2004. Energy Information Administration, U.S. Department of Energy [online]. Available: http://www.eia.doe.gov/cneaf/electricity/page/eia906_920.html [accessed Aug. 17, 2006].

EIA (Energy Information Administration). 2004c. Electricity Sales and Revenue, Office of Coal, Nuclear, Electric, and Alternative Fuels, Energy Information Administration, U.S. Department of Energy [online]. Available: http://www.eia.doe.gov/cneaf/electricity/page/sales_revenue.xls [accessed Aug. 17, 2006].

EIA (Energy Information Administration). 2005a. Electric Power Annual 2004. DOE/EIA-0348(2004). Office of Coal, Nuclear, Electric and Alternate Fuels, U.S. Department of Energy, Washington, DC [online]. Available: http://www.eia.doe.gov/cneaf/electricity/epa/epa_sum.html [accessed July 17, 2006].

EIA (Energy Information Administration). 2005b. Assumptions to the Annual Energy Outlook 2005 with Projections to 2025. Energy Information Administration, U.S. Department of Energy [online]. Available: http://tonto.eia.doe.gov/FTPROOT/forecasting/0554(2005).pdf [accessed May 15, 2007].

EIA (Energy Information Administration). 2006a. Annual Energy Outlook 2006 with Projections to 2030. DOE/EIA-0383(2006). Energy Information Administration, Office of Integrated Analysis and Forecasting, U.S. Department of Energy, Washington, DC. February 2006 [online]. Available: http://www.eia.doe.gov/oiaf/aeo [accessed Aug. 17, 2006].

EIA (Energy Information Administration). 2006b. International Energy Annual 2004. Energy Information Administration, U.S. Department of Energy, Washington, DC. July 2006 [online]. Available: http://www.eia.doe.gov/iea/ [accessed Aug. 17, 2006].

EIA (Energy Information Administration). 2006c. U.S. Carbon Dioxide Emissions from Energy Sources 2005 Flash Estimate. Office of Integrated Analysis and Forecasting, EI-81, Energy Information Administration, U.S. Department of Energy. June 2006 [online]. Available: http://www.eia.doe.gov/oiaf/1605/flash/flash.html [accessed Oct. 26, 2006].

EIA (Energy Information Administration). 2006d. Annual Energy Review. DOE/EIA-0384(2005). July 2006 [online]. Available: http://www.eia.doe.gov/emeu/aer/elect.html [accessed Sept. 9, 2006].

Elliott, D.L., C.G. Holliday, W.R. Barchet, H.P. Foote, and W.F. Sandusky. 1986. Wind Energy Resource Atlas of the United States. DOE/CH 10093-4. Golden, CO: Solar Energy Research Institute. October 1986 [online]. Available: http://rredc.nrel.gov/wind/pubs/atlas/ [accessed Sept. 19, 2006].

E.ON Netz. 2005. Wind Report 2005. E.ON Netz GmbH, Bayreuth, Germany [online]. Available: http://www.eon-netz.com/Ressources/downloads/EON_Netz_Windreport2005_eng.pdf [accessed Sept. 19, 2006].

E.ON Netz. 2006. Data and Facts Relating to Wind Power in Germany, Supplement to the "Wind Report 2005" E.ON Netz GmbH, Bayreuth, Germany [online]. Available: http://www.eon-netz.com/EONNETZ_eng.jsp [accessed Sept. 19, 2006].

EPA (U.S. Environmental Protection Agency). 1992. Framework for Ecological Risk Assessment. EPA/630/R-92/001. Risk Assessment Forum, U.S. Environmental Protection Agency, Washington, DC. February 1992 [online]. Available: http://risk.lsd.ornl.gov/homepage/FRMWRK_ERA.PDF [accessed June 5, 2006].

EPA (U.S. Environmental Protection Agency). 2004. Guidance on State Implementation Plan (SIP) Credits for Emission Reductions from Electric-Sector Energy Efficiency or Renewable Energy Measures. Office of Air and Radiation, U.S. Environmental Protection Agency. August 2004 [online]. Available: http://www.epa.gov/ttn/naaqs/ozone/eac/gm040805_eac_energy-efficiency.pdf [accessed Aug. 17, 2006].

EPA (U.S. Environmental Protection Agency). 2005. Air Emissions Trends—Continued Progress Through 2004, Complete Tables of National Emissions Totals [online]. Available: http://epa.gov/airtrends/2005/pdfs/detailedtable.xls [accessed Aug. 17, 2006].

EPA (U.S. Environmental Protection Agency). 2006a. Controlling Power Plant Emissions: Overview. U.S. Environmental Protection Agency [online]. Available: http://www.epa.gov/mercury/control_emissions/index.htm [accessed July 11, 2006].

EPA (U.S. Environmental Protection Agency). 2006b. eGRID: Emissions and Generation Resource Integrated Database [online]. Available: http://www.epa.gov/cleanenergy/egrid/index.htm [accessed Aug. 17, 2006].

EPA (U.S. Environmental Protection Agency). 2006c. Six Common Air Pollutants. Office of Air and Radiation, U.S. Environmental Protection Agency [online]. Available: http://www.epa.gov/air/urbanair/6poll.html [accessed March 19, 2006].

EPA (U.S. Environmental Protection Agency). 2006d. National Pollutant Discharge Elimination System. U.S. Environmental Protection Agency [online]. Available: http://cfpub.epa.gov/npdes/ [accessed January 30, 2007].

Erickson, W.P., G.D. Johnson, M.D. Strickland, and K. Kronner. 2000. Avian and Bat Mortality Associated with the Vansycle Wind Project, Umatilla County, Oregon 1999 Study Year. Final Report. Prepared by WEST, Inc., Cheyenne, WY, for Umatilla County Department of Resource Services and Development, Pendleton, OR. February 7, 2000 [online]. Available: http://www.west-inc.com/reports/vansyclereportnet.pdf [accessed May 23, 2006].

Erickson, W., G. Johnson, M. Stickland, D. Young, Jr., K. Sernka, and R. Good. 2001. Avian Collisions with Wind Turbines: A Summary of Existing Studies and Comparisons to Other Sources of Avian Collision Mortality in the United States. Washington, DC: Resolve, Inc. August 2001 [online]. Available: http://www.west-inc.com/reports/avian_collisions.pdf [accessed Sept. 13, 2006].

Erickson, W., G. Johnson, D. Young, D. Strickland, R. Good, M. Bourassa, K. Bay, and K. Sernka. 2002. Synthesis and Comparison of Baseline Avian and Bat Use, Raptor Nesting and Mortality Information from Proposed and Existing Wind Developments. Prepared by Western EcoSystems Technology, Inc., Cheyenne, WY, for Bonneville Power Administration, Portland, OR. December 2002 [online]. Available: http://www.bpa.gov/Power/pgc/wind/Avian_and_Bat_Study_12-2002.pdf [accessed May 18, 2006].

Erickson, W.P., J. Jeffrey, K. Kronner, and K. Bay. 2003a. Stateline Wind Project Wildlife Monitoring Annual Report: Results for the Period July 2001–December 2002. Prepared for FPL Energy, Stateline Technical Advisory Committee, Oregon Department of Energy, by Western Ecosystems Technology, Inc., Cheyenne, WY.

Erickson, W.P., K. Kronner, and B. Gritski. 2003b. Nine Canyon Wind Power Project, Avian and Bat Monitoring Report: September 2002-August 2003. Prepared for Nine Canyon Technical Advisory Committee and Energy Northwest, by West, Inc., Cheyenne, WY and Northwest Wildlife Consultants, Inc., Pendleton, OR. October 2003 [online]. Available: http://www.west-inc.com/reports/nine_canyon_monitoring_final.pdf [accessed May 23, 2006].

Erickson, W.P., J. Jeffrey, K. Kronner, and K. Bay. 2004. Stateline Wind Project Wildlife Monitoring Report: July 2001–December 2003. Prepared for FPL Energy, Stateline Technical Advisory Committee, Oregon Department of Energy, by Western EcoSystems Technology, Inc., Cheyenne, WY and Walla Walla, WA; and Northwest Wildlife Consultants, Inc., Pendleton, OR. December 2004 [online]. Available: http://www.west-inc.com/reports/swp_final_dec04.pdf [accessed May 23, 2006].

Erickson, W.P., G.D. Johnson, and D.P. Young. 2005. A summary and comparison of bird mortality from anthropogenic causes with an emphasis on collisions. Pp. 1029-1042 in Bird Conservation Implementation and Integration in the Americas: Proceedings of the Third International Partners in Flight Conference: March 20-24, 2002, Asilomar, CA, C.J. Ralph and T.D. Rich, eds. General Technical Report PSW-GTR-191. Albany, CA: U.S. Department of Agriculture Forest Service, Pacific Southwest Research Station. June 2005 [online]. Available: http://www.fs.fed.us/psw/publications/documents/psw_gtr191/Asilomar/pdfs/1029-1042.pdf [accessed May 21, 2007].

Estep, J. 1989. Avian Mortality at Large Wind Energy Facilities in California: Identification of a Problem. CEC P700-89-001. Sacramento, CA: California Energy Commission.

Estrada, A., and R. Coates-Estrada. 2002. Bats in continuous forest, forest fragments and in an agricultural mosaic habitat-island at Los Tuxtlas, Mexico. Biol. Conserv. 103(2):237-245.

Evans, W.R. 1994. Nocturnal flight call of Bicknell's Thrush. Wilson Bull. 106(1):55-61.

Evans, W.R., and D.K. Mellinger. 1999. Monitoring grassland birds in nocturnal migration. Stud. Avian Biol. 19(1):219-229.

Evans, W.R., and M. O'Brien. 2002. Flight Calls of Migratory Birds (Eastern North American Landbirds) [CD-ROM]. Old Bird, Inc., Ithaca, NY.

Evans, W.R., and K.V. Rosenberg. 1999. Acoustic monitoring of night-migrating birds: A progress report. In Strategies of Bird Conservation: The Partners in Flight Planning Process, Proceedings of the 3rd Partners in Flight Workshop; October 1-5, 1995; Cape May, NJ, R. Bonney, D.N. Pashley, R.J. Cooper, and L. Niles, eds. Proceedings RMRS-P-16. Ogden, UT: U.S. Department of Agriculture, Forest Service, Rocky Mountain Research Station [online]. Available: http://birds.cornell.edu/pifcapemay/evans_rosenberg.htm [accessed Dec. 28, 2006].

EWEA (European Wind Energy Association). 2006. 20 Years After Chernobyl: Wind Power Established as the Safe, Clean and Cheap Option. European Wind Energy Association. Press Releases: April 24, 2006 [online]. Available: http://www.ewea.org/index.php?id=60&no_cache=1&tx_ttnews[tt_news]=107&tx_ttnews[backPid]=1&cHash=2502aa89fd [accessed Aug. 24, 2006].

FAA (Federal Aviation Administration). 2007. Obstruction Marking and Lighting. Advisory Circular AC 70/7460-1K. U.S. Department of Transportation, Federal Aviation Administration [online]. Available: https://www.oeaaa.faa.gov/oeaaa/external/content/AC70_7460_1K.pdf [accessed May 24, 2007].

Fahrig, L. 2003. Effects of habitat fragmentation on biodiversity. Annu. Rev. Ecol. Evol. Syst. 34:487-515.

Farnsworth, A. 2005. Flight calls and their value for future ornithological studies and conservation research. Auk 122(3):733-746.

Farnsworth, A., and I.J. Lovette. 2005. Evolution of nocturnal flight calls in migrating woodwarblers: Apparent lack of morphological constraints. J. Avian Biol. 36(4):337-347.

Farnsworth, A., S.A. Gauthreaux, Jr., and D. van Blaricom. 2004. A comparison of nocturnal call counts of migrating birds and reflectivity measurements on Doppler radar. J. Avian Biol. 35(4):365-369.

Fenton, M.B. 1990. The foraging behavior and ecology of animal-eating bats. Can. J. Zool. 68(3):411-422.

Fenton, M.B. 2000. Choosing the "correct" bat detector. Acta Chiropterol. 2(2):215-224.

Fenton, M.B. 2004. Aerial-feeding bats: Getting the most out of echolocation. Pp. 350-354 in Echolocation in Bats and Dolphins, J.A. Thomas, C.F. Moss, and M. Vater, eds. Chicago, IL: University of Chicago Press.

Fenton, M.B., and D.W. Thomas. 1985. Migrations and dispersal of bats (*Chiroptera*). Contrib. Mar. Sci. 27(Suppl.):409-424.

FERC (Federal Energy Regulatory Commission). 2005. About FERC [online]. Available: http://www.ferc.gov/for-citizens/about-ferc.asp [accessed Sept. 18, 2006].

Fiedler, J.K. 2004. Assessment of Bat Mortality and Activity at Buffalo Mountain Windfarm, Eastern Tennessee. M.S. Thesis, University of Tennessee, Knoxville, TN.

Fiedler, J.K., T.H. Henry, C.P. Nicholson, and R.D. Tankersley. 2007. Results of bat and bird mortality at the expanded Buffalo Mountain wind farm, 2005. Tennessee Valley Authority, Knoxville, Tennessee, USA.

Fingersh, L.J. 2004. Optimization of Utility-Scale Wind-Hydrogen-Battery Systems. Preprint: To be presented at the World Energy Renewable Congress VIII, August 29-September 3, 2004, Denver, CO. NREL/CP-500-36117. National Renewable Energy Laboratory, Golden, CO. July 2004 [online]. Available: http://www.nrel.gov/docs/fy04osti/36117.pdf [accessed Sept. 19, 2006].

Fleming, T.H., and P. Eby. 2003. Ecology of bat migration. Pp. 156-208 in Bat Ecology, T.H. Kunz and M.B. Fenton, eds. Chicago, IL: University of Chicago Press.

Florence, J. 2006. Global Wind Power Expands in 2006. Earth Policy Institute. June 28, 2006 [online]. Available: http://www.earth-policy.org/Indicators/Wind/2006.htm [accessed Sept. 27, 2006].

Flowers, L. 2006. Wind Energy Update. Wind Powering America. PowerPoint Presentation. National Renewable Energy Laboratory, U.S. Department of Energy [online]. Available: http://www.eere.energy.gov/windandhydro/windpoweringamerica/pdfs/wpa/wpa_update.pdf [accessed Sept. 19, 2006].

Forman, R.T.T., D. Sperling, J.A. Bissonette, A.P. Clevenger, C.C. Cutshal, V.H. Dale, L. Fahrig, R. France, C.R. Coldman, K. Heanue, J.A. Jones, W.J. Swanson, T. Turrentine, and T.C. Winter. 2003. Road Ecology: Science and Solutions. Washington, DC: Island Press.

Frank, J.D., T.H. Kunz, J. Horn, C. Cleveland, and S. Petronio. 2003. Advanced infrared detection and image processing for automated bat censusing. Pp. 261-271 in Infrared Technology and Applications XXIX: 21-25 April, 2003, Orlando, FL, B.F. Andresen and G.F. Fulop, eds. Proceedings of SPIE—the International Society for Optical Engineering Vol. 5074. Bellingham, WA: SPIE.

Freemark, C., and B. Collins. 1992. Landscape ecology of birds breeding in temperate forest fragments. Pp. 443-454 in Ecology and Conservation of Neotropical Migrant Landbirds, J.M. Hagan and D.W. Johnston, eds. Washington, DC: Smithsonian Institution Press.

Fristrup, K., and A. Dhondt. 2001. The Impact of Wind Turbines on Birds in New York State. Prepared for Niagara Mohawk Power Corporation, Syracuse, NY.

Fujita, M.S., and T.H. Kunz. 1984. Pipistrellus subflavus. Mamm. Species 228:1-6.

Furlonger, C.L., H.J. Dewar, and M.B. Fenton. 1987. Habitat use by foraging insectivorous bats. Can. J. Zool. 65(2):284-288.

Gaines, W.L., A.L. Lyons, J.F. Lehmkuhl, and K.J. Raedeke. 2005. Landscape evaluation of female black bear habitat effectiveness and capability in the North Cascades, Washington. Biol. Conserv. 125(4):411-425.

Gannes, L.Z., D.M. O'Brien, and C. Martinez Del Rio. 1997. Stable isotopes in animal ecology: Assumptions, caveats, and a call for more laboratory experiments. Ecology 78(4):1271-1276.

Gannon, W.L., and R.E. Sherwin. 2004. Are acoustic detectors a "silver bullet" for assessing habitat use by bats? Pp. 38-45 in Bat Echolocation Research: Tools, Techniques, and Analysis, R.M. Brigham, E.K.V. Kalko, G. Jones, S. Parsons, and H.J.G.A. Limpens, eds. Austin, TX: Bat Conservation International.

GAO (U.S. General Accountability Office). 2004. Renewable Energy: Wind Power's Contribution to Electric Power Generation and Impact on Farms and Rural Communities. GAO-04-756. U.S. General Accountability Office, Washington, DC [online]. Available: http://www.gao.gov/new.items/d04756.pdf [accessed Sept. 26, 2006].

GAO (U.S. Government Accountability Office). 2005. Wind Power: Impacts on Wildlife and Government Responsibilities for Regulating Development and Protecting Wildlife. GAO-05-9006. U.S. Government Accountability Office, Washington, DC. September 2005 [online]. Available: http://www.gao.gov/new.items/d05906.pdf [accessed May 18, 2006].

Gauthreaux, S.A., Jr. 1969. A portable ceilometer technique for studying low-level nocturnal migration. Bird Banding 40(4):309-320.

Gauthreaux, S.A., Jr. 1970. Weather radar quantification of bird migration. BioScience 20(1):17-20.

Gauthreaux, S.A., Jr. 1980. Direct Visual and Radar Methods for the Detection, Quantification, and Prediction of Bird Migration. Special Publication No. 2. Department of Zoology, Clemson University, Clemson, SC. 67 pp.

Gauthreaux, S.A., Jr. 1985a. Radar, Electro-Optical, and Visual Methods of Studying Bird Flight near Transmission Lines. EPRI EA-4120. Palo Alto, CA: Electric Power Research Institute.

Gauthreaux, S.A., Jr. 1985b. An avian migration mobile research laboratory: Hawk migration study. Pp. 339-346 in Proceedings of Hawk Migration Conference IV, M. Harwood, ed. Medford, MA: Hawk Migration Association of North America.

Gauthreaux, S.A., Jr. 1991. The flight behavior of migrating birds in changing wind fields: Radar and visual analyses. Am. Zool. 31(1):187-204.

Gauthreaux, S.A., Jr., and C.G. Belser. 1998. Displays of bird movements on the WSR-88D: Patterns and quantification. Weather Forecast. 13(2):453-464.

Gauthreaux, S.A., Jr., and C.G. Belser. 1999. "Reply" to displays of bird movements on the WSR-88D: Patterns and quantification. Weather Forecast. 14(6):1041-1042.

Gauthreaux, S.A., Jr., and C.G. Belser. 2003a. Radar ornithology and biological conservation. Auk 120(2):266-277.

Gauthreaux, S.A., Jr., and C.G. Belser. 2003b. Bird movements on Doppler weather surveillance radar. Birding 35(6):616-628.

Gauthreaux, S.A., Jr., and C.G. Belser. 2005. Radar ornithology and the conservation of migratory birds. Pp. 871-875 in Bird Conservation Implementation and Integration in the Americas: Proceedings of the Third International Partners in Flight Conference, March 20-24, 2002, Asilomar, CA, Vol. 2, C.J. Ralph and T.D. Rich, eds. Gen. Tech. Rep. PSW-GTR-191. U.S. Department of Agriculture, Forest Service, Pacific Southwest Research Station, Albany, CA. June 2005 [online]. Available: http://www.fs.fed.us/psw/publications/documents/psw_gtr191/Asilomar/pdfs/871-875.pdf [accessed June 13, 2006].

Gauthreaux, S.A., Jr., and C.G. Belser. 2006. Effects of artificial night lighting on migratory birds. Pp. 67-93 in Ecological Consequences of Artificial Night Lighting, C. Rich and T. Longcore, eds. Washington, DC: Island Press.

Gauthreaux, S.A., Jr., and J.W. Livingston. 2006. Monitoring bird migration with a fixed-beam radar and a thermal imaging camera. J. Field Ornithol. 77(3):319-328.

Gauthreaux, S.A., Jr., C.G. Belser, and D. van Blaricom. 2003. Using a network of WSR 88-D weather surveillance radars to define patterns of bird migration at large spatial scales. Pp. 335-346 in Avian Migration, P. Berthold, E. Gwinner, and E. Sonnenschein, eds. Berlin: Springer.

Gilbert, R.O. 1987. Statistical Methods for Environmental Pollution Monitoring. New York: Van Nostrand Reinhold.

Gill, J.P., M. Townsley, and G.P. Mudge. 1996. Review of the Impacts of Wind Farms and Other Aerial Structures upon Birds. Scottish Natural Heritage Review No. 21. Edinburgh: Scottish Natural Heritage.

Gipe, P. 2002. Design as if people matter: Aesthetic guidelines for a wind power future. Pp. 173-214 in Wind Power in View: Energy Landscapes in a Crowded World, M.J. Pasqualetti, P. Gipe, and R.W. Righter, eds. San Diego, CA: Academic Press.

Gipe, P. 2003. Wind Energy "Best Practice" Guides-Wresting Standards From Conflict [online]. Available: http://www.wind-works.org/articles/BestPractice.html [accessed Sept. 15, 2006].

Glover, K.M., K.R. Hardy, T.G. Konrad, W.N. Sullivan, and A.S. Michaels. 1966. Radar observations of insects in free flight. Science 154(3752):967-972.

Goldberg, M., K. Sinclair, and M. Milligan. 2004. Job and Economic Development Impact (JEDI) Model: A User-Friendly Tool to Calculate Economic Impacts from Wind Projects. NREL/CP-500-35953. National Renewable Energy Laboratory, Golden, CO. March 2004 [online]. Available: http://www.eere.energy.gov/windandhydro/windpoweringamerica/pdfs/35953_jedi.pdf [accessed Sept. 6, 2006].

Goosem, M. 1997. Internal fragmentation: The effects of roads, highways, and power line clearings on movements and mortality of rainforest vertebrates. Pp. 241-255 in Tropical Forest Remnants: Ecology, Management, and Conservation of Fragmented Communities, W.F. Laurance and R.O. Bierregaard, eds. Chicago, IL: University of Chicago Press.

Gow, G. 2003. An Adaptable Approach to Short-Term Wind Forecasting [online]. Available: http://www.awea.org/policy/regulatory_policy/transmission_documents/System_Operations/2003_Gow_Forecasting%20Overview.pdf [accessed April 23, 2007].

Graber, R.R. 1968. Nocturnal migration in Illinois: Different points of view. Wilson Bull. 80(1):36-71.

Graber, R.R., and W.W. Cochran. 1959. An audio technique for the study of nocturnal migration of birds. Wilson Bull. 71(3):220-236.

Graber, R.R., and S.S. Hassler. 1962. The effectiveness of aircraft-type (APS) radar in detecting birds. Wilson Bull. 74(4):367-380.

Grady, D. 2004. Public Attitudes Toward Wind Energy in Western and Eastern North Carolina: A Systematic Survey. Efficient North Carolina Conference, Boone, NC. March 4, 2004 [online]. Available: http://www.energy.appstate.edu/docs/wnc_enc_present.pdf [accessed Oct. 17, 2006].

Gramling, R., and W.R. Freudenburg. 1992. Opportunity-threat, development, and adaptation: Toward a comprehensive framework for social impact assessment. Rural Sociol. 57(2):216-234.

Green, N.B., and T.K. Pauley. 1987. Amphibians and Reptiles in West Virginia. Pittsburgh, PA: University of Pittsburgh Press.

Green, R.H. 1979. Sampling Design and Statistical Methods for Environmental Biologists. New York: Wiley.

Greenberg, C.H. 2001. Response of reptile and amphibian communities to canopy gaps created by wind disturbance in the southern Appalachians. Forest Ecol. Manage. 148(1-3):135-144.

Greenhall, A.M., and J.L. Paradiso. 1968. Bats and Bat Banding. Resource Pub. 72. Washington, DC: U.S. Department of the Interior, Fish and Wildlife Service, Bureau of Sports Fisheries and Wildlife.

Griffin, D.R. 1958. Listening in the Dark: The Acoustic Orientation of Bats and Men. New Haven, CT: Yale University Press.

Griffin, D.R. 1970. Migrations and homing of bats. Pp. 233-264 in Biology of Bats, Vol. II, W.A. Wimsatt, ed. New York: Academic Press.

Griffin, D.R. 1971. The importance of atmospheric attenuation for the echolocation of bats (*Chiroptera*). Anim. Behav. 19(1):55-61.

Griffin, D.R. 1972. Nocturnal bird migration in opaque clouds. Pp. 169-188 in Animal Orientation and Navigation, S.R. Galler, K. Schmidt-Koenig, G.J. Jacobs, and R.E. Belleville, eds. NASA SP-262. Washington, DC: National Aeronautics and Space Administration.

Griffin, D.R. 2004. The past, and future of bat detectors. Pp. 6-9 in Bat Echolocation Research: Tools, Techniques, and Analysis, R.M. Brigham, E.K.V. Kalko, G. Jones, S. Parsons, and H.J.G.A. Limpens, eds. Austin, TX: Bat Conservation International.

Griffin, D.R., F.A. Webster, and C.R. Michael. 1960. The echolocation of flying insects by bats. Anim. Behav. 8(3-4):141-154.

Grindal, S.D., and R.M. Brigham. 1998. Short term effects of small-scale habitat disturbance on activity by insectivorous bats. J. Wildlife Manage. 62(3):996-1003.

Grindal, S.D., and R.M. Brigham. 1999. Impacts of forest harvesting on habitat use by foraging insectivorous bats at different spatial scales. Ecoscience 6(1):25-34.

Gruver, J.C. 2002. Assessment of Bat Community Structure and Roosting Habitat Preferences for the Hoary Bat (*Lasiurus cinereus*) Near Foote Creek Rim, Wyoming. M.S. Thesis, University of Wyoming, Laramie, WY.

Gumbert, M.W., J.M. O'Keefe, and J.R. MacGregor. 2002. Roost fidelity in Kentucky. Pp. 143-152 in The Indiana Bat: Biology and Management of an Endangered Species, A. Kurta and J. Kennedy, eds. Austin, TX: Bat Conservation International.

GWEC (Global Wind Energy Council). 2006. Global Wind 2005 Report [online]. Available: http://www.gwec.net/fileadmin/documents/Publications/GWEC-Global_Wind_05_Report_low_res_01.pdf [accessed Dec. 18, 2006].

Hagan, J.M., and A.L. Meehan. 2002. The effectiveness of stand-level and landscape-level variables for explaining bird occurrence in an industrial forest. For. Sci. 48(2):231-242.

Hagler Bailly Consulting, Inc. 1995. The New York State Externalities Cost Study. Dobbs Ferry, NY: Oceana Publications.

Hairston, N.G. 1987. Community Ecology and Salamander Guilds. Cambridge: Cambridge University Press.

Hall, G.A. 1983. West Virginia Birds. Publication No. 7. Pittsburgh, PA: Carnegie Museum of Natural History.

Hamilton, W.J., III. 1962. Evidence concerning the function of nocturnal call notes of migratory birds. Condor 64(5):390-401.

Hammond, J., R. Keeney, and H. Raiffa. 2002. Smart Choices: A Practical Guide to Making Better Decisions. Boston, MA: Harvard Business School Press.

Harmata, A.R., K.M. Podruzny, J.R. Zelenak, and M.L. Morrison. 1999. Using marine surveillance radar to study bird movements and impact assessment. Wildlife Soc. Bull. 27(1):44-52.

Harmata, A.R., G.R. Leighty, and E.L. O'Neil. 2003. A vehicle-mounted radar for dual-purpose monitoring of birds. Wildlife Soc. Bull. 31(3):882-886.

Hassinger, J.D. 2005. Conservation Management Plan for *Neotoma magister*. State Wildlife Grant Final Report. Pennsylvania Game Commission, Harrisburg, PA.

Hassinger, J.D., C.M. Butchkoski, and D.R. Diefenbach. 2005. Conservation Management Plan for *Neotoma magister*. State Wildlife Grant Progress Report. Pennsylvania Game Commission, Harrisburg, PA.

Haskell, D.G. 2000. Effects of forest roads on macroinvertebrate soil fauna of the southern Appalachian Mountains. Conserv. Biol. 14(1):57-63.

Hathaway, A., D. Jacobsen, and C. High. 2005. Model State Implementation Plan (SIP) Documentation for Wind Energy Purchase in States with Renewable Energy Set-Aside. Subcontract Report NREL/SR-500-38075. National Renewable Energy Laboratory, Golden, CO. May 2005 [online]. Available: http://www.nrel.gov/docs/fy05osti/38075.pdf [accessed Sept. 20, 2006].

Hawrot, R.Y., and J.M. Hanowski. 1997. Avian Assessment Document. Avian Population Analysis for Wind Power Generation Regions—012. NRRI/TR-97-23. Center for Water and the Environment, Natural Resources Research Institute, Duluth, MN. 14 pp.

Hayes, J.P. 1997. Temporal variation in activity of bats and the design of echolocation-monitoring studies. J. Mammal. 78(2):514-524.

Hayes, J.P. 2000. Assumptions and practical considerations in the design and interpretation of echolocation-monitoring studies. Acta Chiropterol. 2(2):225-236.

Hayes, J.P. 2003. Habitat ecology and conservation of bats in western coniferous forests. Pp. 81-119 in Mammal Community Dynamics: Management and Conservation in the Coniferous Forests of Western North America, C.J. Zabel and R.G. Anthony, eds. Cambridge: Cambridge University Press.

Hayes, J.P., and J.C. Gruver. 2000. Vertical stratification of bat activity in an old-growth forest in western Washington. Northwest Sci. 74(2):102-108.

Hayes, J.P., and S. Loeb. 2007. The influences of forest management on bats in North America. Pp. 206-235 in Bats in Forests: Conservation and Management, M.J. Lacki, J.P. Hayes, and A. Kurta, eds. Baltimore, MD: John Hopkins University Press.

Hecklau, J. 2005. Visual Characteristics of Wind Turbines. Presentation at the Technical Considerations in Siting Wind Developments: Research Meeting on December 1-2, 2005, Washington, DC [online]. Available: http://www.nationalwind.org/events/siting/presentations/hecklau-visual_characteristics.pdf [accessed Sept. 19, 2006].

Hedges, L.V. 1986. Statistical issues in the meta-analysis of environmental studies. Pp. 261-283 in 1986 Proceedings of Biopharmaceutical Section: The Annual Meeting of the American Statistical Association, August 15-18, 1986, Chicago, IL. Washington, DC: American Statistical Association.

Hedges, L.V., and I. Olkin. 1985. Statistical Methods for Meta-Analysis. Orlando, FL: Academic Press.

Hendrickson, C.T., A. Horvath, S. Joshi, and L.B. Lave. 1998. Economic input-output models for environmental life-cycle assessment. Environ. Sci. Technol. 32(7):184A-191A.

Hendrickson, C.T., L.B. Lave, and H.S. Matthews, eds. 2006. Environmental Life Cycle Assessment of Goods and Services: An Input-Output Approach. Washington, DC: Resources for the Future.

HIDLNR (Hawaii Department of Land and Natural Resources). 2006. Federal and State Governments Approve Conservation Plan for Maui Wind Farm. News Release: February 8, 2006 [online]. Available: http://www.hawaii.gov/dlnr/chair/pio/HtmlNR/06-N021.htm [accessed Sept. 13, 2006].

High, C.J., and K.M. Hathaway. 2006. Avoided Air Emissions from Electric Power Generation at Three Potential Wind Energy Projects in Virginia. Resource Systems Group, Inc., White River Junction, VT. July 5, 2006 [online]. Available: http://www.chesapeakeclimate.org/doc/VA-Wind%20Report.doc [accessed Sept. 20, 2006].

Hill, S.B., and D.H. Clayton. 1985. Wildlife after Dark: A Review of Nocturnal Observation Techniques. Occasional Papers 17. Minneapolis: James Ford Bell Museum of Natural History, University of Minnesota.

Hobson, K.A. 1999. Tracing origins and migration of wildlife using stable isotopes: A review. Oecologia 120(3):314-326.

Hobson, K.A., and L.I. Wassenaar. 2001. Isotopic delineation of North American migratory wildlife population: Loggerhead shrikes. Ecol. Appl. 11(5):1545-1553.

Hodgkison, R., D. Ahmad, S. Balding, T. Kingston, A. Zubaid, and T.H. Kunz. 2002. Capturing bats (*Chiroptera*) in tropical forest canopies. Pp. 160-167 in The Global Canopy Programme Handbook: Techniques of Access and Study in the Forest Roof, A.W. Mitchell, K. Secoy, and T. Jackson, eds. Oxford, UK: Global Canopy Programme.

Hodos, W. 2003. Minimization of Motion Smear: Reducing Avian Collisions with Wind Turbines, Period of Performance: July 12, 1999-August 31, 2002. NREL/SR-500-33249. Prepared for the National Renewable Energy Laboratory, Golden, CO. August 2003 [online]. Available: http://www.nrel.gov/docs/fy03osti/33249.pdf [accessed June 2, 2006].

Hoen, B. 2006. Impacts of Windmills Visibility on Property Values in Madison County, New York. M.S. Thesis, Bard College, Annandale-on-Hudson, NY [online]. Available: http://www.aceny.org/pdfs/misc/Property%20Value%20Study%20Full%20Text5_24_06.pdf [accessed May 22, 2007].

Hogberg, L.K., K.J. Patriquin, and R.M.R. Barclay. 2002. Use by bats of patches of residual trees in logged areas of the boreal forest. Am. Midl. Nat. 148(2):282-288.

Hoover, S.L., and M.L. Morrison. 2005. Behavior of red-tailed hawks in a wind turbine development. J. Wildlife Manage. 69(1):150-159.

Hoppe-Kilpper, M., and U. Steinhauser. 2002. Wind landscapes in the German Milieu. Pp. 94-96 in Wind Power in View: Energy Landscapes in a Crowded World, M.J. Pasqualetti, P. Gipe, and R.W. Righter, eds. San Diego, CA: Academic Press.

Horn, J.W. 2007. Nightly and seasonal behavior of bats in the aerosphere assessed with thermal infrared imaging and NEXRAD Doppler radar. Ph.D. Dissertation. Boston University, Boston, MA.

Horn, J., and E.B. Arnett. 2005. Timing of nightly bat activity with wind turbine blades. Pp. 96-116 in Relationships Between Bats and Wind Turbines in Pennsylvania and West Virginia: An Assessment of Fatality Search Protocols, Patterns of Fatality, and Behavioral Interactions with Wind Turbines, E.B. Arnett, W.P. Erickson, J. Kerns, and J. Horn, eds. Final Report. Prepared for the Bats and Wind Energy Cooperative, by Bat Conservation International, Austin, TX. June 2005 [online]. Available: http://www.batcon.org/wind/BWEC2004finalreport.pdf [accessed May 25, 2006].

Horn, J.W., E.B. Arnett, and T.H. Kunz. In press. Behavioral responses of bats to operating wind turbines. J. Wildlife Manage.

Hötker, H., K.M. Thomsen, and H. Köster. 2004. Auswirkungen regenerativer Energiegewinnung auf die biologische Vielfalt am Beispiel der Vögel und der Fledermäuse-Fakten, Wissenslücken, Anforderungen an die Forschung, ornithologische Kriterien zum Ausbau von regenerativen Energiegewinnungsformen. Gefördert vom Bundesamt für Naturschutz; Förd. Nr. Z1.3-684 11-5/03. Michael-Otto-Institut, NABU, Bonn. December 2004 [online]. Available: http://www.wind-energie.de/fileadmin/dokumente/Themen_A-Z/Vogelschutz/Studie_nabu_VoegelRegEnergien.pdf [accessed May 26, 2006].

Howe, R.W., T.C. Erdman, and K.D. Kruse. 1995. Potential Avian Mortality at Wind Generation Towers in Southeastern Brown County. Richter Museum Special Report No. 4. University of Wisconsin-Green Bay.

Howe, R.W., W. Evans, and A.T. Wolf. 2002. Effects of Wind Turbines on Birds and Bats in Northeastern Wisconsin. Prepared by University of Wisconsin-Green Bay, for Wisconsin Public Service Corporation and Madison Gas and Electric Company, Madison, WI. November 21, 2002 [online]. Available: http://psc.wi.gov/apps/erf_share/view/viewdoc.aspx?docid=35200 [accessed May 23, 2006].

Howell, J.A. 1995. Remote Sensing of Wildlife Habitat for Inventory and Long Term Monitoring: A California Perspective. Conference for Remote Sensing and Environmental Monitoring for the Sustained Development of the Americas, March 1995, San Juan, Puerto Rico.

Howell, J.A. 1997. Bird mortality at rotor swept area equivalents, Altamont Pass and Montezuma Hills, California. T. Western Sec. Wildlife Soc. 33:24-29.

Howell, J.A., and J. Noone. 1992. Examination of Avian Use and Mortality at a U.S. Windpower, Wind Energy Development Site, Montezuma Hills, Solano County, California: Final Report. Fairfield, CA: Solano County, Department of Environmental Management. 41 pp.

Howell, J.C., A.R. Laskey, and J.T. Tanner. 1954. Bird mortality at airport ceilometers. Wilson Bull. 66(3):207-215.

Hoying, K.M., and T.H. Kunz. 1998. Variation in size at birth and post-natal growth in the insectivorous bat *Pipistrellus subflavus* (Chiroptera: Vespertilionidae). J. Zool. 245(1):15-27.

Humphrey, S.R. 1975. Nursery roosts and community diversity of Nearctic bats. J. Mammal. 56(2):321-346.

Humphrey, S.R., A.R. Richter, and J.B. Cope. 1977. Summer habitat and ecology of the endangered Indiana bat, *Myotis sodalis*. J. Mammal. 58(3):334-346.

Hunt, G., and T. Hunt. 2006. The Trend of Golden Eagle Territory Occupancy in the Vicinity of the Altamont Pass Wind Resource Area: 2005 Survey. Pier Final Project Report. CEC-500-2006-056. Prepared for California Energy Commission, Public Interest Energy Research Program, Sacramento, CA, June 2006 [online]. Available: http://www.energy.ca.gov/2006publications/CEC-500-2006-056/CEC-500-2006-056.PDF [accessed May 22, 2007].

Hunt, W.G. 2002. Golden Eagles in a Perilous Landscape: Predicting the Effects of Mitigation for Wind Turbine Blade-Strike Mortality. Consultant Report. P500-97-4033F. Prepared for California Energy Commission, Public Interest Energy Research Program, Sacramento, CA, by University of California, Santa Cruz, CA. July 2002 [online]. Available: http://www.energy.ca.gov/reports/2002-11-04_500-02-043F.PDF [accessed May 24, 2006].

Hüppop, O., J. Dierschke, K.M. Exo, E. Fredrich, and R. Hill. 2006. Bird migration studies and potential collision risk with offshore wind turbines. Ibis 148(1):90-109.

Hutchinson, J.T., and M.J. Lacki. 2001. Possible microclimate benefits of roost site selection in the red bat, *Lasiurus borealis*, in mixed mesophytic forests of Kentucky. Can. Field Nat. 115(2):205-209.

IEA (International Energy Agency). 2006. Energy Policies of IEA Countries: Denmark 2006 Review. Paris: International Energy Agency, Organization for Economic Co-operation and Development.

IEC (International Electrotechnical Commission). 2002. Wind Turbine Generator Systems-Part 11: Acoustic Noise Measurement Techniques. International Standard IEC 61400-11, 2nd Ed., 2002-12. International Electrotechnical Commission, Geneva.

Ihle, J. 2005. Renewable Power Outlook 2005. RPS-7. Platts Renewable Power Service. February 2005 [online]. Available: http://www.esource.com/members/prc_rps/pdf/rps7.pdf [accessed Sept. 7, 2006].

ISO New England Inc. 2006. 2004 New England Marginal Emission Rate Analysis. Draft February 28, 2006 [online]. Available: http://www.iso-ne.com/committees/comm_wkgrps/relblty_comm/pwrsuppln_comm/mtrls/2006/mar162006/02282006_draft2004mea.pdf [accessed Aug. 17, 2006].

Jacques Whitford Limited. 2005. An Investigation of a New Monitoring Technology for Birds and Bats. Jacques Whitford Project No. ONT50563. Prepared for Suncor Energy Products, Inc., Vision Quest Windelectric-TransAlta's Wind Business, Canadian Hydro Developer's, Inc., and Enbridge Inc., by Jacques Whitford Limited, Markham, Ontario. August 2005 [online]. Available: http://www.mgwindpower.info/EchoTrack%20Final%20Report.pdf [accessed Sept. 27, 2006].

Jain, A.A. 2005. Bird and Bat Behavior and Mortality at a Northern Iowa Windfarm. M.S. Thesis, Iowa State University, Ames, IA. 107 pp.

Johnson, G.D. 2005. A review of bat mortality at wind-energy developments in the United States. Bat Res. News 46(2):45-49.

Johnson, G.D., D.P. Young, Jr., W.P. Erickson, C.E. Derby, M.D. Strickland, R.E. Good, and J.W. Kern. 2000a. Wildlife Monitoring Studies: SeaWest Windpower Project, Carbon County, Wyoming, 1995-1999. Final Report. Prepared by WEST, Inc., Cheyenne, WY, for SeaWest Energy Corporation, San Diego, CA, and Bureau of Land Management, Rawlins District Office, Rawlins, WY. August 9, 2000 [online]. Available: http://www.west-inc.com/reports/fcr_final_baseline.pdf [accessed May 24, 2006].

Johnson, G.D., W.P. Erickson, M.D. Strickland, M.F. Shepherd, and D.A. Shepherd. 2000b. Avian Monitoring Studies at the Buffalo Ridge, Minnesota Wind Resource Area: Results of a 4-Year Study. Prepared by WEST, Inc., Cheyenne, WY, for Northern States Power Company, Minneapolis, MN. September 22, 2000 [online]. Available: http://www.west-inc.com/reports/avian_buffalo_ridge.pdf [accessed May 24, 2006].

Johnson, G.D., D.P. Young, W.P. Erickson, M.D. Strickland, R.E. Good, and P. Becker. 2001. Avian and Bat Mortality Associated with the Initial Phase of the Foote Creek Rim Windpower Project, Carbon County, Wyoming: November 3, 1998-October 31, 2000. Prepared for SeaWest Windpower, Inc., San Diego, CA, and Bureau of Land Management, Rawlins District Office, Rawlins, WY, by Western EcoSystems Technology, Inc., Cheyenne, WY. 32 pp.

Johnson, G.D., W.P. Erickson, M.D. Strickland, M.F. Shepherd, D.A. Shepherd, and S.A. Sarappo. 2002. Collision mortality of local and migrant birds at a large-scale wind power development on Buffalo Ridge, Minnesota. Wildlife Soc. Bull. 30(3):879-887.

Johnson, G.D., W.P. Erickson, M.D. Strickland, M.F. Shepherd, and S.A. Sarappo. 2003a. Mortality of bats at a large-scale wind power development at Buffalo Ridge, Minnesota. Am. Midl. Nat. 150(2):332-342.

Johnson, G.D., W.P. Erickson, J. White, and R. McKinney. 2003b. Avian and Bat Mortality During the First Year of Operation at the Klondike Phase I Wind Project, Sherman County, Oregon. Prepared by WEST, Inc., Cheyenne, WY, for Northwestern Wind Power, Goldendale, WA. March 2003 [online]. Available: http://www.west-inc.com/reports/klondike_final_mortality.pdf [accessed May 24, 2006].

Johnson, G.D., M.K. Perlik, W.P. Erickson, and M.D. Strickland. 2004. Bat activity, composition and collision mortality at a large wind plant in Minnesota. Wildlife Soc. Bull. 32(4):1278-1288.

Johnston, L., E. Hausman, A. Sommer, B. Biewald, T. Wolf, D. Schlissel, A. Roschelle, and D. White. 2006. Climate Change and Power: Carbon Dioxide Emissions Costs and Electricity Resource Planning. Synapse Energy Economics, Cambridge, MA. June 8, 2006 [online]. Available: http://www.synapse-energy.com/Downloads/SynapsePaper.2006-06.Climate-Change-and-Power.pdf [accessed Sept. 21, 2006].

Jones, G., and K.E. Barlow. 2004. Cryptic species of echolocating bats. Pp. 345-349 in Echolocation in Bats and Dolphins, J.A. Thomas, C.F. Moss, and M. Vater, eds. Chicago, IL: University of Chicago Press.

Jones, G., N. Vaughan, D. Russo, L.P. Wickramasinghe, and S. Harris. 2004. Designing bat activity surveys using time expansion and direct sampling of ultrasound. Pp. 83-89 in Bat Echolocation Research: Tools, Techniques, and Analysis, R.M. Brigham, E.K.V. Kalko, G. Jones, S. Parsons, and H.J.G.A. Limpens, eds. Austin, TX: Bat Conservation International.

Jones, K., A. Purvis, and J. Gittleman. 2003. Biological correlates of extinction risks in bats. Am. Nat. 161:601-614.

Kalcounis, M.C., K.A. Hobson, R.M. Brigham, and K.R. Hecker. 1999. Bat activity in the boreal forest: Importance of stand type and vertical strata. J. Mammal. 80(2):673-682.

Kalko, E.K.V. 2004. Neotropical leaf-nosed bats (*Phyllostomidae*): "Whispering" bats or candidates for acoustic survey? Pp. 63-69 in Proceedings of a Workshop on Identification and Acoustic Monitoring of Bats, M. Brigham, G. Jones, and E.K.V. Kalko, eds. Austin, TX: Bat Conservation International.

Kaplan, S., and B.J. Garrick. 1981. On the quantitative definition of risk. Risk Anal. 1(1):11-27.

Kapos, V., E. Wandelli, J.L. Camargo, and G. Ganade. 1997. Edge-related changes in environ-
    ment and plant responses due to forest fragmentation in central Amazonia. Pp. 33-44
    in Tropical Forest Remnants: Ecology, Management, and Conservation of Fragmented
    Communities, W.F. Laurance and R.O. Bierregaard, eds. Chicago, IL: University of
    Chicago Press.
Karlsson, J. 1983. Interactions between Birds and Aerogenerators: Result Report 1977-1982
    [in Swedish]. Lund, Sweden: Lund University.
Keith, G., D. White, and B. Biewald. 2002. The OTC Emission Reduction Workbook 2.1
    Description and User's Manual. Prepared for the Ozone Transport Commission, by
    Synapse Energy Economics, Inc., Cambridge, MA. October 15, 2002 [online]. Available:
    http://www.otcair.org/download.asp?FID=69&Fcat=Documents&Fview=Reports&Ffile
    =Workbook%202.1%20Manual.pdf [accessed Sept. 20, 2006].
Keith, G., B. Biewald, A. Sommer, P. Henn, and M. Breceda. 2003. Estimating the Emis-
    sion Reduction Benefits of Renewable Electricity and Energy Efficiency in North
    America: Experience and Methods. Prepared for the Commission for Environmental
    Cooperation. September 22, 2003 [online]. Available: http://www.synapse-energy.com/
    Downloads/SynapseReport.2003-09.CEC.Emission-Reduction-Benefits-Renewables-and-
    EE-Estimates.03-18.pdf [accessed Sept. 21, 2006].
Keith, G., B. Biewald, and D. White. 2004. Evaluating Simplified Methods of Estimating
    Displaced Emissions in Electric Power Systems: What Works and What Doesn't. Pre-
    pared for the Commission for Environmental Cooperation. November 4, 2006 [online].
    Available: http://www.synapse-energy.com/Downloads/SynapseReport.2004-11.CEC.
    Evaluating-Simplified-Methods-of-Estimating-Displaced-Emissions.04-62.pdf [accessed
    Sept. 21, 2006].
Kelly, J.F., and D.M. Finch. 1998. Tracking migrant songbirds with stable isotopes. Trends
    Ecol. Evol. 13(2):48-49.
Kelly, J.F., V. Atudorei, Z.D. Sharp, and D.M. Finch. 2002. Insights into Wilson's Warbler
    migration from analyses of hydrogen stable-isotope ratios. Oecologia 130(2):216-221.
Kelly, J.F., K.C. Ruegg, and T.B. Smith. 2005. Combining isotopic and genetic markers to
    identify breeding origins of migrant birds. Ecol. Appl. 15(5):1487-1494.
Kempthorne, O. 1966. The Design and Analysis of Experiments. New York: John Wiley and
    Sons.
Kerlinger, P. 1982. The migration of common loons through eastern New York. Condor
    84(1):79-100.
Kerlinger, P. 1995. How Birds Migrate, 1st Ed. Mechanicsburg, PA: Stackpole Books.
    228 pp.
Kerlinger, P. 1997. A Study of Avian Fatalities at the Green Mountain Power Corporation's
    Searsburg, Vermont, Windpower Facility 1997. Prepared for Vermont Department of
    Public Service, Green Mountain Power Corporation, National Renewable Energy Labo-
    ratory and Vermont Environmental Research Associates. 12 pp.
Kerlinger, P. 2000. Avian Mortality at Communication Towers: A Review of Recent Litera-
    ture, Research, and Methodology. Prepared for U.S. Fish and Wildlife Service, Office of
    Migratory Bird Management, by Curry & Kerlinger LLC, Cape May Point, NJ. March
    2000 [online]. Available: http://training.fws.gov/library/Pubs9/avian_mortality00.pdf [ac-
    cessed May 22, 2006].
Kerlinger, P. 2002. An Assessment of the Impacts of Green Mountain Power Corporation's
    Wind Power Facility on Breeding and Migrating Birds in Searsburg, Vermont: July 1996-
    July 1998. NREL/SR-500-28591. Prepared for Vermont Public Service, Montpelier, VT.
    U.S. Department of Energy—National Renewable Energy Laboratory, Golden, CO.
    March 2002 [online]. Available: http://www.nrel.gov/docs/fy02osti/28591.pdf [accessed
    July 14, 2006].

Kerlinger, P., and J. Dowdell. 2003. Breeding Bird Survey for the Flat Rock Wind Power Project, Lewis County, New York. Prepared for Atlantic Renewable Energy Corporation.

Kerlinger, P., and S.A. Gauthreaux, Jr. 1985. Flight behavior of raptors during spring migration in south Texas studied with radar and direct visual observations. J. Field Ornithol. 56(4):397-402.

Kerlinger, P., and J. Kerns. 2003. FAA Lighting of Wind Turbines and Bird Collisions. Presentation at the National Wind Coordinating Committee—Wildlife Working Group Meeting: How Is Biological Significance Determined When Assessing Possible Impacts? November 17-18, 2003, Washington, DC [online]. Available: http://www.nationalwind.org/events/wildlife/2003-2/presentations/Kerlinger.pdf [accessed June 14, 2006].

Kerlinger, P., and F.R. Moore. 1989. Atmospheric structure and avian migration. Pp. 109-141 in Current Ornithology, Vol. 6, D.M. Power, ed. New York: Plenum Press.

Kerlinger, P., R. Curry, L. Culp, A. Jain, C. Wilkerson, B. Fischer, and A. Hasch. 2006. Post-Construction Avian and Bat Fatality Monitoring Study for the High Winds Wind Power Project Solano County, California: Two Year Report. Prepared for FPL Energy, High Winds LLC, by Curry and Kerlinger, LLC, Cape May Point, NJ.

Kerns, J., and P. Kerlinger. 2004. A Study of Bird and Bat Collision Fatalities at the Mountaineer Wind Energy Center, Tucker County, West Virginia: Annual Report for 2003. Prepared for FPL Energy and Mountaineer Wind Energy Center Technical Review Committee. February 14, 2004 [online]. Available: http://www.responsiblewind.org/docs/MountaineerFinalAvianRpt3-15-04PKJK.pdf [accessed May 22, 2006].

Kerns, J., W.P. Erickson, and E.B. Arnett. 2005. Bat and bird fatality at wind energy facilities in Pennsylvania and West Virginia. Pp. 24-95 in Relationships Between Bats and Wind Turbines in Pennsylvania and West Virginia: An Assessment of Fatality Search Protocols, Patterns of Fatality, and Behavioral Interactions with Wind Turbines, E.B. Arnett, ed. Final Report. Prepared for the Bats and Wind Energy Cooperative, by Bat Conservation International, Austin, TX. June 2005 [online]. Available: http://www.batcon.org/wind/BWEC2004finalreport.pdf [accessed May 25, 2006].

Kirby Mountain. 2006. Model Large Wind Energy Ordinance. Kirby Mountain, June 07, 2006 [online]. Available: http://kirbymtn.blogspot.com/2006_06_01_kirbymtn_archive.html [accessed Sept. 14, 2006].

Klazura, G.E., and D.A. Imy. 1993. A description of the initial set of analysis products available from the NEXRAD WSR-88D System. Bull. Am. Meteorol. Soc. 74(7):1293-1311.

Knapp, S.M., C.A. Haas, D.G. Harpole, and R.L. Kirkpatrick. 2003. Initial effects of clearcutting and alternative silvicultural practices on terrestrial salamander abundance. Conserv. Biol. 17(3):752-762.

Knust, R., P. Dalhoff, J. Gabriel, J. Heuers, O. Hüppop, and H. Wendeln. 2003. Investigations to Avoid and Reduce Possible Impacts of Wind Energy Parks on the Marine Environment in the Offshore Areas of North and Baltic Sea. Final Report. Alfred-Wegener-Institut für Polar- und Meeresforschung, Bremerhaven, and Umweltbundesamt, Berlin. March 15, 2003 [online]. Available: http://www.umweltdaten.de/publikationen/fpdf-l/2686.pdf [accessed June 14, 2006].

Koford, R., A. Jain, G. Zenner, and A. Hancock. 2004. Avian Mortality Associated with the Top of Iowa Wind Farm. Progress Report: Calendar Year 2003. Iowa State University, Ames, IA. February 28, 2004 [online]. Available: http://www.ohiowind.org/ohiowind/page.cfm?pageID=2011 [accessed May 24, 2006].

Kragh, J., D. Theofiloyiannakos, H. Klug, T. Osten, B. Andersen, N. van der Borg, S. Ljunggren, O. Fegeant, R.J. Whitson, J. Bass, D. Englich, C. Eichenlaub, and R. Weber. 1999. Noise Immission from Wind Turbines. Contractor Report No. ETSU W/13/00503/REP. National Engineering Laboratory [online]. Available: http://www.greatplacetowork.gov.uk/renewables/publications/pdfs/W1300503.pdf [accessed Dec. 19, 2006].

Krebs, C.J. 1989. Ecological Methodology. New York: Harper & Row.

Krusic, R.A., M. Yamasaki, C.D. Neefus, and P.J. Pekins. 1996. Bat habitat use in White Mountain National Forest. J. Wildlife Manage. 60(3):625-631.

KSEC (Kansas Energy Council). 2004. Guidelines and Standards for the Siting of Wind-Energy Development in Kansas and the Flint Hills, August 30, 2004 [online]. Available: http:// www.kansasenergy.org/KEC/SitingGuidelines(KEC).pdf [accessed Sept. 18, 2006].

KSREWG (The Kansas Renewable Energy Working Group). 2003. Siting Guidelines for Wind-power Projects in Kansas. The Kansas Renewable Energy Working Group, Environmental and Siting Committee [online]. Available: http://www.naseo.org/energy_sectors/wind/ kansas_siting_guidelines.pdf [accessed May 24, 2007].

Kunz, T.H. 1982a. Lasionycteris noctivagans. Mamm. Species 173:1-5.

Kunz, T.H., ed. 1982b. Ecology of Bats. New York: Plenum Press.

Kunz, T.H. 1982c. Roosting ecology of bats. Pp. 1-55 in Ecology of Bats, T.H. Kunz, ed. New York: Plenum Press.

Kunz, T.H., ed. 1988. Ecological and Behavioral Methods for the Study of Bats. Washington, DC: Smithsonian Institution Press. 533 pp.

Kunz, T.H. 2003. Censusing bats: Challenges, solutions, and sampling biases. Pp. 9-20 in Monitoring Trends in Bat Populations of the United States and Territories: Problems and Prospects, T.J. O'Shea and M.A. Bogan, eds. Biological Resources Discipline, Information and Technology Report USGS/BRD/ITR-2003-003. U.S. Geological Survey [online]. Available: http://www.fort.usgs.gov/products/publications/21329/21329.pdf [accessed June 9, 2006].

Kunz, T.H. 2004. Foraging habits of North American insectivorous bats. Pp. 13-25 in Bat Echolocation Research: Tools, Techniques, and Analysis, R.M. Brigham, E.K.V. Kalko, G. Jones, S. Parsons, and H.J.G.A. Limpens, eds. Austin, TX: Bat Conservation International.

Kunz, T.H., and A. Kurta. 1988. Methods of capturing and holding bats. Pp. 1-30 in Ecological and Behavioral Methods for the Study of Bats, T.H. Kunz, ed. Washington, DC: Smithsonian Institution Press.

Kunz, T.H., and L.F. Lumsden. 2003. Ecology of cavity and foliage roosting bats. Pp. 3-89 in Bat Ecology, T.H. Kunz and M.B. Fenton, eds. Chicago, IL: University of Chicago Press.

Kunz, T.H., and S. Parsons, eds. In press. Ecological and Behavioral Methods for the Study of Bats, 2nd Ed. Baltimore, MD: Johns Hopkins University Press.

Kunz, T.H., G.R. Richards, and C.R. Tidemann. 1996. Capturing small volant mammals. Pp. 157-164 in Measuring and Monitoring Biological Diversity: Standard Methods for Mammals, D.E. Wilson, F.R. Cole, J.D. Nichols, R. Rudran, and M.S. Foster, eds. Washington, DC: Smithsonian Institution Press.

Kunz, T., A. Zubaid, and G.F. McCracken, eds. 2006. Functional and Evolutionary Ecology of Bats. New York: Oxford University Press.

Kunz, T.H., E.B. Arnett, W.P. Erickson, A.R. Hoar, G.D. Johnson, R.P. Larkin, M.D. Strickland, R.W. Thresher, and M.D. Tuttle. 2007. Ecological impacts of wind energy development on bats: Questions, research needs and hypotheses. Front. Ecol. Environ.

Kunz, T.H., R. Hodgkison, and C. Weisse. In press a. Methods of capturing and holding bats. In Ecological and Behavioral Methods for the Study of Bats, 2nd Ed., T.H. Kunz, and S. Parsons, eds. Baltimore, MD: Johns Hopkins University Press.

Kunz, T.H., E.B. Arnett, B.M. Cooper, W.P. Erickson, R.P. Larkin, T. Mabee, M.L. Morrison, M.D. Strickland, and J.M. Szewczak. In press b. Methods and metrics for assessing impacts of wind energy development on nocturnally-active birds and bats. J. Wildlife Manage.

Kurta, A. 1982. Flight patterns of Eptesicus fuscus and Myotis lucifugus over a stream. J. Mammal. 63(2):335-337.

Kurta, A., and J. Kennedy, eds. 2002. The Indiana Bat: Biology and Management of an Endangered Species. Austin, TX: Bat Conservation International.

Kurta, A., K.J. Williams, and R. Mies. 1996. Ecological, behavioural, and thermal observa-
tions of a peripheral population of Indiana bats (Myotis sodalis). Pp. 102-117 in Bats and
Forests Symposium, October 19-21, 1995, Victoria, British Columbia, Canada, R.M.R.
Barclay and R.M. Brigham, eds. Working Paper 23. Victoria, BC: British Columbia,
Ministry of Forests Research Branch.

Kurta, A., S.W. Murray, and D.H. Miller. 2002. Roost selection and movements across
the summer landscape. Pp. 118-129 in The Indiana Bat: Biology and Management of
an Endangered Species, A. Kurta and J. Kennedy, eds. Austin, TX: Bat Conservation
International.

Laake, J.L. 1992. Catch-per-unit-effort models: An application to an elk population in Colorado. Pp. 44-55 in Wildlife 2001: Populations, D.R. McCullough and R.H. Barrett, eds.
London: Elsevier Applied Science.

Laake, J.L., S.T. Buckland, D.R. Anderson, and K.P. Burnham. 1993. DISTANCE User's
Guide. Version 2.0. Colorado Cooperative Fish and Wildlife Research Unit, Colorado
State University, Fort Collins, CO. 72 pp.

Lack, D., and G.C. Varley. 1945. Detection of birds by radar. Nature 156:446.

Lancia, R.A., W.L. Kendall, K.H. Pollock, and J.D. Nichols. 2005. Estimating the number of
animals in wildlife populations. Pp. 106-133 in Techniques for Wildlife Investigations
and Management, 6th Ed., C.E. Braun, ed. Bethesda, MD: Wildlife Society.

Larkin, R.P. 1991. Flight speeds observed with radar, a correction: Slow "birds" are insects.
Behav. Ecol. Sociobiol. 29(3):221-224.

Larkin, R.P. 2005a. Review of strengths, weaknesses, and application of tools. Pp. 79-80 in
Proceedings of the Onshore Wildlife Interactions with Wind Developments: Research
Meeting V: November 3-4, 2004, Lansdowne, VA, S. Savitt Schwartz, ed. National
Wind Coordinating Committee [online]. Available: http://www.nationalwind.org/events/
wildlife/2004-2/proceedings.pdf [accessed May 22, 2007].

Larkin, R.P. 2005b. Radar techniques for wildlife biology. Pp. 448-464 in Techniques for
Wildlife Investigations and Management, 6th Ed., C.E. Braun, ed. Bethesda, MD: Wildlife Society.

Larkin, R.P. 2006. Migrating bats interacting with wind turbines: What birds can tell us. Bat
Res. News 47(2):23-32.

Larkin, R.P., and B.A. Frase. 1988. Circular paths of birds flying near a broadcasting tower
in cloud. J. Comp. Psychol. 102(1):90-93.

Larkin, R.P., W.R. Evans, and R.H. Diehl. 2002. Nocturnal flight calls of Dickcissels and Doppler radar echoes over south Texas in spring. J. Field Ornithol. 73(1):2-8.

Larom, D. 2002. Auditory communication, meteorology, and the Unwelt. J. Comp. Psychol.
116(2):133-136.

Larom, D., M. Garstang, K. Payne, R. Raspet, and M. Lindeque. 1997. The influence of
surface atmospheric conditions on the range and area reached by animal vocalizations.
J. Exp. Biol. 200(3):421-431.

Larsen, J.K., and J. Madsen. 2000. Effects of wind turbines and other physical elements on
field utilization by pink-footed geese (Anser brachyrhynchus): A landscape perspective.
Landscape Ecol. 15(8):755-764.

Larson, D.J., and J.P. Hayes. 2000. Variability in sensitivity of Anabat II detectors and a
method of calibration. Acta Chiropterol. 2(2):209-213.

Laurance, W.F. 1997. Hyper-disturbed parks: Edge effects and the ecology of isolated rain-
forest reserves in Tropical Australia. Pp. 71-84 in Tropical Forest Remnants: Ecology,
Management, and Conservation of Fragmented Communities, W.F. Laurance and R.O.
Bierregaard, eds. Chicago, IL: University of Chicago Press.

Laurance, W.F., and M.A. Cochrane. 2001. Special section: Synergistic effects in fragmented landscapes. Conserv. Biol. 15(6):1488-1489.

LaVal, R.K., and M.L. LaVal. 1980. Ecological Studies and Management of Missouri Bats with Emphasis on Cave-Dwelling Species. Terrestrial Series No. 8. Jefferson City, MO: Missouri Department of Conservation.

Law, B.S., J. Anderson, and M. Chidel. 1999. Bat communities in a fragmented forest landscape on the south-west slopes of New South Wales, Australia. Biol. Conserv. 88(3):333-345.

Lawrence, B.D., and J.A. Simmons. 1982. Measurements of atmospheric attenuation at ultrasonic frequencies and the significance for echolocation by bats. J. Acoust. Soc. Am. 71(3):585-590.

Leddy, K.L., K.F. Higgins, and D.E. Naugle. 1999. Effects of wind turbines on upland nesting birds in Conservation Reserve Program grasslands. Wilson Bull. 111(1):100-104.

Lee, R., A. Krupnick, and D. Burtraw. 1995. Estimating Externalities of Electric Fuel Cycles: Analytical Methods and Issues, Estimating Externalities of Coal Fuel Cycles. Washington, DC: McGraw-Hill/Utility Data Institute.

Lenzen, M., and J. Munksgaard. 2002. Energy and CO_2 life-cycle analyses of wind turbines—review and applications. Renew. Energ. 26(3):339-362.

Lewis, T. 1970. Patterns of distribution of insects near a windbreak of tall trees. Ann. Appl. Biol. 65:213-220.

Liechti, F., B. Bruderer, and H. Parpoth. 1995. Quantification of nocturnal bird migration by moonwatching: Comparison with radar and infrared observations. J. Field Ornithol. 66(4):457-468.

Liechti, F., P. Dieter, and S. Komenda-Zehnder. 2003. Nocturnal bird migration in Mauritania—first results. J. Ornithol. 144(4):445-450.

Limpens, H.J.G.A., and K. Kapteyn. 1991. Bats, their behaviour and linear landscape elements. Myotis 29:63-71.

Limpens, H.J.G.A., and G.F. McCracken. 2004. Choosing a bat detector: Theoretical and practical aspects. Pp. 28-37 in Bat Echolocation Research: Tools, Techniques, and Analysis, R.M. Brigham, E.K.V. Kalko, G. Jones, S. Parsons, and H.J.G.A. Limpens, eds. Austin, TX: Bat Conservation International.

Lincoln, F.C., S.R. Peterson, and J.L. Zimmerman. 1998. Migration of Birds. Circular 16. U.S. Department of the Interior, U.S. Fish and Wildlife Service, Washington, DC [online]. Available: http://www.npwrc.usgs.gov/resource/birds/migratio/migratio.htm [accessed June 16, 2006].

Link, W.A., and J.R. Sauer. 2002. A hierarchical analysis of population change with application to cerulean warblers. Ecology 83(10):2832-2840.

Litton, R.B., Jr. 1979. Descriptive approaches to landscape analysis. Pp. 77-87 in Proceedings of Our National Landscape: A Conference on Applied Techniques for Analysis and Management of the Visual Resource, April 23-25, 1979, Incline Village, Nevada, G.H. Elsner and R.C. Smardon, eds. General Technical Report PSW 35. U.S. Department of Agriculture, Forest Service, Pacific Southwest Forest and Range Experiment Station, Berkeley, CA.

Lloyd, P., T.E. Martin, R.L. Redmond, U. Langner, and M.M. Hart. 2005. Linking demographic effects of habitat fragmentation across landscapes to continental source-sink dynamics. Ecol. Appl. 15(5):1504-1514.

Loeb, S.C., F.H. Tainter, and E. Cazares. 2000. Habitat associations of hypogeous fungi in the Southern Appalachians: Implications for the endangered northern flying squirrel (*Glaucomys sabrinus coloratus*). Am. Midl. Nat. 144(2):286-296.

LoGiudice, K. 2003. Trophically transmitted parasites and the conservation of small populations: Raccoon roundworm and the imperiled Allegheny woodrat. Conserv. Biol. 17(1):258-266.

Longcore, T., C. Rich, and S.A. Gauthreaux. 2005. Scientific Basis to Establish Policy Regulating Communications Towers to Protect Migratory Birds: Response to Avatar Environmental, LLC, Report Regarding Migratory Bird Collisions with Communications Towers, WT Docket No. 03-187, Federal Communications Commission Notice of Inquiry. Prepared for American Bird Conservancy, Defenders of Wildlife, Forest Conservation Council, The Humane Society of the United States, by Land Protection Partners, Los Angeles, CA, and Clemson University, Clemson, SC. February 14, 2005 [online]. Available: http:// www.abcbirds.org/policy/towers/lpptowerkill.pdf [accessed May 24, 2006].

Lowery, G.H., Jr. 1951. A quantitative study of the nocturnal migration of birds. Univ. Kansas Publ. Museum Nat. Hist. 3(2):361-472.

Lowery, G.H., Jr., and R.J. Newman. 1955. Direct studies of nocturnal bird migration. Pp. 238-263 in Recent Studies in Avian Biology, A. Wolfson, ed. Urbana: University of Illinois Press.

Lowery, G.H., Jr., and R.J. Newman. 1966. A continent wide view of bird migration on four nights in October. Auk 83(4):547-586.

Lycoming County. 2005. Zoning Ordinance. Lycoming County, Pennsylvania [online]. Available: http://www.seda-cog.org/lyco/cwp/view.asp?a=3&Q=416972&lycoRNavrad9590 1=l [accessed Sept. 13, 2006].

Lynch, K. 1960. The Image of the City. Cambridge: MIT Press.

Lynch, K. 1971. Site Planning. Cambridge: MIT Press.

Mabee, T.J., and B.A. Cooper. 2002. Nocturnal Bird Migration at the Stateline and Vansycle Wind Energy Projects, 2000–2001. Prepared for CH2MHILL and FPL Energy Vansycle, LLC, Juno Beach, FL, by ABR, Inc., Forest Grove, OR. 16 pp.

Mabee, T.J., and B.A. Cooper. 2004. Nocturnal bird migration in northeastern Oregon and southeastern Washington. Northwest. Nat. 85(2):39-47.

Mabee, T.J., B.A. Cooper, and J.H. Plissner. 2004. Radar Study of Nocturnal Bird Migration at the Proposed Mount Storm Wind-Power Development, West Virginia, Fall 2003. Final Report. Prepared for Western EcoSystems Technology, Inc., Cheyenne, WY, and NedPower US LLC, Chantilly, VA, by ABR, Inc., Forest Grove, OR. March 2004 [online]. Available: http://www.abrinc.com/news/Publications_Newsletters/Mt.%20Storm%20Ra dar%20Study%20of%20Nocturnal%20Bird%20Migration,%20WV,%20Fall%20200 3.pdf [accessed June 13, 2006].

Mabee, T.J., B.A. Cooper, J.H. Plissner, and D.P. Young. 2006. Nocturnal bird migration over an Appalachian ridge at a proposed wind power project. Wildlife Soc. Bull. 34(3):682-690.

Macaulay Library. 2006. Sound and Video Catalog. Macaulay Library, Cornell Lab of Ornithology [online]. Available: http://www.birds.cornell.edu/macaulaylibrary/ [accessed June 28, 2006].

MADEM (Massachusetts Department of Environmental Management). 1982. Massachusetts Landscape Inventory: A Survey of the Commonwealth's Scenic Areas. Boston: Massachusetts Department of Environmental Management.

Maine Land Use Regulatory Commission. 1997. Comprehensive Land Use Plan for Areas Within the Jurisdiction of the Maine Land Use Regulation Commission [online]. Available: http://www.maine.gov/doc/lurc/reference/clup.html [accessed Sept. 15, 2006].

Manly, B.F.J. 1991. Randomization and Monte Carlo Methods in Biology, 1st. Ed. New York: Chapman and Hall.

Manly, B.F.J. 1997. Randomization, Bootstrap and Monte Carlo Methods in Biology, 2nd Ed. London: Chapman and Hall.

Manly, B.F.J. 2001. Statistics for Environmental Science and Management. Boca Raton, FL: Chapman and Hall/CRC.

Manly, B.F.J., L.L. McDonald, and D.L. Thomas. 1993. Resource Selection by Animals: Sta-tistical Design and Analysis for Field Studies, 1st Ed. London: Chapman and Hall.

Manly, B.F.J., L.L. McDonald, and G.W. Garner. 1996. Maximum likelihood estimation for the double-count method with independent observers. J. Agric. Biol. Envir. S. 1(2):170-189.

Manville, A.M., II. 2001. The ABCs of avoiding bird collisions at communication towers: Next steps. Pp. 85-103 in Avian Interactions with Utility and Communication Structures: Proceedings of a workshop held in Charleston, SC, December 2-3, 1999, R.L. Carlton, ed. Concord, CA: Electric Power Research Institute.

Marra, P.P., K.A. Hobson, and R.T. Holmes. 1998. Linking winter and summer events in a migratory bird by using stable-carbon isotopes. Science 282(5395):1884-1886.

Marsh, D.M., and N.G. Beckman. 2004. Effects of forest roads on the abundance and activity of terrestrial salamanders. Ecol. Appl. 14(6):1882-1891.

Marsh, D.M., G.S. Milam, N.P. Gorham, and N.G. Beckman. 2005. Forest roads as partial barriers to terrestrial salamander movement. Conserv. Biol. 19(6):2004-2008.

Martí, R. 1995. Bird/wind turbine investigations in southern Spain. Pp. 48-52 in Proceedings of the National Avian-Wind Power Planning Meeting, Denver, Colorado, 20-21 July 1994. King City, Ontario: LGL Ltd, Environmental Research Associates [online]. Avail-able: http://www.nationalwind.org/publications/wildlife/avian94/avian94-03.htm#Bird/Wind%20Turbine%20Investigations%20in%20Southern%20Spain [accessed May 24, 2006].

Martin, W.H. 1993. Reproduction of the timber rattlesnake (Crotalus horridus) in the Ap-palachian Mountains. J. Herpetol. 27(2):133-143.

Mathusa, P., and E. Hogan. 2006. Cost estimates for electric generation technologies. Pp. 87-88 in Alternatives to the Indian Point Energy Center for Meeting New York Electric Power Needs. Washington, DC: The National Academies Press.

Matocq, M.D., and F.X. Villablanca. 2001. Low genetic diversity in an endangered species: Recent or historic pattern? Biol. Conserv. 98(1):61-68.

Mattsson, B.J., and G.J. Niemi. 2006. Factors influencing predation on ovenbird (Seiurus aurocapilla) nests in northern hardwoods: Interactions across spatial scales. Auk 123(1):82-96.

McCracken, G.F. 1996. Bats aloft: A study of high-altitude feeding. Bats 14(3):7-10.

McCracken, G.F. 2003. Estimates of population sizes in summer colonies of Brazilian free-tailed bats (Tadarida brasiliensis). Pp. 21-30 in Monitoring Trends in Bat Populations of the United States and Territories: Problems and Prospects, T.J. O'Shea and M.A. Bogan, eds. Biological Resources Discipline, Information and Technology Report USGS/BRD/ITR-2003-003. U.S. Geological Survey [online]. Available: http://www.fort.usgs.gov/products/publications/21329/21329.pdf [accessed May 25, 2006].

McCracken, G.F., and J.K. Westbrook. 2002. Bat patrol. Natl. Geogr. 201(4):114-123.

McCrary, M.D., R.L. McKernan, W.D. Wagner, and R.W. Schreiber. 1983. Nocturnal Avian Migration Assessment of the San Gorgonio Wind Resource Study Area, Spring 1982. Report 83-RD-108. Prepared for Southern California Edison Company, Research and Development Division, Rosemead, CA.

McCrary, M.D., R.L. McKernan, W.D. Wagner, and R.E. Landry. 1984. Nocturnal Avian Mi-gration Assessment of the San Gorgonio Wind Resource Study Area, Fall 1982. Report 84-RD-11. Prepared for Southern California Edison Company, Research and Develop-ment, Rosemead, CA. 87 pp.

McCrary, M.D., R.L. McKernan, and R.W. Schreiber. 1986. San Gorgonio Wind Resource Area: Impacts of Commercial Wind Turbine Generators on Birds, 1985 Data Report. Prepared for Southern California Edison Company, Rosemead, CA. 33 pp.

McGill University. 2000. Migrations of birds and insects. Radar Meteorology at McGill. McGill University, Montreal, Quebec, CA [online]. Available: http://www.radar.mcgill. ca/bird_migration.html [accessed Oct. 4, 2006].

McKinney, M.L. 1997. Extinction vulnerability and selectivity: Combining ecological and paleontological views. Annu. Rev. Ecol. Syst. 28:495-516.

MDDNR (Maryland Department of Natural Resources). 2003. Rare, Threatened, and Endangered Animals of Maryland. Maryland Department of Natural Resources, Wildlife and Heritage Service, Natural Heritage Program, Annapolis, MD. December 12, 2003 [online]. Available: http://dnrweb.dnr.state.md.us/download/rteanimals.pdf [accessed Oct. 16, 2006].

MDDNR (Maryland Department of Natural Resources). 2006. Maryland's Power Plant Research Program (PPRP). Maryland Department of Natural Resources, Annapolis, MD [online]. Available: http://www.dnr.state.md.us/bay/pprp/pp_brochure.html [accessed Sept. 14, 2006].

MDPSC (Maryland Public Service Commission). 1997. Title 20 Subtitle 79 Applications Concerning the Construction or Modification of Generating Stations and Overhead Transmission Lines [online]. Available: http://esm.versar.com/pprp/licensing/regs/pscregs7901. html#defs [accessed Sept. 13, 2006].

MD Windpower TAG (Maryland Windpower Technical Advisory Group). 2006. Windpower Technical Advisory Group-Introduction. Maryland's Power Plant Research Program [online]. Available: http://esm.versar.com/pprp/windpower/default.htm [accessed Sept. 15, 2006].

Mearns, E.A. 1898. A study of the vertebrate fauna of the Hudson Highlands, with observations on the Mollusca, Crustacea, Lepidoptera, and the flora of the region. B. Am. Mus. Nat. Hist. 10:303-352.

MEDEP (Maine Department of Environmental Protection). 2003. Chapter 315: Assessing and Mitigating Impacts to Existing Scenic and Aesthetic Uses (06-096) [online]. Available: http://www.maine.gov/sos/cec/rules/06/096/096c315.doc [accessed April 17, 2007].

Menzel, J.M. 2003. An Examination of the Habitat Requirements of the Endangered Virginia Northern Flying Squirrel (*Glaucomys sabrinus fuscus*) by Assessing Nesting Sites, Habitat Use and the Development of a Habitat Model. Ph.D. Dissertation, West Virginia University, Morgantown, WV. 122 pp.

Menzel, J.M., M.A. Menzel, Jr., J.C. Kilgo, W.M. Ford, J.W. Edwards, and G.F. McCracken. 2005a. Effect of habitat and foraging height on bat activity in the coastal plain of South Carolina. J. Wildlife Manage. 69(1):235-245.

Menzel, J.M., W.M. Ford, M.A. Menzel, T.C. Carter, J.E. Gardner, J.D. Garner, and J.E. Hofmann. 2005b. Summer habitat use and home-range analysis of the endangered Indiana bat. J. Wildlife Manage. 69(1):430-436.

Menzel, J.M., W.M. Ford, J.W. Edwards, and L.J. Ceperley. 2006. A Habitat Model for Virginia Northern Flying Squirrel (*Glaucomys sabrinus fuscus*) in the Central Appalachian Mountains. Research Paper NE-729. U.S. Department of Agriculture, Forest Service, Northeastern Research Station, Newton Square, PA. 14 pp [online]. Available: http:// www.fs.fed.us/ne/newtown_square/publications/research_papers/pdfs/2006/ne_rp729.pdf [accessed Dec. 20, 2006].

Menzel, M.A., T.C. Carter, and D.M. Krishon. 1995. Roosting, Foraging, and Habitat Use by Bats of Sapelo Island, Georgia. Georgia Department of Natural Resources, Athens, GA.

Menzel, M.A., T.C. Carter, B.R. Chapman, and J. Laerm. 1998. Quantitative comparison of tree roosts used by red bats (*Lasiurus borealis*) and Seminole bats (*L. seminolus*). Can. J. Zool. 76(4):630-634.

Menzel, M.A., T.C. Carter, J.M. Menzel, W.M. Ford, and B.R. Chapman. 2002. Effects of group selection silviculture in bottomland hardwoods on the spatial activity patterns of bats. For. Ecol. Manage. 162(2-3):209-218.

Merriam, G., M. Kozakiewicz, E. Tsuchiya, and K. Hawley. 1989. Barriers as boundaries for metapopulations and demes of *Peromyscus leucopus* in farm landscapes. Landscape Ecol. 2(4):227-235.

MIDLEG (Michigan Department of Labor and Economic Growth). 2005. Michigan Siting Guidelines for Wind Energy Systems. Energy Office, Michigan Department of Labor and Economic Growth. December 14, 2005 [online]. Available: http://www.michigan.gov/documents/Wind_and_Solar_Siting_Guidlines_Draft_5_96872_7.pdf [accessed Sept. 13, 2006].

Mielke, M., D. Soctomah, M. Marsden, and W. Ciesla. 1986. Decline and Mortality of Red Spruce in West Virginia. Report No. 86-4. U.S. Department of Agriculture, Forest Service, Forest Pest Management, Methods Application Group, Fort Collins, CO.

Millspaugh, J.J., and J.M. Marzluff, eds. 2001. Radio Tracking and Animal Populations. San Diego, CA: Academic Press.

MIT (Massachusetts Institute of Technology). 2003. The Future of Nuclear Power: An Interdisciplinary MIT Study. Massachusetts Institute of Technology [online]. Available: http://web.mit.edu/nuclearpower/pdf/nuclearpower-full.pdf [accessed May 15, 2007].

Mitchell, D. 2001. Spring and fall diet of the endangered West Virginia northern flying squirrel (*Glaucomys sabrinus fuscus*). Am. Midl. Nat. 146(2):439-443.

Mitchell, M.S, and R.A. Powell. 2003. Response of black bears to forest management in the southern Appalachian Mountains. J. Wildlife Manage. 67(4):692-705.

Morrison, M.L., W.M. Block, M.D. Strickland, and W.L. Kendall. 2001. Wildlife Study Design. New York: Springer.

MSU (Michigan State University). 2004. Land Use and Zoning Issues Related to Site Development for Utility Scale Wind Turbine Generators. Michigan State University [online]. Available: http://web1.msue.msu.edu/cdnr/otsegowindflicker.pdf [accessed Sept. 6, 2006].

Mukhin, A. 2004. Night movements of young reed warblers (*Acrocephalus scirpaceus*) in summer: Is it postfledging dispersal? Auk 121(1):203-209.

Munn, C.A. 1991. Tropical canopy netting and shooting lines over tall trees. J. Field Ornithol. 62(4):454-463.

Musgrave, R.S., and M.A. Stein. 1993. State Wildlife Laws Handbook: Center for Wildlife Law at the Institute of Public Law, University of New Mexico, Albuquerque, NM. Rockville, MD: Government Institutes.

Nardi, R.R., and J.H. Daniels, Jr. 2005a. Wind Energy Easements. Windustry, Minneapolis, MN [online]. Available: http://www.windustry.org/opportunities/WindEasements.pdf [accessed Sept. 6, 2006].

Nardi, R.R., and J.H. Daniels, Jr. 2005b. Wind Energy Easement and Lease Agreements: Best Practices and Recommenadtions. Windustry, Minneapolis, MN [online]. Available: http://www.windustry.org/Easements/BestPractices&Policy_Sept2005.pdf [accessed Oct. 16, 2006].

NCDC (National Climatic Data Center). 2006. NCDC Radar Resources. National Climatic Data Center, U.S. Department of Commerce [online]. Available: http://www.ncdc.noaa.gov/oa/radar/radarresources.html [accessed July 3, 2006].

NEPA Task Force (National Environmental Policy Task Force). 2003. Modernizing NEPA Implementation, Report to the Council on Environmental Quality. September 2003 [online]. Available: http://ceq.eh.doe.gov/ntf/report/finalreport.pdf [accessed Dec. 27, 2006].

NESCAUM (Northeast States for Coordinated Air Management). 2002. Greenhouse Gas Case Studies: Exelon Corporation Wind Power Purchases. May 6, 2002 [online]. Available: http://www.nescaum.org/projects/greenhouse-gas-early-action-demonstration-project/ greenhouse-gas-case-studies/windpower.doc [accessed Sept. 21, 2006].

NESCAUM (Northeast States for Coordinated Air Management). 2003. Mercury Emissions from Coal-Fired Power Plants: The Case for Regulatory Action. October 2003 [online]. Available: http://www.nescaum.org/documents/rpt031104mercury.pdf/ [accessed Sept. 21, 2006].

Neuweiler, G. 2000. The Biology of Bats. New York: Oxford University Press.

Newman, R.J., and G.H. Lowery, Jr. 1959. The changing seasons: A summary of the 1959 spring migration and its geographic background. Audubon Field Notes 13(4):346-352.

Nicholson, C.P. 2003. Buffalo Mountain Windfarm, Bird and Bat Mortality Monitoring Report: October 2001-September 2002. Tennessee Valley Authority, Knoxville, TN. February 2003 [online]. Available: http://psc.wi.gov/apps/erf_share/view/viewdoc. aspx?docid=35049%20 [accessed May 24, 2006].

Nielsen, A. 2003. Shadow Flicker Briefing. Prepared for Zilkha Renewable Energy, Portland, OR, by Wind Engineers, Inc. November 20, 2003 [online]. Available: http://www.efsec. wa.gov/wildhorse/apl/Exhibits PDF/E09_Shadow Flicker Briefing Memo.pdf [accessed Sept. 6, 2006].

Nielsen, F.B. 2002. A formula for success in Denmark. Pp. 115-132 in Wind Power in View: Energy Landscapes in a Crowded World, M.J. Pasqualetti, P. Gipe, and R.W. Righter, eds. San Diego: Academic Press.

Nisbet, I.C.T. 1959. Calculation of flight directions of birds observed crossing the face of the moon. Wilson Bull. 17(3):237-243.

Nisbet, I.C.T. 1963a. Estudio de la migracion sobre el disco lunar (I). Ardeola 8:5-17.

Nisbet, I.C.T. 1963b. Quantitative study of migration with 23-centimetre radar. Ibis 105(4):435-460.

Nohara, T.J., P. Weber, A. Premji, C. Kranor, S. Gauthreaux, M. Brand, and G. Key. 2005. Affordable avian radar surveillance systems for natural resource management and BASH applications. Pp. 10-15 in IEEE 2005 International Radar Conference, May 9-12, 2005, Arlington, VA [online]. Available: http://ieeexplore.ieee.org/xpl/tocresult. jsp?isnumber=30939 [accessed June 15, 2006].

Nol, E., C.M. Francis, and D.M. Burke. 2005. Using distance from putative source woodlots to predict occurrence of forest birds in putative sinks. Conserv. Biol. 19(3):836-844.

Northrop, Devine, & Tarbell, Inc. 1995a. New England Wind Energy Station, Spring 1994 Nocturnal Migration Study Report. Prepared for Kenetech Windpower, Inc., Portland, ME.

Northrop, Devine, & Tarbell, Inc. 1995b. New England Wind Energy Station, Fall 1994 Nocturnal Migration Study Report. Prepared for Kenetech Windpower, Inc., Portland, ME.

NRC (National Research Council). 1985. Oil in the Sea: Inputs, Fates, and Effects. Washington, DC: National Academy Press.

NRC (National Research Council). 1986. Ecological Knowledge for Environmental Problem Solving: Concepts and Case Studies. Washington, DC: National Academy Press.

NRC (National Research Council). 1994. Science and Judgment in Risk Assessment. Washington, DC: National Academy Press.

NRC (National Research Council). 2000. Waste Incineration and Public Health. Washington, DC: National Academy Press.

NRC (National Research Council). 2001. Climate Change and Science: An Analysis of Some Key Questions. Washington, DC: National Academy Press.

NRC (National Research Council). 2002. New Tools for Environmental Protection: Education, Information, and Voluntary Measures, T. Dietz and P.C. Stern, eds. Washington, DC: National Academy Press.

NRC (National Research Council). 2003. Cumulative Environmental Effects of Oil and Gas Activities on Alaska's North Slope. Washington, DC: The National Academies Press.

NRC (National Research Council). 2004. Atlantic Salmon in Maine. Washington, DC: The National Academies Press.

NRC (National Research Council). 2005. Assessing and Managing the Ecological Impacts of Paved Roads. Washington, DC: The National Academies Press.

NRC (National Research Council). 2006. New Source Review for Stationary Sources of Pollution. Washington, DC: The National Academies Press.

NREL (National Renewable Energy Laboratory). 2003. MidAtlantic Regional Wind High Resolution. GIS Coverages, Dynamic Maps, GIS Data and Analysis Tools [online]. Available: http://www.nrel.gov/gis/data_analysis.html [accessed March 29, 2006].

NREL (National Renewable Energy Laboratory). 2006a. Projected Benefits of Federal Energy Efficiency and Renewable Energy Programs, FY 2007 Budget Request. NREL/TP-620-39684. National Renewable Energy Laboratory, Energy Efficiency and Renewable Energy, U.S. Department of Energy. March 2006 [online]. Available: http://www1.eere.energy.gov/ba/pdfs/39684.pdf [accessed Sept. 19, 2006].

NREL (National Renewable Energy Laboratory). 2006b. Wind Deployment System (WinDS) Model. National Renewable Energy Laboratory [online]. Available: http://www.nrel.gov/analysis/winds/ [accessed Sept. 21, 2006].

NWCC (National Wind Coordinating Committee). 2001. Guidelines for Assessing the Economic Development Impacts of Wind Power. National Wind Coordinating Committee, Washington, DC. October 2001 [online]. Available: http://www.nationalwind.org/publications/economic/guidelines.pdf [accessed Sept. 6, 2006].

NWCC (National Wind Coordinating Committee). 2002. Permitting of Wind Energy Facilities: A Handbook Revised 2002. August 2002 [online]. Available: http://www.nationalwind.org/publications/siting/permitting2002.pdf [accessed Sept. 12, 2006].

NYSDEC (New York State Department of Conservation). 2005. SEQR Handbook [online]. Available: http://www.dec.ny.gov/permits/6188.html [accessed June 14, 2007].

NYSERDA (New York State Energy Research and Development Authority). 2005a. Overview of the SEQR Process. New York State Energy Research and Development Authority, Albany, NY. October 2005 [online]. Available: http://www.powernaturally.org/Programs/Wind/toolkit/17_overviewSEQRprocess.pdf [accessed Sept. 11, 2006].

NYSERDA (New York State Energy Research and Development Authority). 2005b. Assessing and Mitigating Visual Impacts. New York State Energy Research and Development Authority, Albany, NY. October 2005 [online]. Available: http://www.powernaturally.org/Programs/Wind/toolkit/6_visualimpactupfront.pdf [accessed May 22, 2007].

Odum, W.E. 1982. Environmental degradation and the tyranny of small decision. BioScience 32(9):728-729.

OFCM (Office of the Federal Coordinator for Meteorological Services and Supporting Research). 2006. Doppler Radar Meteorological Observations, Part A. System Concepts, Responsibilities, and Procedures. Federal Meteorological Handbook No. 11. FCM-H11A-2006. Office of the Federal Coordinator for Meteorological Services and Supporting Research, U.S. Department of Commerce/National Oceanic and Atmospheric Administration, Washington, DC. April 2006 [online]. Available: http://www.ofcm.gov/fmh11/fmh11parta/pdf/FMH11PTA-04-2006.pdf [accessed June 13, 2006].

Old Bird, Inc. 2005. Avian Night Flight Call. Old Bird, Inc., Ithaca, NY [online]. Available: http://www.oldbird.org/ [accessed June 28, 2006].

Oregon Department of Environmental Quality. 2006. Noise Control Regulations Chapter 340, Division 35. Oregon State Archive, August 2006 [online]. Available: http://arcweb.sos. state.or.us/rules/OARs_300/OAR_340/340_035.html [accessed Sept. 6, 2006].

Orloff, S., and A. Flannery. 1992. Wind Turbine Effects on Avian Activity, Habitat Use, and Mortality in Altamont Pass and Solano County Wind Resource Areas, 1989-1991. Final Report. P700-92-001. Prepared for Planning Departments of Alameda, Contra Costa and Solano Counties and the California Energy Commission, Sacramento, CA, by BioSystems Analysis, Inc., Tiburon, CA. March 1992.

Orloff, S., and A. Flannery. 1996. A Continued Examination of Avian Mortality in the Altamont Pass Wind Resource Area. Consultant Report. P700-96-004CN. Prepared for California Energy Commission, Sacramento, CA, by BioSystems Analysis, Inc., Tiburon, CA. August 1996.

Ortega, Y.K., and D.E. Capen. 1999. Effects of forest roads on habitat quality for ovenbirds in forested landscape. Auk 116(4):937-946.

Ortega, Y.K., and D.E. Capen. 2002. Roads as edges: Effects on birds in forested landscapes. Forest Sci. 48(2):381-390.

Osborn, R.G., C.D. Dieter, K.F. Higgins, and R.E. Usgaard. 1998. Bird flight characteristics near wind turbines in Minnesota. Am. Midl. Nat. 139(1):29-38.

O'Shea, T.J., and M.A. Bogan, eds. 2003. Monitoring Trends in Bat Populations of the United States and Territories: Problems and Prospects. Biological Resources Discipline, Information and Technology Report USGS/BRD/ITR-2003-003. U.S. Geological Survey [online]. Available: http://www.fort.usgs.gov/products/publications/21329/21329.pdf [accessed June 9, 2006].

O'Shea, T.J., M.A. Bogan, and L.E. Ellison. 2003. Monitoring trends in bat populations of the United States and territories: Status of the science and recommendations for the future. Wildlife Soc. Bull. 31(1):16-29.

O'Shea, T.J., L.E. Ellison, and T.R. Stanley. 2004. Survival estimation in bats: Historical overview, critical appraisal, and suggestions for new approaches. Pp. 297-336 in Sampling Rare or Elusive Species: Concepts, Designs, and Techniques for Estimating Population Parameters, W.L. Thompson, ed. Washington, DC: Island Press.

OTC (Ozone Transport Commission). 2002. OTC Emission Reduction Workbook 2.1. Ozone Transport Commission. November 12, 2002 [online]. Available: http://www.otcair.org/document.asp?fview=Report# [accessed Sept. 21, 2006].

Pacca, S., and A. Horvath. 2002. Greenhouse gas emissions from building and operating electric power plants in the Upper Colorado River Basin. Environ. Sci. Technol. 36(14):3194-3200.

PADCNR (Pennsylvania Department of Conservation and Natural Resources). 2006. Wind Farm Permitting and Regulation in Pennsylvania. Pennsylvania Wind Farms and Wildlife Collaborative Resource Center [online]. Available: http://www.dcnr.state.pa.us/wind/resources.aspx [accessed Dec. 27, 2006].

Paetkau, D. 2003. An empirical exploration of data quality in DNA-based population inventories. Mol. Ecol. 12(6):1375-1387.

PAGC (Pennsylvania Game Commission). 2006. Endangered Species. Pennsylvania Game Commission [online]. Available: http://www.pgc.state.pa.us/pgc/cwp/view.asp?a=458&q=150321 [accessed Oct. 16, 2006].

Palmer, J.F. 1983. Visual quality and visual impact assessment. Pp. 263-284 in Social Impact Assessment Methods, K. Finsterbusch, L.G. Llewellyn, and C.P. Wolf, eds. Beverly Hills, CA: Sage Publications.

Palmer, J.F. 1997. Public Acceptance Study of the Searsburg Wind Power Project: Year One Post-Construction. Prepared by James F. Palmer, Clinton Solutions, Fayetteville, NY, for Vermont Environmental Research Associates, Inc., Waterbury Center, VT, and Green Mountain Power Corporation, South Burlington, VT. December 1997 [online]. Available: http://www3.digitalfrontier.com/essential_wc5/wind/images/photos/searsburg/pub_acceptance_study.pdf [accessed Sept. 6, 2006].

Parsons, S., and M.K. Obrist. 2004. Recent methodological advances in the recording and analysis of chiropteran biosonar signals in the field. Pp. 468-477 in Echolocation in Bats and Dolphins, J.A. Thomas, C. Moss, and M. Vater, eds. Chicago: University of Chicago Press.

Pasqualetti, M. 2005. Visual Impacts Introduction. Pp. 2 in Proceedings: NWCC Technical Considerations in Siting Wind Developments, NWCC Meeting, December 1-5, 2005, Washington, DC, S. Savitt Schwartz, ed. National Wind Coordinating Committee Staff c/o RESOLVE, Inc., Washington, DC. March 2006 [online]. Available: http://www.nationalwind.org/events/siting/proceedings.pdf [accessed May 23, 2007].

Pasqualetti, M.J., P. Gipe, and R.W. Righter, eds. 2002. Wind Power in View: Energy Landscapes in a Crowded World. San Diego: Academic Press.

Pasqualetti, M.J., R. Righter, and P. Gipe. 2004. History of wind energy. Pp. 419-433 in Encyclopedia of Energy, Vol. 6, C. Cleveland, ed. Amsterdam: Elsevier.

Patterson, J.W., Jr. 2005. Development of Obstruction Lighting Standards for Wind Turbine Farms. DOT/FAA/AR-TNO5/50. U.S. Department of Transportation, Federal Aviation Administration. November 2005 [online]. Available: http://www.tc.faa.gov/its/worldpac/techrpt/artn05-50.pdf [accessed Sept. 13, 2006].

Patton-Mallory, M. 2006. The Role of the Federal Government and Federal Lands in Fueling Renewable and Alternative Energy. Testimony to the Subcommittee on Energy and Mineral Resources, Committee on Resources, United States House of Representatives. April 6, 2006 [online]. Available: http://www.fs.fed.us/congress/109/house/oversight/patton-mallory/040606.html [accessed Oct. 10, 2006].

Pauley, T.K. 1981. The range and distribution of the Cheat Mountain salamander, Plethodon nettingi. Proc. WV Acad. Sci. 53(2):31-35.

Pavey, C.R. 1998. Habitat use by the eastern horseshoe bat, Rhinolophus megaphyllus, in a fragmented woodland mosaic. Wildlife Res. 25(5):489-498.

PAWWG (Pennsylvania Wind Working Group). 2006. Model Ordinance for Wind Energy Facilities in Pennsylvania. March 21, 2006 [online]. Available: http://www.pawindenergynow.org/pa/Model_Wind_Ordinance_Final_3_21_06.pdf [accessed Sept. 19, 2006].

Pedden, M. 2006. Analysis: Economic Impacts of Wind Applications in Rural Communities: June 18, 2004-January 31, 2005. Subcontract Report NREL/SR-500-39009. National Renewable Energy Laboratory, Golden, CO. January 2006 [online]. Available: http://www.nrel.gov/docs/fy06osti/39099.pdf [accessed Sept. 6, 2006].

Pedersen, M.B., and E. Poulsen. 1991. Impact of a 90m/2MW Wind Turbine on Birds: Avian Responses to the Implementation of the Tjaereborg Wind Turbine at the Danish Wadden Sea [in Danish]. Kalö, Denmark: Miljöministeriet, Danmarks Miljöundersögelser (Danske Vildtundersogelser 47:34-44).

Peterson, B.S., and H. Nohr. 1989. Consequences of Minor Wind Mills for Bird Fauna. Ornis Consult, Copenhagen.

Petranka, J.W., and C.K. Smith. 2005. A functional analysis of streamside habitat use by southern Appalachian salamanders: Implications for riparian forest management. For. Ecol. Manag. 210(1-3):443-454.

Petranka, J.W., M.E. Eldridge, and K.E. Haley. 1993. Effects of timber harvesting on southern Appalachain salamanders. Conserv. Biol. 7(2):363-370.

Pettersson, L. 2004. Time expansion: Analysis capabilities and limitations and field design. Pp. 91-94 in Bat Echolocation Research: Tools, Techniques, and Analysis, R.M. Brigham, E.K.V. Kalko, G. Jones, S. Parsons, and H.J.G.A. Limpens, eds. Austin, TX: Bat Conservation International.

Phillips, J.F. 1994. The Effects of the Windfarm on the Upland Breeding Bird Communities of Bryn Titli, Mid-Wales: 1993-1994. Royal Society for the Protection of Birds, the Welsh Office, Bryn Aderyn, The Bank, Newton, Powys, UK.

Pierpont, N. 2006. Wind Turbine Syndrome. Testimony before the New York State Legislature Energy Committee. March 7, 2006 [online]. Available: http://www.ninapierpont.com/pdf/Wind_turbine_syndrome,_NYS_Energy_Committee_3-7-06.pdf [accessed Sept. 6, 2006].

Pierson, E.D. 1998. Tall trees, deep holes, and scarred landscapes: Conservation biology of North American bats. Pp. 309-325 in Bat Biology and Conservation, T.H. Kunz and P.A. Racey, eds. Washington, DC: Smithsonian Institution Press.

Piorkowski, M. 2006. Breeding Bird Habitat Use and Turbine Collisions of Birds and Bats Located at a Wind Farm in Oklahoma Mixed-Grass Prairie. M.S. Thesis, Oklahoma State University, Stillwater, OK.

Plissner, J.H., T.J. Mabee, and B.A. Cooper. 2006. A Radar and Visual Study of Nocturnal Bird and Bat Migration at the Proposed Highland New Wind Development Project, Virginia, Fall 2005. Final Report. Prepared for Highland New Wind Development, LLC, Harrisonburg, VA, by ABR, Inc.—Environmental Research & Services, Forest Grove, OR. January 2006 [online]. Available: http://esm.versar.com/pprp/windpower/Highland-VA-Radar-Study2006.pdf [accessed Sept. 29, 2006].

PJM. 2005a. PJM Load Forecast Report. The Load Analysis Subcommittee, PJM Interconnection, Inc. February 2005 [online]. Available: http://www.pjm.com/planning/res-adequacy/downloads/2005-load-forecast-report.pdf [accessed Sept. 21, 2006].

PJM. 2005b. Rules and Procedures for Determination of Generating Capability. PJM Manual 21, Revision 04. Prepared by PJM System Planning Department. August 15, 2005 [online]. Available: http://www.pjm.com/contributions/pjm-manuals/pdf/m21v04.pdf [accessed Sept. 21, 2006].

PJM. 2006a. 2005 State of the Market Report. Market Monitoring Unit, PJM Interconnection, Inc. March 8, 2006 [online]. Available: http://www.pjm.com/markets/market-monitor/som.html [accessed Sept. 21, 2006].

PJM. 2006b. PJM Interconnection Generator Attribute Tracking System [online]. Available: http://gats.pjm-eis.com/myModule/rpt/myrpt.asp [accessed August 17, 2006].

Podolsky, R.H. 2003. Method and Article of Manufacture for Determining Probability of Avian Collision. U.S. Patent Filing #92717353USPL.

Porneluzi, P.A., and J. Faaborg. 1999. Season-long fecundity, survival, and viability of ovenbirds in fragmented and unfragmented landscapes. Conserv. Biol. 13(5):1151-1161.

Ports, M.A., and P.V. Bradley. 1996. Habitat affinities of bats from northeastern Nevada. Great Basin Nat. 56(1):48-53.

Poupart, G.J. 2003. Wind Farms Impact on Radar Aviation Interests-Final Report. FES W/14/00614/00/REP. DTI PUB URN 03/1294. U.K. Department of Trade and Industry [online]. Available: http://www.dti.gov.uk/files/file18038.pdf [accessed April 27, 2007].

Power, M., D. Tilman, J. Estes, B. Menge, W. Bond, L.S. Mills, G. Daily, J.C. Castilla, J. Lubchenco, and R. Paine. 1996. Challenges in the quest for keystones. BioScience 46(8):609-620.

Preatoni, D.G., M. Nodari, R. Chirichella, G. Tosi, L.A. Wauters, and A. Martinoli. 2005. Identifying bats from time-expanded recordings of search calls: Comparing classification methods. J. Wildlife Manage. 69(4):1601-1614.

Priestley, T. 2006. Wind power visual impact assessment: Practical issues and links to research. Pp. 23-27 in Proceedings: NWCC Technical Considerations in Siting Wind Developments, NWCC Meeting, December 1-5, 2005, Washington, DC, S. Savitt Schwartz, ed. National Wind Coordinating Committee Staff c/o RESOLVE, Inc., Washington, DC. March 2006 [online]. Available: http://www.nationalwind.org/events/siting/proceedings. pdf [accessed May 23, 2007].

Publicover, D. 2004. A Methodology for Assessing Conflicts Between Windpower Development and Other Land Uses. AMC Technical Report 04-2. Appalachian Mountain Club, Research Department, Gorham, NH. May 2004 [online]. Available: http://www.outdoors. org/pdf/upload/Windpower-Siting-Project-Report.pdf [accessed Sept. 13, 2006].

Puechmaille, S.J., and E.J. Petit. 2007. Empirical evaluation of non-invasive capture-mark-recapture estimation of population size based on a single sampling session. J. Appl. Ecol. 44(4):843-852.

Purvis, A., J.L. Gittleman, G. Cowlishaw, and G.M. Mace. 2000. Predicting extinction risk in declining species. P. Roy. Soc. Lond. B Bio. 267(1456):1947-1952.

Quang, P.X., and E.F. Becker. 1996. Line transect sampling under varying conditions with applications to aerial surveys. Ecology 77(4):1297-1302.

Quang, P.X., and E.F. Becker. 1997. Combining line transect and double count sampling techniques for aerial surveys. J. Agric. Biol. Envir. S. 2(2):229-242.

Ralph, C.J., G.R. Geupel, P. Pyle, T.E. Martin, and D.F. DeSante. 1993. Handbook of Field Methods for Monitoring Landbirds. GTR-PSW-GTR-144. U.S. Department of Agriculture, Forest Service, Pacific Southwest Research Station, Albany, CA [online]. Available: http://www.fs.fed.us/psw/publications/documents/gtr-144/ [accessed Sept. 29, 2006].

Reed, R.A., J. Johnson-Barnard, and W.A. Baker. 1996. Contribution of roads to forest fragmentation in the Rocky Mountains. Conserv. Biol. 10(4):1098-1106.

Renevey, B. 1981. Study of the wing-beat frequency of night-migrating birds with a tracking radar. Rev. Suisse Zool. 88:875-886.

Reynolds, D.S. 2006. Monitoring the potential impact of a wind development site on bats in the Northeast. J. Wildlife Manage. 70(5):1219-1227.

Reynolds, R.T., J.M. Scott, and R.A. Nussbaum. 1980. A variable circular-plot method for estimating bird numbers. Condor 82(3):309-313.

RGGI (Regional Greenhouse Gas Initiative). 2006. Regional Greenhouse Gas Initiative Model Rule. August 18, 2006 [online]. Available: http://www.rggi.org/docs/model_rule_8_15_06.pdf [accessed Sept. 21, 2006].

Richardson, W.J. 1972. Autumn migration and weather in eastern Canada: A radar study. Am. Birds 26:10-17.

Richardson, W.J. 1978. Reorientation of nocturnal landbird migrants over the Atlantic Ocean near Nova Scotia in autumn. Auk 95(4):717-732.

Richardson, W.J. 1979. Radar techniques for wildlife studies. Pp. 171-179 in Pecora IV, Proceedings of the Symposium: Application of Remote Sensing Data to Wildlife Management. National Wildlife Federation Scientific Technical Series 3. Sioux Falls, SD: National Wildlife Federation.

Richardson, W.J. 1990. Timing of bird migration in relation to weather: Updated review. Pp. 78-101 in Bird Migration: Physiology and Ecophysiology, E. Gwinner, ed. Berlin: Springer-Verlag.

RICS (Royal Institution of Chartered Surveyors). 2004. RICS Housing Market Survey [online]. Available: http://www.rics.org/ [accessed April 23, 2007].

RIDEM (Rhode Island Department of Environmental Management). 1990. The Rhode Island Landscape Inventory: A Survey of the State's Scenic Areas. Providence, RI: Department of Environmental Management.

Riggs, M.R., and K.H. Pollock. 1992. A risk ratio approach to multivariate analysis of survival in longitudinal studies of wildlife populations. Pp. 74-89 in Wildlife 2001: Populations, D.R. McCullough and R.H. Barrett, eds. London: Elsevier Applied Science.

Riley, J.R. 1989. Remote sensing in entomology. Annu. Rev. Entomol. 34:247-271.

Riley, J.R., D.R. Reynolds, and M.J. Farmery. 1983. Observations of the flight behaviour of the armyworm moth, *Spodoptera exempta*, at an emergence site using radar and infra-red optical techniques. Ecol. Entomol. 8:395-418.

Rinehart, J.B., and T.H. Kunz. 2001. Preparation and deployment of canopy mist nets made by Avinet. Bat Res. News 42(3):85-88.

Ringkøbing Amt, Møller og Grønborg, and Carl Bro. 2002. Windmøller På Land: Drejebog For VVM. February 2002 [online]. Available: http://www.windenergy-in-the-bsr.net/download/DrejebogVVMvindmoeller.pdf [accessed Sept. 18, 2006].

Robbins, C.S., D.K. Dawson, and B.A. Dowell. 1989. Habitat Area Requirements of Breeding Forest Birds of the Middle Atlantic States. Wildlife Monographs 103. Washington, DC: Wildlife Society.

Robbins, C.S., J.W. Fitzpatrick, and P.B. Hamel. 1992. A warbler in trouble: *Dendroica cerulea*. Pp. 549-562 in Ecology and Conservation of Neotropical Migrant Landbirds, J.M. Hagan and D.W. Johnston, eds. Washington, DC: Smithsonian Institution Press.

Roble, S.M. 2006. Natural Heritage Resources of Virginia: Rare Animal Species. Natural Heritage Technical Report 06-10. Virginia Department of Conservation and Recreation, Division of Natural Heritage, Richmond, VA. 44 pp.

Roman, J., and S.R. Palumbi. 2003. Whales before whaling in the North Atlantic. Science 301(5632):508-510.

Rosenstock, S.S., D.R. Anderson, K.M. Giesen, T. Leukering, and M.F. Carter. 2002. Landbird counting techniques: Current practices and an alternative. Auk 119(1):46-53.

Rossell, C.R., Jr., B. Gorsira, and S. Patch. 2005. Effects of white-tailed deer on vegetation structure and woody seedling composition in three forest types on the Piedmont Plateau. For. Ecol. Manag. 210(1-3):415-424.

Royle, J.A., and D.R. Rubenstein. 2004. The role of species abundance in determining the breeding origins of migratory birds using stable isotopes. Ecol. Appl. 14(6):1780-1788.

Rubenstein, D.R., and K.A. Hobson. 2004. From birds to butterflies: Animal movement patterns and stable isotopes. Trends Ecol. Evol. 19(5):256-263.

Rubenstein, D.R., C.P. Chamberlain, R.T. Holmes, M.P. Ayres, J.R. Waldbauer, G.R. Graves, and N.C. Tuross. 2002. Linking breeding and wintering ranges of a migratory songbird using stable isotopes. Science 295(5557):1062-1065.

Russell, A.L., and G.F. McCracken. 2006. Population genetic structuring of very large populations: The Brazilian free-tailed bat *Tadarida brasiliensis*. Pp. 227-247 in Functional and Evolutionary Ecology of Bats, T.H. Kunz, A. Zubaid, and G.F. McCracken, eds. New York: Oxford University Press.

Russell, A.L., R.A. Medellin, and G.F. McCracken. 2005. Genetic variation and migration in the Mexican free-tailed bat (*Tadarida brasiliensis mexicana*). Mol. Ecol. 14(7):2207-2222.

Russell, K.R., and S.A. Gauthreaux, Jr. 1998. Use of weather radar to characterize movements of roosting purple martins. Wildlife Soc. Bull. 26(1):5-16.

Russell, K.R., D.S. Mizrahi, and S.A. Gauthreaux, Jr. 1998. Large-scale mapping of purple martin pre-migratory roosts using WSR-88D weather surveillance radar. J. Field Ornithol. 69(2):316-325.

Sabol, B.M., and M.K. Hudson. 1995. Technique using thermal infrared-imaging for estimating populations of gray bats. J. Mammal. 76(4):1242-1248.

Sauer, J.R., J.E. Hines, and J. Fallon. 2005. The North American Breeding Bird Survey, Results and Analysis 1966-2004. Version 2005.2. USGS Patuxent Wildlife Research Center, Laurel, MD [online]. Available: http://www.mbr-pwrc.usgs.gov/bbs/ [accessed June 16, 2006].

Sawyer, H., R.M. Nielson, F. Lindzey, and L.L. McDonald. 2006. Winter habitat selection of mule deer before and during development of a natural gas field. J. Wildlife Manage. 70(2):396-403.

Schleede, G.R. 2003. Wind Energy Economics in West Virginia. Minnesotans for Sustainability. January 20, 2003 [online]. Available: http://www.mnforsustain.org/windpower_schleede_west_virginia_part2.htm [accessed May 18, 2006].

Schmidt, U., and G. Joermann. 1986. The influence of acoustical interferences on echolocation in bats. Mammalia 50(3):379-389.

Schulze, M.D., N.E. Seavy, and D.F. Whitacre. 2000. A comparison of the phyllostomid bat assemblages in undisturbed neotropical forest and in forest fragments of a slash-and-burn farming mosaic in Peten, Guatemala. Biotropica 32(1):174-184.

Schüz, E., P. Berthold, E. Gwinner, and H. Oelke. 1971. Grundriss der Vogelzugskunde, 2nd Ed. Berlin: P. Parey.

SDGFP (South Dakota Department of Game, Fish and Parks). 2005. Siting Guidelines for Wind Power Projects in South Dakota [online]. Available: http://nathist.sdstate.edu/SD-BWG/Subpages/windGuidelines.pdf [accessed Sept. 12, 2006].

Shoenfeld, P.S. 2004. Suggestions Regarding Avian Mortality Extrapolation. Prepared for the Mountaineer Wind Energy Center Technical Review Committee, by P.S. Shoenfeld, West Virginia Highlands Conservancy, Davis, WV [online]. Available: http://www.wvhighlands.org/Birds/SuggestionsRegardingAvianMortalityExtrapolation.pdf [accessed May 23, 2007].

Short, W., N. Blair, P. Denholm, and D. Heimiller. 2006. Modeling High-Penetration Wind Scenarios: WinDS Model Results. Presentation to Wind Powering America Summit, June 8, 2006 [online]. Available: http://www.eere.energy.gov/windandhydro/windpoweringamerica/pdfs/workshops/2006_summit/short.pdf [accessed Sept. 7, 2006].

Siemers, B.M., K. Beedholm, C. Dietz, I. Dietz, and T. Ivanova. 2005. Is species identity, sex, age or individual quality conveyed by echolocation call frequency in European horseshoe bats? Acta Chiropterol. 7(2):259-274.

Skalski, J.R., and D.S. Robson. 1992. Techniques for Wildlife Investigations: Design and Analysis of Capture Data. San Diego, CA: Academic Press.

Smallwood, K.S., and C.G. Thelander. 2004. Developing Methods to Reduce Bird Mortality in the Altamont Pass Wind Resource Area. Final Report. P500-04-052. Prepared for California Energy Commission, Public Interest Energy Research Program, Sacramento, CA, by BioResources Consultants, Ojai, CA. August 2004 [online]. Available: http://www.energy.ca.gov/pier/final_project_reports/500-04-052.html [accessed May 25, 2006].

Smallwood, K.S., and C.G. Thelander. 2005. Bird Mortality at the Altamont Pass Wind Resource Area: March 1998-September 2001. Subcontract Report NREL/SR-500-36973. Prepared for National Renewable Energy Laboratory, Golden, CO, by BioResource Consultants, Ojai, CA. August 2005 [online]. Available: http://www.nrel.gov/docs/fy05osti/36973.pdf [accessed June 22, 2006].

Smardon, R.C., J.F. Palmer, and J.P. Felleman. 1986. Foundations for Visual Project Analysis. New York: Wiley.

Smith, C., E. Demeo, and S. Smith. 2006. Integrating Wind Generation into Utility Systems. North American Windpower. September 2006 [online]. Available: http://www.uwig.org/WindIntegration-NAW92006.pdf [accessed May 16, 2007].

Smith, H.J., and H. Kunreuther. 2001. Mitigation and benefits measures as policy tools for siting potentially hazardous facilities: Determinants of effectiveness and appropriateness. Risk Anal. 21(2):371-382.

SonoBat. 2005. Software for Bat Call Analysis [online]. Available: http://www.sonobat.com/ [accessed June 28, 2006].

Speakman, J.R., and P.A. Racey. 1991. No cost of echolocation for bats in flight. Nature 350: 421-423.

Spencer, H.J., C. Palmer, and K. Parry-Jones. 1991. Movements of fruit-bats in eastern Australia, determined by using radio-tracking. Wildlife Res. 18(4):463-468.

Stanton, C. 2005. Visual Impacts: UK and European Perspectives. Presentation at the Technical Considerations in Siting Wind Developments: Research Meeting on December 1-2, 2005, Washington, DC [online]. Available: http://www.nationalwind.org/events/siting/presentations/stanton-visual_impacts.pdf [accessed April 23, 2007].

State of Oregon. 2006. The Siting Process for Energy Facilities [online]. Available: http://www.oregon.gov/ENERGY/STING/process.shtml [accessed Sept. 11, 2006].

Stemer, D. 2002. A Roadmap for PIER Research on Avian Collisions with Wind Turbines in California. P500-02-070F. Sacramento, CA: California Energy Commission [online]. Available: http://www.energy.ca.gov/reports/2002-12-24_500-02-070F.PDF [accessed May 18, 2006].

Sterzinger, G., F. Beck, and D. Kostiuk. 2003. The Effect of Wind Development on Local Property Values. Renewable Energy Policy Project, Washington, DC [online]. Available: http://www.crest.org/articles/static/1/binaries/wind_online_final.pdf [accessed April 23, 2007].

Stewart-Oaten, A. 1986. The Before-After/Control-Impact-Pairs Design for Environmental Impact. Prepared for Marine Review Committee, Inc., Encinitas, CA. June 20.

Stilz, W.P., and H.U. Schnitzler. 2005. Estimation of the Acoustical Range of Bat Echolocation for Extended Targets [online]. Available: http://www.biosonarlab.uni-tuebingen.de/public/distance/distance_abstract.pdf [accessed May 21, 2007].

Strbac, G. 2002. Quantifying the System Costs of Additional Renewables in 2020, A Report to the Department of Trade and Industry [online]. Available: http://www.dti.gov.uk/files/file21352.pdf [accessed March 27, 2007].

Suh, S., M. Lenzen, G.J. Treloar, H. Hondo, A. Horvath, G. Huppes, O. Joiliet, U. Klann, W. Krewitt, Y. Moriguchi, J. Munksgaard, and G. Norris. 2004. System boundary selection in life-cycle inventories using hybrid approaches. Environ. Sci. Technol. 38(3):657-664.

Suter, G.W., II, R.A. Efroymson, B.E. Sample, and D.S. Jones. 2000. Ecological Risk Assessment for Contaminated Sites. Boca Raton, FL: Lewis Press.

Sutherland, W.J., I. Newton, and R. Green. 2004. Bird Ecology and Conservation: A Handbook of Techniques. New York: Oxford University Press.

Swezicak, J.M. 2004. Advanced analysis techniques for identifying bat species. Pp. 121-126 in Bat Echolocation Research: Tools, Techniques, and Analysis, R.M. Brigham, E.K.V. Kalko, G. Jones, S. Parsons, and H.J.G.A. Limpens, eds. Austin, TX: Bat Conservation International.

Swihart, R.K., and N.A. Slade. 1984. Road crossing in Sigmodon hispidus and Microtus ochrogaster. J. Mammal. 65(2):357-360.

Thayer, R.L., and H.A. Hansen. 1988. Wind on the land. Landscape Archit. 78(2):69-73.

Thayer, R.L., and H.A. Hansen. 1989. Pp. 17-19 in Consumer Attitude and Choice in Local Energy Development. Davis, CA: Center for Design Research, Department of Environmental Design, University of California-Davis.

Thelander, C.G., and L. Rugge. 2000. Avian Risk Behavior and Fatalities at the Altamont Wind Resource Area: March 1998 to February 1999. NREL/SR-500-27545. Prepared for the National Renewable Energy Laboratory, Golden, CO, by BioResource Consultants, Ojai, CA. May 2000 [online]. Available: http://www.nrel.gov/docs/fy00osti/27545.pdf [accessed May 25, 2006].

Thomas, J.A., C.F. Moss, and M. Vater, eds. 2004. Echolocation in Bats and Dolphins. Chicago, IL: The University of Chicago Press.

Thompson, W.L., G.C. White, and C. Gowan. 1998. Monitoring Vertebrate Populations. San Diego, CA: Academic Press.

Tidemann, C.R., and J.E. Nelson. 2004. Long-distance movements of the grey-headed flying fox (Pteropus poliocephalus). J. Zool. 263(2):141-146.

Trombulak, S.C., and C.A. Frissell. 2000. Review of ecological effects of roads on terrestrial and aquatic communities. Conserv. Biol. 14(1):18-30.

Tucker, V.A. 1996. A mathematical model of bird collisions with wind turbine rotors. J. Sol. Energ. T ASME 118(4):253-262.

Turton, S.M., and H.J. Freiburger. 1997. Edge and aspect effects on the microclimate of a small tropical forest remnant on the Atherton Tableland, Northeast Australia. Pp. 45-54 in Tropical Forest Remnants: Ecology, Management, and Conservation of Fragmented Communities, W.F. Laurance and R.O. Bierregaard, eds. Chicago, IL: University of Chicago Press.

Tuttle, M.D. 1976. Collecting techniques. Pp. 71-88 in Biology of bats of the new world family *Phyllostomatidae*, Part I, R.J. Baker, D.C. Carter, and J.K. Jones, Jr., eds. Special Pub. No. 10. Lubbock, TX: Texas Tech Press.

Tuttle, M.D., and D. Stevenson. 1977. An analysis of migration as a mortality factor in the gray bat based on public recoveries of banded bats. Am. Midl. Nat. 97(1):235-240.

Udevitz, M.S., and K.H. Pollock. 1992. Change-in-ratio methods for estimating population size. Pp. 90-101 in Wildlife 2001: Populations, D.R. McCullough and R.H. Barrett, eds. London: Elsevier Science.

Underwood, A.J. 1994. On beyond BACI: Sampling designs that might reliably detect environmental disturbances. Ecol. Appl. 4(1):3-15.

Underwood, A.J. 1997. Experiments in Ecology: Their Logical Design and Interpretation Using Analysis of Variance. Cambridge: Cambridge University Press.

UNEP/EUROBATS. 2006. 1991-2006. EUROBATS Celebrates Its 15th Anniversary. EUROBATS Publication Series, No 1. United Nations Environment Programme (UNEP)/The Agreement on the Conservation of Population of European Bats, Bonn, Germany [online]. Available: http://www.eurobats.org/publications/publication%20series/Eurobats_15th_Anniversary_INTERNET.pdf [accessed Dec. 18, 2006].

UNFCCC (United Nations Framework Convention on Climate Change). 2006. Consolidated Baseline Methodology for Grid-Connected Electricity Generation from Renewable Sources. ACM0002/Version 06. Clean Development Mechanism–Executive Board, United Nations Framework Convention on Climate Change [online]. Available: http://cdm.unfccc.int/UserManagement/FileStorage/CDMWF_AM_BW759ID58S-T5YEEV6WUCN5744MN763 [accessed Sept. 21, 2006].

University of Chicago. 2004. The Economic Future of Nuclear Power. University of Chicago. August 2004 [online]. Available: http://www.anl.gov/Special_Reports/NuclEconAug04.pdf [accessed May 15, 2007].

USFS (U.S. Forest Service). 1974. National Forest Landscape Management, Vol. 2, Chapter 1. The Visual Managements System. Agriculture Handbook No. 462. U.S. Department of Agriculture, Forest Service, Washington, DC.

USFS (U.S. Forest Service). 1979. Proceedings of Our National Landscape: A Conference on Applied Techniques for Analysis and Management of the Visual Resource, April 23-25, 1979, Incline Village, NV. General Technical Report PSW-35. U.S. Department of Agriculture, Forest Service, Pacific Southwest Forest and Range Experiment Station, Berkeley, CA [online]. Available: http://www.fs.fed.us/psw/publications/documents/psw_gtr035/psw_gtr035.pdf [accessed May 23, 2007].

USFS (U.S. Forest Service). 1995. Landscape Aesthetics: A Handbook for Scenery Management. Agriculture Handbook No. 701. U.S. Department of Agriculture, Forest Service, Washington, DC.

USFWS (U.S. Fish and Wildlife Service). 2002a. Migratory Bird Mortality: Many Human-Caused Threats Afflict our Bird Populations. U.S. Fish and Wildlife Service, Division of Migratory Bird Management, Arlington, VA. January 2002 [online]. Available: http://www.fws.gov/birds/mortality-fact-sheet.pdf [accessed May 16, 2007].

USFWS (U.S. Fish and Wildlife Service). 2002b. Birds of Conservation Concern 2002. U.S. Fish and Wildlife Service, Division of Migratory Bird Management, Arlington, VA. December 2002 [online]. Available: http://www.fws.gov/migratorybirds/reports/bcc2002.pdf [accessed Oct. 2, 2006].

USFWS (U.S. Fish and Wildlife Service). 2003. Interim Guidelines to Avoid and Minimize Wildlife Impacts from Wind Turbines. U.S. Department of the Interior, U.S. Fish and Wildlife Service, Washington, DC [online]. Available: http://www.fws.gov/habitatconservation/wind.pdf [accessed June 2, 2006].

USFWS (U.S. Fish and Wildlife Service). 2006. West Virginia Northern Flying Squirrel: *Glaucomys sabrinus fuscus*. U.S. Fish and Wildlife Service, Northeast Region, Hadley, MA [online]. Available: http://www.fws.gov/northeast/pdf/flyingsq.pdf [accessed Oct. 17, 2006].

Usgaard, R.E., D.E. Naugle, R.G. Osborn, and K.F. Higgins. 1997. Effects of wind turbines on nesting raptors at Buffalo Ridge in southwestern Minnesota. Proc. S. Dakota Acad. Sci. 76:113-117.

UWIG (Utility Wind Integration Group). 2006. Utility Wind Integration State of the Art. Prepared by Utility Wind Integration Group, Reston, VA. May 2006 [online]. Available: http://www.uwig.org/UWIGWindIntegration052006.pdf [accessed Sept. 21, 2006].

VADEQ (Virginia Department of Environmental Quality). 2006. Department of Environmental Quality—Letter Enclosing Additional Information (Part 1 of 3) [online]. Available: http://docket.scc.state.va.us:8080/CyberDocs/Libraries/Default_Library/Common/frameviewdsp.asp?doc=63377&lib=CASEWEBP%5FLIB&mimetype=application%2Fpdf&rendition=native [accessed Oct. 9, 2006].

Valladares, G., A. Salvo, and L. Cagnolo. 2006. Habitat fragmentation effects on trophic processes of insect-plant food webs. Conserv. Biol. 20(1):212-217.

van den Berg, G.P. 2004. Effects of the wind profile at night on wind turbine sound. J. Sound Vib. 277(4-5):955-970.

van den Berg, G.P. 2006. The Sound of High Winds: The Effect of Atmospheric Stability on Wind Turbine Sound and Microphone Noise. Ph.D. Dissertation, University of Groningen [online]. Available: http://dissertations.ub.rug.nl/faculties/science/2006/g.p.van.den.berg/?FullItemRecord=ON [accessed Sept. 7, 2006].

VASCC (Virginia State Corporation Commission). 2005. Highland New Wind Development. Document List for Case Number: PUE-2005-00101 [online]. Available: http://docket.scc.virginia.gov:8080/vaprod/DOCUMENTS.ASP?MATTER_NO=121126 [accessed Sept. 21, 2006].

VASCC (Virginia State Corporation Commission). 2006a. Filing Requirements in Support of Applications for Authority to Construct and Operate an Electric Generating Facility [online]. Available: http://leg1.state.va.us/000/reg/TOC20005.HTM#C0302 [accessed Sept. 21, 2006].

VASCC (Virginia State Corporation Commission). 2006b. Highland New Wind Development Case Number PUE-2005-00101 (all filings, notices and comments related to the case) [online]. Available: http://docket.scc.virginia.gov:8080/vaprod/main.asp [accessed Sept. 13, 2006].

Vaughn, C.R. 1985. Birds and insects as radar targets: A review. Proc. IEEE 73(2):205-227.

Vauk, G. 1990. Biological and Ecological Study of the Effects of Construction and Operation of Wild Power Sites [in German]. Jahrgang/Sonderheft, Enbericht. Norddeutsche Naturschutzakademie, Germany.

Veilleux, J.P., and S.L. Veilleux. 2004. Intra-annual and interannual fidelity to summer roost areas by female eastern pipistrelles, *Pipistrellus subflavus*. Am. Midl. Nat. 152(1):196-200.

Veilleux, J.P., J.O. Whitaker, Jr., and S.L. Veilleux. 2004. Reproductive stage influences roost use by tree roosting female eastern pipistrelles, *Pipistrellus subflavus*. Ecoscience 11(2):249-256.

Verboom, B., and K. Spoelstra. 1999. Effects of food abundance and wind on the use of tree lines by an insectivorous bat, *Pipistrellus pipistrellus*. Can. J. Zool. 77(9):1393-1401.

Vermont Commission on Wind Energy Regulatory Policy. 2004. Findings and Recommendations. Prepared per Executive Order 04-04. December 15, 2004 [online]. Available: http://publicservice.vermont.gov/energy-efficiency/ee_files/wind/WindCommissionFinal-Report-12-15-04.pdf [accessed Sept. 11, 2006].

Vermont Division for Historic Preservation. 2007. Project Review: Criteria for Evaluating the Effect of Telecommunications Facilities on Historic Resources [online]. Available: http://historicvermont.org/programs/evaluatingcelltowers.pdf [accessed April 23, 2007].

Vilà, C., I.R. Amorim, J.A. Leonard, D. Posada, J. Castroviejo, F. Petrucci-Fonseca, K.A. Crandall, H. Ellegren, and R.K. Wayne. 1999. Mitochondrial DNA phylogeography and population history of the grey wolf *Canis lupus*. Mol. Ecol. 8(12):2089-2103.

Villard, M.A., M.K. Trzcinski, and G. Merriam. 1999. Fragmentation effects on forest birds: Relative influence of woodland cover and configuration on landscape occupancy. Conserv. Biol. 13(4):774-783.

Vissering, J. 2001. Wind Energy and Vermont's Scenic Landscape: A Discussion Based on the Woodbury Stakeholders Workshop [online]. Available: http://publicservice.vermont.gov/energy-efficiency/ee_files/wind/vissering_report.pdf#search=%22Quechee%20Analysis%22 [accessed Oct. 10, 2006].

von Hensen, F. 2004. Thought and working hypotheses on the bat compatibility of wind energy plants [in German]. Nyctalus 9(5):427-436.

Vonhof, M.J. 1996. Roost-site preferences of big brown bats (*Eptesicus fuscus*) and silver-haired bats (*Lasionycteris noctivagans*) in the Pend d'Oreille Valley in southern British Columbia. Pp. 62-80 in Bats and Forests Symposium, R.M.R. Barclay and R.M. Brigham, eds. Working Paper 23/1996. Research Program, B.C. Ministry of Forests, Victoria, Canada.

VTANR (Vermont Agency of Natural Resources). 2004. Wind Energy and other Renewable Energy Development on ANR Lands. December 2004 [online]. Available: http://www.vermontwindpolicy.org/finalpol.pdf#search=%22State%20of%20Vermont%20commercial%20wind%20energy%22 [accessed Sept. 21, 2006].

VTANR (Vermont Agency of Natural Resources). 2006. Act 250 Statute: Title 10: Conservation and Development Chapter 151: State Land Use and Development Plans [online]. Available: http://www.nrb.state.vt.us/lup/statute.htm [accessed Sept. 27, 2006].

VTPSB (Vermont Public Service Board). 2003. Title 30: Public Service. 30 V.S.A. § 202. Electrical Energy Planning [online]. Available: http://www.leg.state.vt.us/statutes/fullsection.cfm?Title=30&Chapter=005&Section=00202 [accessed Sept. 29, 2006].

VTPSB (Vermont Public Service Board). 2006. Citizens' Guide to the Vermont Public Service Board's Section 248 Process. Vermont Public Service Board, Montpelier, VT [online]. Available: http://www.state.vt.us/psb/document/Citizens_Guide_to_248.pdf [accessed Sept. 11, 2006].

Waits, L.P. 2004. Using noninvasive genetic sampling to detect and estimate abundance of rare wildlife species. Pp. 211-228 in Sampling Rare or Elusive Species: Concepts, Designs, and Techniques for Estimating Population Parameters, W.L. Thompson, ed. Washington, DC: Island Press.

Waits, L.P., and P.L. Leberg. 2000. Biases associated with population estimation using molecular tagging. Anim. Conserv. 3(3):191-200.

Walsh, A.L., and S. Harris. 1996. Foraging habitat preferences of vespertilionid bats in Britain. J. Appl. Ecol. 33(3):508-518.

Walter, W.D., D.M. Leslie, Jr., and J.A. Jenks. 2004. Response of Rocky Mountain Elk (*Cervus elaphus*) to Wind-Power Development in Southwestern Oklahoma. Presentation at the 11th Annual Meeting of the Wildlife Society, September 19, 2004, Calgary, Alberta, Canada.

WBCSD and WRI (World Business Council for Sustainable Development and World Resources Institute). 2005. The Green House Gas Protocol for Project Accounting, World Business Council for Sustainable Development, Geneva, Switzerland and World Resources Institute, Washington, DC [online]. Available: http://pdf.wri.org/ghg_project_accounting.pdf [accessed Sept. 21, 2006].

Weakland, C.A., and P.B. Wood. 2005. Cerulean warbler (*Dendroica cerulea*) microhabitat and landscape-level habitat characteristics in southern West Virginia. Auk 122(2):497-508.

Weathers, K.C., M.L. Cadenasso, and S.T.A. Pickett. 2001. Forest edges as nutrient and pollutant concentrators: Potential synergisms between fragmentation, forest canopies and the atmosphere. Conserv. Biol. 15(6):1506-1514.

Webster, M.S., P.P. Marra, S.M. Haig, S. Bensch, and R.T. Holmes. 2002. Links between worlds: Unraveling migratory connectivity. Trends Ecol. Evol. 17(2):76-83.

Weed, D.L. 2007. Weight of evidence: A review of concept and methods. Risk Anal. 25(6): 1545-1557.

Wenny, D.G., R.L. Clawson, J. Faaborg, and S.L. Sheriff. 1993. Population density, habitat selection and minimum area requirements of three forest-interior warblers in central Missouri. Condor 95(4):968-979.

WEST (Western EcoSystems Technology). 2006. Diablo Winds Wildlife Monitoring Progress Report: March 2005-February 2006. Technical Report. Prepared for Alameda County and FPL Energy, by WEST, Inc., Cheyenne, WY.

West, E.W., and U. Swain. 1999. Surface activity and structure of a hydrothermally-heated maternity colony of the little brown bat, *Myotis lucifugus*, in Alaska. Can. Field Nat. 113(3):425-429.

Westbrook, J.K., and W.W. Wolf. 1998. Migratory flights of bollworms, *Helicoverpa zea* (Boddie), indicated by Doppler weather radar. Pp. 354-355 in Preprints-Weather Data Requirements for Integrated Pest Management: 23rd Conference on Agricultural and Forest Meteorology, 2nd Urban Environment Symposium, and 13th Conference on Biometeorology and Aerobiology, November 2-6, 1998, Albuquerque, NM. Boston: American Meteorological Society.

Wethington, T.A., D.M. Leslie, Jr., M.S. Gregory, and M.K. Wethington. 1996. Prehibernation habitat use and foraging activity by endangered Ozark big-eared bats (*Plecotus townsendii ingens*). Am. Midl. Nat. 135(2):218-230.

Whitaker, J.O., Jr. 1998. Life history and roost switching in six summer colonies of eastern pipistrelles in buildings. J. Mammal. 79(2):651-659.

Whitaker, J.O., Jr., V. Brack, and J.B. Cope. 2002. Are bats in Indiana declining? P. Indiana Acad. Sci. 1:95-106.

White, J.G. 2002. Oregon's Siting Process for Large Wind Energy Facilities. Presentation at the Windpower 2002 Conference, June 2-5, 2002, OR [online]. Available: http://www.oregon.gov/ENERGY/SITING/docs/WindSite.PDF [accessed Sept. 14, 2006].

WIDNR (Wisconsin Department of Natural Resources). 2004. Considering Natural Resource Issue in Windfarm Siting in Wisconsin: A Guidance. Wisconsin Department of Natural Resources. July 2004 [online]. Available: http://www.dnr.state.wi.us/org/es/science/energy/wind/guidelines.pdf [accessed Sept. 13, 2006].

Wilkinson, G.S., and T.H. Fleming. 1996. Migration and evolution of lesser-long-nosed bats *Leptonycteris curasoae*, inferred from mitochondrial DNA. Mol. Ecol. 5(3):329-339.

Wilkinson, G.S., and J.M. South. 2002. Life history, ecology and longevity of bats. Aging Cell 1(2):124-131.

Williams, T.C., J.M. Williams, J.M. Teal, and J.W. Kanwisher. 1972. Tracking radar studies of bird migration. Pp. 115-128 in Animal Orientation and Navigation, S.R. Galler, K. Schmidt-Koenig, G.J. Jacobs, and R.E. Belleville, eds. NASA SP-262. Washington, DC: National Aeronautics and Space Administration.

Williams, T.C., L.C. Ireland, and J.M. Williams. 1973. High altitude flights of the free-tailed bat, *Tadarida brasiliensis*, observed with radar. J. Mammal. 54(4):807-821.

Williams, T.C., J.M. Williams, L.C. Ireland, and J.M. Teal. 1977. Autumnal bird migration over the western North Atlantic Ocean. Am. Birds 31:251-267.

Williams, T.C., J.M. Williams, P.G. Williams, and P. Stokstad. 2001. Bird migration through a mountain pass studied with high resolution radar, ceilometers, and census. Auk 118(2):389-403.

Winhold, L., A. Kurta, and R. Foster. 2005. Are red bats (*Lasiurus borealis*) declining in southern Michigan? Bat Res. News 46(4):229.

Winkelman, J.E. 1985. Impact of medium-sized wind turbines on birds: A survey on flight behavior, victims, and disturbance. Neth. J. Agr. Sci. 33:75-78.

Winkelman, J.E. 1989. Birds at a Wind Park near Urk: Bird Collision Victims and Disturbance of Wintering Ducks, Geese and Swans [in Dutch]. RIN-Rapport 89/15. Arnhem: Rijksinstituut voor Natuurbeheer.

Winkelman, J.E. 1990. Disturbance of Birds by the Experimental Wind Park near Oosterbierum (Fr.) During Building and Partly Operative Situations [1984-1989] [in Dutch]. RIN-Rapport 90/9. Arnhem: Rijksinstituut voor Natuurbeheer.

Winkelman, J.E. 1992a. The Impact of the SEP Wind Park near Oosterbierum (Fr.), the Netherlands, on Birds, Vol. 4: Disturbance [in Dutch]. RIN-report 92/5. Arnhem: DLO-Instituut voor Bos-en Natuuronderzoek.

Winkelman, J.E. 1992b. The Impact of the SEP Wind Park near Oosterbierum (Fr.), the Netherlands, on Birds, Vol. 2: Nocturnal Collision Risks [in Dutch]. RIN-report 92/3. Arnhem: DLO-Instituut voor Bos-en Natuuronderzoek.

Winkelman, J.E. 1992c. The Impact of the SEP Wind Park near Oosterbierum (Fr.), the Netherlands, on Birds, Vol. 1: Collision Victims [in Dutch]. RIN-report 92/2. Arnhem: DLO-Instituut voor Bos-en Natuuronderzoek.

Winkelman, J.E. 1995. Bird/wind turbine investigations in Europe. Pp. 43-47 in Proceedings of the National Avian-Wind Power Planning Meeting, July 20-21, 1994, Denver, CO. Prepared by LGL Ltd., Environmental Research Associates, King City, Ontario. February 24, 1995.

Wirsing, A.J., T.D. Steury, and D.L. Murray. 2002. A demographic analysis of a southern snowshoe hare population in a fragmented habitat: Evaluating the refugium model. Can. J. Zool. 80(1):169-177.

Wolsink, M. 1990. The siting problem: Wind power as a social dilemma. Pp. 725-729 in European Community Wind Energy Conference: Proceedings of an International Conference, September 10-14, 1990, Madrid, Spain, W. Palz, ed. Bedford, England: H.S. Stephens.

Wood, P.B., J.P. Duguay, and J.V. Nichols. 2005. Cerulean warbler use of regenerated clearcuts and two-age harvests. Wildlife Soc. Bull. 33(3):851-858.

Wood, P.B., S.B. Bosworth, and R. Dettmers. 2006. Cerulean warbler abundance and occurrence relative to large-scale edge and habitat characteristics. Condor 108(1):154-165.

WVDNR (West Virginia Department of Natural Resources). 2003. Rare and Threatened and Endangered Species. West Virginia Department of Natural Resources [online]. Available: http://www.wvdnr.gov/wildlife/endangered.shtm [accessed Sept. 21, 2006].

WVPSC (West Virginia Public Service Commission). 2005. Rules Governing Siting Certificates for Exempt Wholesale Generators. 150CSR30 [online]. Available: http://www.wvmcre.org/links/150-03-psc_siting_regs.pdf [accessed Sept. 14, 2006].

WVPSC (West Virginia Public Service Commission). 2006a. Commission Final Order Hearing Cancel for 8/9/2006: Case 05-1740-E-CS, July 24, 2006 [online]. Available: http://www.psc.state.wv.us/webdocket/default.htm [accessed Oct. 11, 2006].

WVPSC (West Virginia Public Service Commission). 2006b. Vision Statement of the Public Service Commission [online]. Available: http://www.psc.state.wv.us/missionstatement.htm [accessed Sept. 29, 2006].

WWEA (World Wind Energy Association). 2006. Worldwide Wind Energy Boom in 2005 [online]. Available: http://www.wwindea.org/home/index.php?option=com_content&task=view&id=13&Itemid=40 [accessed Sept. 27, 2006].

Wyman, R.L. 1998. Experimental assessment of salamanders as predators of detrital food webs: Effects on invertebrates, decomposition and the carbon cycle. Biodivers. Conserv. 7(5):641-650.

Young, D.P., Jr., and W. Erickson. 2006. Wildlife issue solutions: What have marine radar surveys taught us about avian risk assessment? Presentation at the Wildlife Workgroup Research Meeting VI, National Wind Coordinating Collaborative, November 14-16, 2006, San Antonio, TX [online]. Available: http://www.nationalwind.org/events/wildlife/2006-3/presentations/birds/young-marine_radar.pdf [accessed May 23, 2007].

Young, D.P., Jr., G.D. Johnson, W.P. Erickson, M.D. Strickland, R.E. Good, and P. Becker. 2001. Avian and Bat Mortality Associated with the Initial Phase of the Foote Creek Rim Windpower Project, Carbon County, Wyoming: November 1998-October 31, 2000. Prepared for SeaWest Windpower, Inc., San Diego, CA, and Bureau of Land Management, Rawlins District Office, Rawlins, WY, by Western EcoSystems Technology, Inc., Cheyenne, WY. 32 pp.

Young, D.P., Jr., W.P. Erickson, J.D. Jeffrey, K.J. Bay, R.E. Good, and B.G. Lack. 2003a. Avian and Sensitive Species Baseline Study Plan and Final Report TPC Combine Hills Turbine Ranch, Umatilla County, Oregon. Prepared for Eurus Energy America Corporation, San Diego, CA and Aeropower Services, Inc., Portland, OR, by Western EcoSystems Technology, Inc., Cheyenne, WY. March 2003.

Young, D.P., Jr., W.P. Erickson, M.D. Strickland, R.E. Good, and K.J. Sernka. 2003b. Comparison of Avian Responses to UV-Light-Reflective Paint on Wind Turbines: July 1999-December 2000. NREL/SR-500-32840. Prepared for National Renewable Energy Laboratory, Golden, CO, by Western EcoSystems Technology, Inc., Cheyenne, WY. January 2003 [online]. Available: http://www.west-inc.com/reports/fcr_nrel.pdf [accessed June 22, 2006].

Young, D.P., Jr., W.P. Erickson, R.E. Good, M.D. Strickland, and G.D. Johnson. 2003c. Avian and Bat Mortality Associated with the Initial Phase of the Foote Creek Rim Windpower Project, Carbon County, Wyoming: November 1998-June 2002. Prepared for Pacific Corp, Inc., Portland, OR, SeaWest Windpower, Inc., San Diego, CA, and Bureau of Land Management, Rawlins District Office, Rawlins, WY, by Western EcoSystems Technology, Inc., Cheyenne, WY. January 10, 2003 [online]. Available: http://www.west-inc.com/reports/fcr_final_mortality.pdf [accessed May 25, 2006].

Young, D.P., Jr., M.D. Strickland, W.P. Erickson, K. Bay, R. Canterbury, T. Mabee, B. Cooper, and J. Plissner. 2004. Baseline Avian Studies Mount Storm Wind Power Project, Grant County, West Virginia: May 2003-March 2004. Prepared for NedPower Mount Storm, LLC, Chantilly, VA, by Western EcoSystems Technology, Inc., Cheyenne, WY. April 23, 2004 [online]. Available: http://www.west-inc.com/reports/mount_storm_final.pdf [accessed June 2, 2006].

Young, D.P., Jr., J.D. Jeffrey, W.P. Erickson, K. Bay, K. Kronner, B. Gritski, and J. Baker. 2005. Combine Hills Turbine Ranch Wildlife Monitoring First Annual Report: March 2004-March 2005. Prepared for Eurus Energy America Corporation, Umatilla County, and the Combine Hills Technical Advisory Committee.

Zehnder, S., and L. Karlsson. 2001. Do ringing numbers reflect true migratory activity of nocturnal migrants? J. Ornithol. 142(2):173-183.

Zehnder, S., S. Åkesson, F. Liechti, and B. Bruderer. 2001. Nocturnal autumn bird migration at Falsterbo, South Sweden. J. Avian Biol. 32(3):239-248.

Zeiss, C., and J. Atwater. 1989. Property value guarantees for waste facilities. J. Urban Plan. Dev. 115(3):123-134.

Zimmerman, G.S., and W.E. Glanz. 2000. Habitat use by bats in eastern Maine. J. Wildlife Manage. 64(4):1032-1040.

Zube, I., and L. Mills, Jr. 1976. Cross cultural explorations in landscape perception. Pp. 162-169 in Studies in Landscape Perception, E.H. Zube, ed. Publication No. R-76-1. Amherst: Institute for Man and Environment, University of Massachusetts.

Appendixes

APPENDIX
A

About the Authors

Paul Risser *(chair)* is Chair and Chief Operating Officer, University Research Cabinet of the University of Oklahoma. Until recently, he was Chancellor of the Oklahoma State System of Higher Education, where he led a state system comprised of 25 state colleges and universities, 9 constituent agencies, and 1 higher education center. Dr. Risser led two universities as President, Oregon State University (1996-2002) and Miami University of Ohio (1993-1996). Dr. Risser's research specialties are the flow of energy and materials through grassland and forested ecosystems, the effects of climate on plant community productivity, and landscape ecology. He has served as chair of the National Research Council (NRC) Board of Environmental Studies and Toxicology and as a member of numerous NRC committees. He is a fellow of the American Association for the Advancement of Science and the American Academy of Arts and Sciences. He now chairs the Science Committee of the Smithsonian National Museum of Natural History, Washington, DC. Chancellor Risser previously served as Secretary-General of the Scientific Committee on Problems of the Environment, Paris, France, and as Program Director for Ecosystem Studies at the National Science Foundation. Dr. Risser has also served as President of three scientific organizations, the American Institute of Biological Sciences, the Ecological Society of America, and the Association of Southwestern Naturalists. He received his Ph.D. in botany and soils from the University of Wisconsin-Madison.

Ingrid Burke is a Professor in the Department of Forest, Rangeland, and Watershed Stewardship at Colorado State University, where she is also a University Distinguished Teaching Scholar. Her areas of interests are in soil organic matter dynamics, ecosystem ecology, biogeochemistry, regional modeling, and global change, as well as pedagogical techniques. She has served as a member of the NRC Committee to Review EPA's Environmental Monitoring and Assessment Program and as a member of the NRC Board on Environmental Studies and Toxicology. She is an associate editor of *Ecological Applications* and has been on the editorial board of *Ecosystems* and *Forest Ecology and Management*. She is a member of the American Association for the Advancement of Science, the American Institute of Biological Sciences, and the Ecological Society of America. She received her Ph.D. from the University of Wyoming.

Christopher Clark is the I.P. Johnson Director of the Bioacoustics Research Program at the Cornell Laboratory of Ornithology. He also is a Senior Scientist and member of the graduate faculty in the Department of Neurobiology and Behavior at Cornell University. His research interests are in the development and application of passive acoustic techniques for understanding how and why animals communicate and to monitor the health of wildlife populations. He directed the development of Canary and Raven, software programs used by scientists to study the sounds of birds and other animals. He arrived at Cornell in 1987 after seven years at The Rockefeller University as a postdoctoral fellow and assistant professor. He received an M.S. in Engineering and a Ph.D. in Biology from the State University of New York at Stony Brook.

Mary English is a Research Leader at the Energy, Environment and Resources Center at the University of Tennessee, Knoxville. Her current research interests include land-use and growth-management planning at the local and state levels, political and economic conditions for sustainable consumption, and participatory processes for environmental decision making. Her research has focused on ways in which environmental decision-making processes can be improved. Dr. English is a member of the Tennessee Air Pollution Control Board. She also has served on the NRC's Board on Radioactive Waste Management and several NRC study committees, including the Committee on Remediation of Buried and Tank Wastes as Vice-Chair and as a member on the Committee on Prioritization and Decision-Making in the Department of Energy-Office of Science and Technology. She received her Ph.D. in Sociology from the University of Tennessee, Knoxville.

Sidney Gauthreaux, Jr., is a Professor of Biological Sciences at Clemson University where he has worked for the last 35 years. His area of expertise is in laboratory and field studies of bird migration, orientation, and navigation. He is one of the pioneers in the use of radar to study bird movements in the atmosphere, and he has been particularly interested in bird migration across the Gulf of Mexico and over the eastern United States. Currently, the Department of the Interior and the Nature Conservancy are using information gathered by Dr. Gauthreaux on important migration stopover areas in an effort to evaluate these areas for habitat protection projects. He is a fellow of several societies including the American Ornithologists Union, the American Association for the Advancement of Science, and the Deutsche Ornithologen-Gesellschaft. Dr. Gauthreaux has served as President of the Animal Behavior Society and Chair of the South Carolina Heritage Trust Advisory Board. He received his Ph.D. in Ornithology from Louisiana State University.

Sherri W. Goodman is General Counsel at the Center for Naval Analyses Corporation, a non-profit research and analysis organization. Ms. Goodman was the Deputy Under Secretary of Defense (Environmental Security) from 1993 to 2001. As the chief environmental, safety, and occupational-health officer for the Department of Defense, Ms. Goodman was responsible for programs including energy efficiency and climate change, cleanup at active and closing bases, compliance with environmental laws, environmental cooperation with foreign militaries, conservation of natural and cultural resources, explosives safety, and pest management. Ms. Goodman has twice received the Department of Defense award for Distinguished Public Service, the Gold Medal from the American Defense Preparedness Association, and EPA's Climate Change Award. She practiced law at the Boston law firm Goodwin Procter. She also served on the staff of the Senate Armed Services Committee, 1987-1990, working for the Chairman, Senator Sam Nunn, where she oversaw the Department of Energy's Defense and Environmental Programs, including nuclear weapons research and development production, waste management, and environmental remediation. Ms. Goodwin is a member of the NRC Board on Environmental Studies and Toxicology. She received her J.D. from the Harvard School of Law, her Masters in Public Policy from Harvard's John F. Kennedy School of Government, and her B.A. from Amherst College.

John Hayes is Chair of the Department of Wildlife Ecology and Conservation at the University of Florida. Before coming to the University of Florida he was a Professor in the Department of Forest Science at Oregon State University, where he also served as the Associate Dean of International Programs for the College of Forestry. His research interests include the influence of forest management and habitat alteration on wildlife populations, the influence of spatial scale on habitat selection, and the ecology and management of bats. Currently, he is co-investigator on a research project exploring approaches to evaluate risk of wind-energy sites to bats. In addition, Dr. Hayes has served as a member of a scientific advisory committee for a research cooperative focused on implications of wind energy on bats, and he has served as a consultant in assessing risks of proposed wind-energy sites for bats. Dr. Hayes received his Ph.D. in Ecology and Evolutionary Biology from Cornell University.

Arpad Horvath is an Associate Professor in the Engineering and Project Management Program in the Department of Civil and Environmental Engineering at the University of California, Berkeley. His research interests are in developing methods and tools for environmental and economic analysis of civil infrastructure systems, primarily for the built environment. His research has focused on the environmental implications of the construction, electronics and various service industries, life-cycle assessment modeling by using environmentally augmented economic input-output analysis, and environmental performance measurement. He is the director of the Consortium on Green Design and Manufacturing, which encourages multidisciplinary research and education on environment and pollution prevention issues. He is also associate editor of the *Journal of Infrastructure Systems*. He received his Ph.D. in Civil Engineering from Carnegie Mellon University.

Thomas H. Kunz is Professor of Biology and Director of the Center for Ecology and Conservation Biology at Boston University, where he has been on the faculty for the past 34 years. His research focuses on the ecology, behavior, evolution, and conservation biology of bats. He is the author or co-author of over 200 publications and is the editor of *Ecology of Bats* (Plenum Press 1982) and *Ecological and Behavioral Methods for the Study of Bats* (Smithsonian Institution Press 1988), and co-editor of *Bat Biology and Conservation* (Smithsonian Institution Press, 1998), *Bat Ecology* (University of Chicago Press 2003), and *Functional and Evolutionary Ecology of Bats* (Oxford University Press, et al. 2006). He is an elected Fellow of the American Association for the Advancement of Science, Past-President of the American Society of Mammalogists, and a recipient of the Gerrit S. Miller Jr. Award (1984) and the C. Hart Merriam Award (2000). He is currently funded by grants from the National Science Foundation and the National Park Service, where his research focuses on assessing the ecological and economic impact of Brazilian free-tailed bats on agroecosystems and the influence of anthropogenic factors on the prevalence of rabies in two common species of North American insectivorous bats. He received his Ph.D. from the University of Kansas.

Lynn Maguire is Professor of the Practice of Environmental Decision Analysis and Director of Professional Studies at the Nicholas School of the Environment and Earth Sciences at Duke University, where she has been since 1982. Dr. Maguire's current research uses a combination of methods from decision analysis, environmental conflict resolution, and social psychology to study environmental decision making. Dr. Maguire focuses on collaborative decision processes where values important to the general public and stakeholders must be combined with technical analysis to determine management strategies. She has applied these methods to management of endangered species, invasive species, multiple use of public forestland, and water quality planning. She was on the editorial board of the journal *Biological Conservation*, and she was also was a member of the board of Governors of the Society of Conservation Biology. She has served as a member on the NRC Committee on Scientific Issues in the Endangered Species Act and on the NRC Lake Ontario-St. Lawrence River review committee. She received a Ph.D. in Wildlife Ecology from Utah State University.

Lance Manuel is an Associate Professor in the Department of Civil Engineering at the University of Texas at Austin. His areas of interest are structural reliability, structural dynamics, probabilistic seismic hazard analysis, and wind engineering. He has worked with Sandia National Laboratories on the statistical analysis of inflow and loads data for wind turbines, on characterization of the spatial coherence in inflow turbulence for wind turbines, and on the development of turbine design loads using inverse reliability techniques. He received his Ph.D. in Civil Engineering from Stanford University.

Erik Lundtang Petersen is the Head of the Wind Energy Department at Risø National Laboratory in Roskilde, Denmark. The research of the department aims to develop new opportunities for industry and society in the exploitation of wind power and to map and alleviate atmospheric aspects of environmental problems. He has worked with the Wind Atlas Analysis and Application Programme, and was principal consultant for the World Meteorological Organization on the Meteorological Information for Development of Renewable Energy project. He is the editor of the journal *Wind Energy* and is a member and founding member of the European Renewable Energy Centres Agency. He has served on a variety of missions for the United Nations Development Program and Danida in the capacity as advisor for wind-energy feasibility projects and was advisor to the Algerian Commissariel National aux Energies Nouvelles. He received his Ph.D. from the Technical University of Denmark.

Dale Strickland is Vice President and Senior Ecologist at Western EcoSystems Technology, Inc. (WEST). His areas of expertise include the design and conduct of wildlife studies, impact and risk assessment, and natural resource damage assessment studies. Prior to his employment with WEST he served as a scientist and administrator with the Wyoming Game and Fish

Department and served on the faculty of the Department of Statistics at the University of Wyoming. He has also taught courses in wildlife management and statistics as a visiting instructor at the University of Wyoming. He contributed to documents for the National Oceanic and Atmospheric Administration for the quantification of injury due to oil spills in Type B Natural Resource Damage Assessments. He authored a chapter in a guidance document on the conduct of research on interactions between birds and wind-energy facilities for the National Wind Coordinating Committee. Dr. Strickland is currently serving as the Executive Director of the Platte River Endangered Species Partnership. He is also currently serving as an Associate Editor for the *Journal of Wildlife Management*. Dr. Strickland received a Ph.D. in Zoology from the University of Wyoming.

Jean Vissering is a landscape architect who has presented and written extensively on the issues of scenic resource evaluation and visual impact assessment and aesthetics within Vermont. Ms. Vissering has worked with wind-energy developers, local communities, and other stakeholders in assessing the impacts of wind-energy projects in Vermont. She has presented at the National Wind Coordinating Committee, and has written a paper for the State of Vermont on the subject of visual aesthetics and wind-energy projects. Ms. Vissering has been a Lecturer at the University of Vermont's School of Natural Resources and Department of Plant and Soil Science. She received a Masters of Landscape Architecture from North Carolina State University.

James Roderick (Rick) Webb is a Senior Scientist with the Department of Environmental Sciences at the University of Virginia, where he is Projects Coordinator of the Shenandoah Watershed Study and the Virginia Trout Stream Sensitivity Study. His primary research focus is the effects of air pollution on streams associated with forested mountain watersheds in the central Appalachian Mountain region. He has served on several cases as an expert witness on aquatic effects of acidic deposition for the U.S. Department of Justice. Previously, he worked with conservation organizations concerned with the direct environmental effects of coal extraction. He represented the Virginia Society of Ornithology on the Virginia Wind Energy Collaborative Environmental Working Group and co-authored a document on land-based wind-energy projects and environmental effects. He received a Masters in Environmental Sciences from the University of Virginia.

Robert Whitmore is a Professor of Wildlife Ecology at West Virginia University where he has been since 1975. His research interests are in conservation ecology, ornithology, interpretive bird studies, and quantitative ecology. He has performed field work and published on birds and bats in the area of the Alleghany Highlands where wind-energy projects are being developed. Within the Appalachian ecosystems, Dr. Whitmore has conducted extensive field research in the habitat types that are involved in wind-energy development. He is an Elected Member of the American Ornithologists Union, as well as a member of the Cooper Ornithological Society, the Wilson Ornithological Society, and the Society of Field Ornithologists. He received a Ph.D. in Zoology from Brigham Young University.

Emission Rates for Electrical Generation

TABLE B-1 Annual Emission Rates for Electrical Generating Units (lb/MWh)

Data Set	Geographic Area	Period	Generation Type	CO_2	NO_x	SO_2
eGrid 2000[a]						
USA	50 U.S. states	2000	System average	1,392	3.0	6.0
MAH States	MD, PA, VA, WV, DC	2000	System average	1,426	3.5	9.7
USA	50 U.S. states	2000	Coal	2,188	4.8	10.9
MAH States	MD, PA, VA, WV, DC	2000	Coal	2,053	5.0	14.1
USA	50 U.S. states	2000	Natural gas	1,187	1.7	0.3
MAH States	MD, PA, VA, WV, DC	2000	Natural gas	878	1.0	0.3
BLM EIS[b]	Western U.S.	Pre-1991	Coal	2,860	15.4	15.4
BLM EIS[b]	Western U.S.	Pre-2001	Natural gas	1,200	0.0	1.3
PJM 2005[c]	PJM grid system	2005	System average	1,292	2.6	8.5
RSG-ERT SIP[d]	MD, PA, WV, VA	Pre-2003	Coal	2,113	5.7	17.7
RSG-ERT VA[e]	VA, WV	2004	Primarily coal	2,037	3.9	5.3
ISO-NE[f]						
2000	New England	2000	Marginal units	1,488	1.9	6.2
2004	New England	2004	Marginal units	1,102	0.5	2.0
2000	New England	2000	System average	913	1.1	3.9
2005	New England	2004	System average	876	0.8	2.3
OTC[g]						
NY	NY	2002	System average	810	1.2	2.7
NE	New England	2002	System average	1,000	1.1	3.3
PJM	PJM grid system	2002	System average	1,180	2.3	8.0
NESCAUM[b]						
Greenpoint	NY	1999	System average	944	1.5	4.4
Exelon	PJM grid system	1998	System average	1,199	2.8	9.0

[a]eGRID 2006.
[b]BLM 2005a.
[c]PJM 2006b.
[d]Hathaway et al. 2005.
[e]High and Hathaway 2006.
[f]ISO New England Inc. 2006.
[g]Keith et al. 2002.
[h]NESCAUM 2002.

NOTE: The committee has not assessed the uncertainty associated with the numbers presented.

TABLE B-2 Wind Resource Database: Standard Version, May 2005[a]

State	Data Source[b]	State	Data Source[b]
Arizona	N/TWS 2003	Nebraska	N/TWS 2005
Alabama	PNL 1987	New Hampshire	N/TWS 2002
Arkansas	PNL 1987	New Jersey	N/TWS 2003
California	N/TWS 2003	New Mexico	N/TWS 2003
Colorado	N/TWS 2003	New York	PNL 1987
Connecticut	N/TWS 2002	North Carolina	N/TWS 2003
Delaware	N/TWS 2003	North Dakota	NREL 2000
Florida	PNL 1987	Ohio	N/TWS 2004
Georgia	PNL 1987	Oklahoma	PNL 1987
Idaho	N/TWS 2002	Oregon	N/TWS 2002
Illinois	NREL 2001	Pennsylvania	N/TWS 2003
Indiana	N/TWS 2004	Rhode Island	N/TWS 2002
Iowa	PNL 1987	South Carolina	PNL 1987
Kansas	PNL 1987	South Dakota	NREL 2000
Kentucky	PNL 1987	Tennessee	PNL 1987
Louisiana	PNL 1987	Texas	PNL 1987
Maine	N/TWS 2002		NREL 2000
Maryland	N/TWS 2003	Vermont	N/TWS 2002
Massachusetts	N/TWS 2002	Virginia	N/TWS 2003
Michigan	N/TWS 2005	Washington	N/TWS 2002
Minnesota	PNL 1987	West Virginia	N/TWS 2003
Mississippi	PNL 1987	Wisconsin	PNL 1987
Missouri	N/TWS 2004	Wyoming	N/TWS 2002
Montana	N/TWS 2002		

[a]Data source and exclusion criteria for U.S. wind potential map coverage provided on March 16, 2006, by National Renewable Energy Laboratory, Golden, CO.

[b]YrSource: Yr = Year validated (1987 to present); Source = PNL, NREL, or N/TWS (NREL with AWS TrueWind).

NOTE: PNL data resolution is 1/4 degree of latitude by 1/3 degree of longitude; each cell has a terrain exposure percent (5% for ridgecrest to 90% for plains) to define base resource area in each cell. Ridgecrest areas have 10% of the area assigned to the next higher power class. NREL data were generated with the WRAMS model and do not account for surface roughness. Resolution is 1 km. Texas includes the Texas mesas study area updated by NREL using WRAMS. N/TWS data was generated by AWS TrueWind and validated by NREL. Resolution is 400 m for the northwest states (WA, OR, ID, MT, and WY) and 200 m everywhere else. These data consider surface roughness in their estimates.

TABLE B-3 Wind Resource Exclusion Database:[a] Criteria for Defining Available Windy Land[b]

Criteria	Data/Comments
Environmental	
(2) 100% exclusion of National Park Service and Fish and Wildlife Service managed lands.	USGS Federal and Indian Lands shapefile, Jan. 2005.
(3) 100% exclusion of federal lands designated as park, wilderness, wilderness study area, national monument, national battlefield, recreation area, national conservation area, wildlife refuge, wildlife area, wild and scenic river, or inventoried roadless area.	USGS Federal and Indian Lands shapefile, Jan. 2005.
(4) 100% exclusion of state and private lands equivalent to criteria 2 and 3, where GIS data are available.	State/GAP land stewardship data management status 1, from Conservation Biology Institute Protected Lands database, 2004.
(8) 50% exclusion of remaining USDA Forest Service (FS) lands (incl. National Grasslands).	USGS Federal and Indian Lands shapefile, Jan. 2005.
(9) 50% exclusion of remaining Dept. of Defense lands.	USGS Federal and Indian Lands shapefile, Jan. 2005.
(10) 50% exclusion of state forest land, where GIS data is available.	State/GAP land stewardship data management status 2, from Conservation Biology Institute Protected Lands database, 2004.
Land Use	
(5) 100% exclusion of airfields, urban, wetland, and water areas.	USGS North America Land Use Land Cover (LULC), version 2.0, 1993; ESRI airports and airfields (2003).
(11) 50% exclusion of non-ridgecrest forest.[c]	Ridgecrest areas defined using a terrain definition script, overlaid with USGS LULC data screened for the forest categories.
Other	
(1) Exclude areas of slope > 20%.	Derived from elevation data used in the wind resource model.
(6) 100% exclude 3 km surrounding criteria 2-5 (except water).	Merged datasets and buffer 3 km.
(7) Exclude resource areas that do not meet a density of 5 km^2 of class 3 or better resource within the surrounding 100 km^2 area.	Focalsum function of class 3+ areas (not applied to 1987 PNL resource data).

[a]Standard Version, last revised Jan. 2004.
[b]Numbered in the order they are applied.
[c]50% exclusions are not cumulative. If an area is non-ridgecrest forest on FS land, it is just excluded at the 50% level one time.

APPENDIX
C

Methods and Metrics for Wildlife Studies

A wide range of methods are available for assessing the ecological influences of wind-energy and aspects of the ecology and behavior of species that may be affected by wind-energy facilities; most of them are reviewed here. For additional information on methods readers are referred to syntheses presented in Anderson et al. (1999), Braun (2005), and Kunz and Parsons (in press).

Key Variables and Monitoring Methods

Researchers have only begun to investigate the ecological impacts of wind-energy facilities, especially impacts on bats. The possibility of large cumulative impacts on bat populations has not previously been considered in siting plans and wind-energy development in the United States, and thus research and monitoring studies are needed to develop predictive models of cumulative effects and to inform decision makers. Understanding of impacts on birds also is limited because of the lack of replication of studies at existing wind-energy facilities, the lack of information in some regions of the country, and inadequate evaluation of predicted impacts following facility construction and operation.

Bat fatalities at wind turbines have been reported at nearly every wind-energy facility where post-construction surveys have been conducted, yet few of these studies have included more than one year of monitoring, and of these none monitored fatalities consistently from spring migration through fall migration at any single site. Moreover, only four studies prior to that of Arnett (2005) used fresh bat carcasses to assess searcher efficiency and/or

279

conducted scavenger-removal experiments to correct estimates for potential biases.

Study Design

The most important element in designing a study is deciding on the study objective. Once the study objective is determined, other essential issues include the following:

- The area of interest,
- Time period of interest,
- Species of interest,
- Potentially confounding variables,
- The time and budget available for the required studies, and
- The magnitude of the impact being evaluated.

The following is a general discussion of methods, metrics, and study design for achieving objectives commonly addressed in the study of wildlife impacts from wind-energy development. For a more detailed discussion of this topic, readers are referred to Green (1979), Underwood (1994), Anderson et al. (1999), Manly (2001), and Morrison et al. (2001). There is no fundamental difference between monitoring and research, but a commonly used criterion for distinguishing them is the duration of study. Monitoring schemes are essentially repeated surveys (Manly 2001) and are usually designed to detect changes and trends in the variable of interest. Because considerations in study design are essentially the same for both monitoring and observational studies, no effort will be made to further discriminate between the two.

Reliable study designs available for environmental impact assessments are limited. The before-after/control impact (BACI) design is commonly used in observational studies (e.g., Stewart-Oaten 1986) and has been considered the optimal impact-study design by Green (1979). As the name implies, this type of study involves the collection of data in the assessment area and a similar (control) area both before and after an impact occurs (Morrison et al. 2001). An effect typically is measured as a change in the difference between estimates of a variable for the control and an assessment area following an impact. Confidence intervals can increase the reliability of an impact estimate when data from more than one control area are available (Underwood 1994). Ideally, control areas should be randomly selected from a population of similar sites (Manly 2001). Study areas within the assessment and control area may be matched to reduce the natural variation common in impact studies (Skalski and Robson 1992), although characteristics of study sites may change in longer-term studies, and thus matching may be unreliable.

When data are lacking before an impact, the control-impact design may be used. This type of study differs from the BACI design only in the lack of pre-impact data. As in the BACI design, if a significant difference is attributed to the impact of a perturbation the assumption is that nothing else could cause a change of that magnitude (Manly 2001). Before-after designs can be used when data from a control area cannot be obtained. A change immediately following an impact is assumed to be a result of the impact and not from some other cause. In the absence of data from control areas, the attribution of cause may be difficult to support, unless the impact is large and easily attributable to the cause. For example, a decline in bird abundance following the construction of a wind-energy facility might be attributed to the facility by finding large numbers of bird carcasses killed by turbines. In the absence of strong corroborative evidence, attributing the change in abundance to the wind-energy plant may be difficult to defend.

The impact-gradient design may be used for quantifying impacts in relatively small assessment areas with homogeneous environments (Anderson et al. 1999; Manly 2001). With this design, an effect is assumed if it appears to be reduced as the distance increases from the source of the impact (Manly 2001). The most important assumption made when using the impact-gradient design is that the environment is homogeneous. Homogeneity is relatively uncommon in the environment and the analysis of data resulting from this study design should take spatial correlation into account (Manly 2001). For example, wind turbines are typically placed on the windiest sites available in a wind-resource area, such as ridge tops. Thus, moderating environmental conditions as a function of distance from the turbines may create subtle differences in the characteristics of the sites that could mask impacts.

Morrison et al. (2001) suggested improving observational studies by using several general approaches to study design that can increase precision without requiring increased replication. Their suggestions include:

• Vary sampling effort (or apply treatments) within homogenous groups of experimental units (blocking).
• Measure non-treatment factors (co-variates) and use analysis of covariance when analyzing the response to a treatment to consider the added influence of variables having a measurable influence on the dependent variable.
• Refine experimental techniques, including greater sampling precision within experimental units (Cochran and Cox 1957; Cox 1958).

Mensurative studies involve making measurements of uncontrolled events at one or more points in space or time with space and time being the only experimental variable or treatment (Morrison et al. 2001). Mensurative studies are most convincing when the impacts are large and it is difficult or impossible to attribute the impact to some other cause. Nevertheless, mensurative studies often are conducted because there is no alternative, and they give more information than no study at all (Manly 2001). A study of impact should not rely on a single response variable, but should use the strongest design possible and accumulate all available evidence in a weight-of-evidence approach (Anderson et al. 1999) when evaluating the existence and magnitude of an impact. Table C-1, taken from Anderson et al. (1999), provides a decision matrix for selecting the appropriate impact-study design.

Methods for Estimating Abundance

Estimating abundance of species at proposed and existing wind-energy sites can be important in assessing the ecological impacts of wind-energy facilities. This section reviews several methods that are appropriate for assessing fatalities and effects of habitat alterations on populations of bats and birds. Direct impacts are fatalities resulting from collisions with wind-turbine blades or turbine monopoles while animals are in flight. Direct impacts may alter sex and age ratios, densities of resident or migratory populations, and survivorship and reproductive success. Indirect impacts include animal, plant, or ecosystem responses to habitat alteration caused by wind-energy facilities; they may include altered foraging behavior, breeding activities, migratory patterns, and demographics. Anderson et al. (1999) provided a detailed discussion of methods and metrics for the study of impacts on birds caused by wind-energy development. While many of these methods and metrics were developed for birds, an improved summary for methods and metrics useful in the study of bats and nocturnally active birds is included in this appendix; a complementary document also is being developed by the National Wind Coordinating Committee (Kunz et al. in press b).

Abundance of some animals can be determined from a census or estimated using line-transect sampling, point-counts, quadrat sampling, and other techniques (Buckland et al. 2001, 2004; Manly 2001; Morrison et al. 2001). Abundance also can be estimated through indirect approaches such as mark-resight and capture-mark-recapture estimation (Skalski and Robson 1992; Amstrup et al. 2005), catch-per-unit-effort (Laake 1992), survival analysis (Riggs and Pollock 1992), and change-in-ratio methods (Udevitz and Pollock 1992).

Censusing wildlife in designated areas or estimating absolute abundance is generally difficult, expensive, and time consuming. Impact-assessment

TABLE C-1 Study-Design Decision Matrix for Observational Studies

Design Options

Study Conditions	Recommended Design	Study Conditions	Potential Design Modification
Pre-impact data possible	BACI	Matching of	Matched
Reference area indicated	BACI	study sites on assessment and reference areas possible	pair, design with BACI
Pre-impact data not possible	Impact-reference	Matching of	Matched
Reference area indicated	Impact-reference	study sites on assessment and reference areas possible	pair, design with impact-reference
Pre-impact data possible Reference area not indicated	Before-after		
Small homogenous area of potential impact	Impact-gradient[a]		

Sampling Plan Options

Sampling Plan	Recommended Use
Haphazard/judgment sampling	Preliminary reconnaissance
Probability-based sampling:	
Simple random sampling	Homogenous area with respect to impact indicators and covariates
Stratified random sampling	Strata well defined and relatively permanent, and study of short duration
Systematic sampling	Heterogeneous area with respect to impact indicators and covariates, and study of long duration

Parameters to Measure

Parameter	Empirical Description
Abundance/relative use	Use per unit area and/or per unit time as an index[b]
Mortality	Carcasses per unit area and/or per unit time
Reproduction	Young per breeding pair of adults
Habitat use	Use as a function of availability
Covariates	Vegetation, topography, structure, distance, species, weather, season, etc.

[a]Impact-gradient design can be used in conjunction with BACI, impact reference, and before-after designs.
[b]Can be summarized by activity/behavior for evaluation of risk.
SOURCE: Anderson et al. 1999. Reprinted with permission; copyright 1999, National Wind Coordinating Committee.

studies often estimate animal use as a surrogate for abundance. Animal use can be estimated by a variety of methods such as counting the animals detected from a given set of observation points, the amount of time spent by individual animals within a survey plot, the number of animals seen moving past a particular point, the number of targets passing through a radar beam, the number of targets within altitude bands, the number of nests present in a given area, the number of animals trapped or netted, the number of calls detected, or the amount of sign (e.g., tracks or scat) recorded within sample plots. Counts are expressed as the number of observations per unit area, per unit time, or both. Estimates of use allow comparisons among defined time periods and areas (Anderson et al. 1999; Hayes and Loeb 2007; Kunz et al. in press a). Comparison of indices such as animal use among studies or sites requires that indices be estimated using similar protocols.

Estimates of use also can assist in the interpretation of fatality data. For example, if two wind-energy facilities are being compared based on fatalities alone, the facility with the greater number of fatalities might be considered to have the greater impact. However, if the facility with more fatalities also has much greater use by the species being killed, then the greater use must be taken into account in any comparison. For example, at a minimum, estimation of use should include the intensity of activity, flight paths, flight heights, and the behavior of the animals of interest.

Monitoring productivity and survivorship may be an alternative to the direct estimation of fatalities and abundance when looking at the cumulative effects of wind-energy development on wildlife populations. The Monitoring Avian Productivity and Survivorship (MAPS) program was designed to accurately assess changes in bird productivity and survivorship in response to environmental changes (DeSante et al. 2001). The MAPS program provides annual and regional indices of post-fledging productivity from the number and proportion of young birds captured, annual and regional estimates of adult survivorship, recruitment in the adult population, and adult population size from capture-recapture data on adult birds. At the local level, Hunt (2002) used radiotelemetry data on golden eagles in the Altamont Pass Wind Resources Area (APWRA) to estimate the population's annual growth rate, which was used to evaluate the effect of wind-energy production on fatalities.

This type of study often can provide more information about the mechanisms of impact than simply evaluating fatalities. For example, while Hunt (2002) concluded that the population of golden eagles had characteristics of a growing population, the confidence intervals around the point estimate of positive growth rate included zero, thus making it impossible to verify whether the population was growing or declining. Hunt (2002) concluded that golden eagle territories were consistently occupied and a sufficient number of non-territorial (floater) eagles existed to re-populate vacant territories, suggesting a relatively healthy population. Nevertheless, the relatively high fatalities attributable to the wind-energy facilities resulted in a population without sufficient floaters to ensure stability, making the population susceptible to future declines should fatalities increase for any reason. It also was clear from Hunt's study that the targeted group of eagles was part of a larger population. Thus, the APWRA may represent a mortality sink for the regional population of golden eagles. Certainly, at the current level of eagle fatalities in the APWRA (Smallwood and Thelander 2004, 2005), the viability of the eagle population depends on adequate immigration from surrounding areas.

The detection, identification, and counting of diurnally active organisms in the lower atmosphere is rather straightforward, despite the lack of standard protocols for making daytime observations at planned or existing wind-energy facilities. The situation at night is more difficult. Several methods for detecting, identifying, and counting birds, bats, and insects in the atmosphere at night have been developed (Hayes and Loeb 2007; Kunz et al. in press a). Table C-2 (modified from Larkin 2005a) provides a summary of current technology with respect to the detection range of the equipment, the ability to identify the type of animal, the ability to provide information on passage rates or density estimates, measurement of the altitude of a target, and cost of the equipment.

When confirmation of the age, sex, and reproductive condition of a species in an area of interest is desirable (as may often be the case during pre-siting and pre-construction surveys), capture is required. Information on species identity, sex, age, and reproductive condition can also be assessed from bats and birds killed by wind turbines. Remote sensing (e.g., radar) can provide information needed to assess risks to bats and birds at larger spatial and temporal scales.

In many cases, using a combination of approaches will be of value as no single method can be used for unambiguously assessing natural populations or the effects of wind turbines on biotic communities. Each approach has its own strengths, limitations, and biases. Investigators should understand the limitations, applicability, and operational considerations of each method before deploying them in the field. Local field guides and taxonomic keys for species identification are essential tools for investigators if they wish to identify the species composition at each locality and the identity of animals that are captured or killed. Use of mitochondrial- and nuclear-DNA sequence data that can be derived by extractions from feathers, hair, and skin of carcasses killed by wind turbines offers the potential for estimating population size of birds and bats (e.g., Waits 2004; Kunz et al. in press a; N.B. Simmons, American Museum of Natural History, personal communication 2006). Moreover, similar DNA-sequence data may be needed to verify the identity of some closely related or cryptic species (e.g., *Myotis* species). In

TABLE C-2 Remote-Sensing Tools for Detecting, Tracking, and
Quantifying Flying Birds, Bats, and Insects

Equipment	Range	Identification[a]
Small marine radar	30 m-6 km with proper siting of unit	+ Bird bats vs. insects – Birds vs. bats straight flight: unknown
Large Doppler surveillance radar (NWS)	10-200 km	+ Can discriminate targets by speed if winds are known + Waterfowl & raptors vs. other birds & bats + Insects slower than songbirds
Thermal infrared	Depends on equipment and cost: $75,000 US unit can detect birds at 3 km	Size but not species + Discriminates birds, insects and foraging bats – Migrating birds & bats
Image intensifier	Good equipment: small birds at 400 m cheap equipment: shorter range	– Cheap equipment: poor + Good equipment: better + Discriminate birds, bats vs. insects nearby
Ceilometer-spotlight	< 400 m	– Poor for small targets – Insects can sometimes be confused with birds & bats
Moon watching	Observer-dependent	+ Skilled observers can identify many types of birds and discriminate birds from bats + Insect contamination rare, butterflies & moths can be identified
Radio tracking	0-2 km	Perfect
Audio microphones for birds	400 m, depends on ambient noise	+ Some nocturnal songbird species + Data include no insects
Ultrasound microphones for bats	< 30 m, depends on humidity	-? Bats may or may not emit sounds + If they do, may be species-specific

[a]+ indicates capability; – indicates a lack of capability
SOURCE: Modified from Larkin 2005a. Modified table reprinted with permission; copyright 2005, Wildlife Society.

addition, voucher specimens of killed animals should be collected and deposited in recognized museum collections for future reference.

An overview of how different equipment and approaches are being used in studies associated with proposed and existing wind-energy facili-

Passage Rates	Height Information	Cost
Good to excellent	Unmodified marine radar antenna in vertical surveillance: yes Parabolic antenna: yes	Specialized, expensive if done correctly
Good in the infrequent cases where a radar siting happens to be opportune	Very coarse with poor low altitude coverage	Data are cheap; skilled labor for analysis
Excellent when altitude of target is known	Coarse when calibrated with vertically pointing radar and then used alone	Expensive if high-quality equipment used
Yes	Same as last	Rather expensive if high-quality equipment used
Yes but light may affect flying animals	Same as last	Inexpensive but labor-intensive
2 days before and 2 days after full moon and with no cloud cover	Very crude	A good telescope of at least 20× is required; labor-intensive
Poor	Crude	High
Only some species call and quantification is assumption-ridden	Microphones: Single: no Arrays: possible	Recording equipment inexpensive, analysis expensive
No, only presence/absence; too many unknowns at present state of knowledge	Some; depends on microphones and placement	Moderate costs

ties, including both remote sensing (including passive acoustic recording, ultrasonic bat detectors, radar, moon-watching, ceilometer, reflectance infrared imaging, thermal infrared imaging, and radiotelemetry) and capture approaches are presented later in this appendix.

Estimating Abundance Using Molecular Markers

Estimates of population size, population structure, genetic diversity, and effective population size are important parameters for assessing life histories of natural populations and for managing endangered and threatened species at risk (Dinsmore and Johnson 2005; Lancia et al. 2005). Estimates of these parameters for both resident and migrating birds and bats are needed to better understand how populations are likely to respond to naturally occurring perturbations and to anthropogenic factors such as global climate change, deforestation, and habitat alteration. Wind-energy development, along with other anthropogenic factors, may have adverse effects on some animal populations by directly causing fatalities and indirectly altering critical nesting, roosting, and foraging habitats. To adequately assess whether fatalities or altered habitats are of biological significance to resident and migrating birds and bats, knowledge of baseline population levels, population structure, and genetic diversity is needed. These parameters can be expected to differ among species, which will be subject to different risks from local and regional environmental factors. For example, species represented by large populations, large genetic diversity, and little spatial breeding structure are likely to be less affected by anthropogenic factors than species represented by small populations, low genetic diversity, and strong spatial breeding structure (Avise 1992, 2004).

Rare and elusive species may be at greatest risk from anthropogenic changes (Thompson et al. 1998). An important challenge for population ecologists has been applying traditional census methods to rare and elusive species (Thompson et al. 1998). For example, for bats, few statistically defensible estimates of population size have been published—and this is especially the case for migratory tree-roosting species (O'Shea and Bogan 2003; O'Shea et al. 2003, 2004). Historically, population estimates of birds and bats have been derived using a variety of methods, including direct counts, point counts, and other estimating procedures such as capture-mark-recapture methods, photographic sampling, probability sampling, maximum likelihood models, and Bayesian methods (Bibby et al. 2000; Braun 2005; Kunz et al. in press a). Direct counts often are not practical, especially for nocturnally active bird and bat species, in part because these animals typically are small, cryptic, or otherwise difficult to census visually using most existing methods, either during daily or nightly emergences from roosts, or during migratory or foraging flights. Relatively recent approaches have been developed to use capture-mark-recapture models where some or all of the assumptions are relaxed; however, these approaches also have limitations in that a proportion of the originally marked individuals must be recaptured. More recently, capture-mark-recapture models have been used to estimate population sizes derived using non-invasive genetic sampling (Waits 2004).

For example, using this approach, Puechmaille and Petit (2007) compared estimates of colony sizes of the lesser horseshoe bat (*Rhinolophus hipposideros*) using DNA extracted from feces with independent estimates based on visual counts conducted during nightly emergence flights. Their results indicate that analysis of DNA extracted from feces can provide accurate estimates of colony size.

Large populations accumulate more genetic diversity and retain this diversity longer than do small populations. At the DNA level, these processes have predictable effects on both levels of genetic diversity and how this diversity is distributed among individuals within populations. Because these effects are predictable, it is possible to estimate long-term effective population size based solely on observed patterns of DNA diversity. If a population changes in size, predictable effects on patterns of diversity occur, and these effects are proportional to that change. Thus, significant declines in population size through time can be documented, although there is some time lag between changes in population size and observable effects on genetic diversity. A conceptual description of the "coalescent" process that results in these effects is provided below. Those interested in more detailed descriptions and applications are referred to Roman and Palumbi (2003), Avise (2004), Russell et al. (2005), and references cited therein.

The variation at any particular gene in a population can be illustrated as a topology ("gene tree") reflecting the historical relationships or genealogy of the gene copies found in different individuals. The number of mutations (i.e., nucleotide substitutions) separating these variable DNA sequences is a function of the demographic history of the population. Because mutations accumulate through time, sequences that diverged longer ago will be separated by a larger number of mutations than those that diverged more recently. If a historically large population remains large, its gene trees will have many "branches" of varying lengths that reflect the accumulation and retention of older and younger mutations. If a large population is reduced in size, its gene tree will be "pruned." That is, genes reflecting both long and short branches will be lost, with the result of less overall diversity. Short branches also will be proportionally fewer in the reduced population because fewer mutations occur, and older ones are less likely to be retained simply because of the smaller population size. Correspondingly, if a population that was historically small expands in size, its gene tree will consist mostly of short branches reflecting the increased occurrence and retention of more recent mutations.

Estimates of population size based on gene diversity have been applied to a variety of animals to investigate patterns of change caused by climatic change or human intervention. For example, the historical population sizes of humpback and fin whales prior to hunting by humans were estimated at approximately 240,000 and 360,000 whales, respectively, contrasted to modern population sizes of 10,000 and 56,000 individuals (Roman and Palumbi 2003). The historical estimate of the effective population size of grey wolves prior to human settlement of North America was approximately 5,000,000, as compared to the current estimate of 173,000 (Vilà et al. 1999). The critically endangered Morro Bay kangaroo rat apparently never had an effective population size greater than about 13,000 (Matocq and Villablanca 2001). For bats, coalescent analysis indicates an expansion of migratory populations of Brazilian free-tailed approximately 3,000 years ago, a date that corresponds with the development of a wetter climate and increased insect availability (Russell et al. 2005; Russell and McCracken 2006). This was apparently followed by an approximately 16-fold decline in estimated population size in more recent times (Russell et al. 2005; Russell and McCracken 2006), perhaps as a consequence of human activity.

For the lesser long-nosed bat (*Leptonycteris curasoae yerbabuenae*), the most recent current estimate of effective population size was 159,000 individuals (Wilkinson and Fleming 1996), although no estimate of historical effective population sizes is available for comparison. These and other estimates of effective population size reflect the current distributional range of a given species. However, data from censuses of local populations also need to be considered when evaluating impacts of anthropogenic factors. For example, current colony sizes of the Brazilian free-tailed bats, determined using thermal infrared-imaging and computer-vision technologies (approximately 400,000), are important biological units that deserve special attention (Frank et al. 2003) apart from estimates of effective population size.

Migratory tree-roosting bats are especially challenging to census, largely because they are solitary and roost in foliage (eastern red bats, western red bats, and hoary bats) or tree cavities (silver-haired bats). Instead of using traditional marking methods, molecular markers could be used to estimate population sizes after identifying individuals from the DNA obtained noninvasively from samples of feces, hair, or skin tissue. As with traditional methods, the reliability of population estimates based on molecular methods depends on certain assumptions. For example, population size can be under- or overestimated if scoring errors are made when the alleles of heterozygous individuals are not amplified during a positive polymerase chain reaction (PCR), or PCR-generated alleles create a slippage artifact during the first cycles of the reaction (Waits and Leberg 2000). Errors of this type can be corrected by repeating the process of genotyping and comparing genotypes to each other (Paetkau 2003).

It is important to understand the extent of population-level structuring, because it may differ markedly among species. For example, population-genetic studies on Brazilian free-tailed bats show high genetic diversity and little population structuring (Russell and McCracken 2006), whereas other species, such as the lesser long-nosed bat, show relatively low genetic diversity and high population structuring. The implications of these and other studies using molecular markers (Avise 1992, 2004) are that different species are subject to different risks from anthropogenic influences, and should be studied to assess whether a given species is more or less at risk from changing environments. Sex ratios, effective population size, and genetic diversity are intimately related. Changes in sex ratios in populations cause changes in effective population size, and when effective population size decreases, populations tend to lose genetic diversity.

Researchers charged with collecting samples of dead and moribund bats at wind-energy facilities can provide valuable data for advancing knowledge about local and migratory populations by recording the date, location, species, sex, age, reproductive condition, and standard external measurements for each individual recovered.

Collecting hair samples from bats and feathers from birds also is useful for analysis of the geographic origin of migrants and residents based on stable-isotope analysis. Ideally, data for stable-isotope analyses should be collected for all species found at each location. When necessary, representative specimens—and especially unidentifiable carcasses—should be collected in their entirety and deposited as voucher specimens with active scientists associated with natural-history museums. Data derived from feathers of birds and hair and wing biopsies from bats killed by wind turbines also offers the potential for identifying closely related or cryptic species (e.g., *Myotis* species). Collaborations with researchers affiliated with natural-history museums and other research laboratories equipped for genetic and stable isotope analysis are important. In the United States, the American Museum of Natural History, New York, serves as a repository for all carcasses and tissues collected from dead bats and birds collected from beneath wind turbines, and the Conservation Genetics Research Center at the University of California at Los Angeles serves as a repository for feather samples for genetic analysis.

Types of Studies: Strengths and Limitations of Different Approaches

Pre-siting Studies

Wind-energy developers spend much time and effort evaluating potential sites prior to investing in developing a particular site (macro-siting). Once a site is selected for development, evaluations are made in an effort to plan how best to develop the site (micro-siting) to optimize electricity production (Anderson et al. 2002). Macro- and micro-siting decisions are extremely important in minimizing the potential impacts of wind-energy facilities on wildlife and other natural resources.

Pre-siting studies will provide more information if they evaluate likely impacts relative to other potentially developable sites, as well as evaluating impacts from an absolute perspective. Studies to address this question are usually short-term and do not qualify as either monitoring or research. They may vary from relatively simple reconnaissance surveys for species and habitat presence or absence to more sophisticated baseline studies and impact and risk assessments.

The elements of a reconnaissance survey for the purpose of comparing sites should include determination of the wildlife species known to use the area based on existing data and literature, the possible presence of species of concern (e.g., federal and state protected species), the presence of habitat that potentially supports species of concern, unique habitat features (e.g., old-growth forest, raptor-nesting sites), and wildlife concerns important to state and federal management agencies. A survey of existing information on wildlife in the area being considered for possible development, one or more seasonally appropriate site visits to examine habitat characteristics for potential occurrence of wildlife species of interest, and visits with knowledgeable agency professionals and local experts all provide valuable sources of information. Beyond the simple ranking of relative importance of each area to wildlife, pre-siting evaluation also should consider the potential for impacts to occur if a wind-energy facility is constructed on a particular site, and possible cumulative impacts, placed in the context of other sites being developed or proposed. The U.S. Fish and Wildlife Service (USFWS 2003) published Interim Voluntary Guidelines that recommend the development of a Potential Impact Index (PII). Although these guidelines still are under review, they describe the PII as a two-step process:

- "Identify and evaluate reference sites within the general geographic area of Wind Resource Areas (WRA) being considered for development of a facility. Reference sites are areas where wind development would result in the maximum negative impact on wildlife, resulting in a high PII score. Reference sites are used to determine the comparative risks of developing other potential sites.
- Evaluate potential development sites to determine risk to wildlife, and rank sites against each other using the highest-ranking reference site as a standard. While high-ranking sites are generally less desirable for wind development, a high rank does not necessarily preclude development of a site, nor does a low rank automatically eliminate the need to conduct pre-development assessments of wildlife use and impact potential."

The reference-area concept described for the PII emphasizes the value of a highly diverse site, such as a wetland or a woodland complex within a grassland community, or a mosaic of grasslands and forests, rather than comparing similar areas. This approach places a relatively high value on species diversity and does not consider whether the species present are at high risk from impact. For example, the approach increases the possibility that areas with a single important species, such as a grassland area with relatively low species diversity but important habitat for a species at risk or of special concern, might actually appear to be a good site for wind-energy development when compared to an area with higher species diversity. Furthermore, the definition of the reference area by the person developing the PII score is highly subjective.

An alternative paradigm for selecting reference areas is to identify those that are similar to the one being proposed for development. If the objective is to predict potential impacts, the reference area or areas should be in similar habitats with comparable wildlife communities where wind-energy facilities already exist. If the objective is to combine pre-development assessments with post-development surveys to estimate possible project impact, then a reference area without a wind-energy facility and similar to the area proposed for development should be chosen for comparison.

Potential impacts resulting from perturbations are often evaluated using a general framework called "Ecological Risk Assessment" (ERA), defined by the U.S. Environmental Protection Agency (EPA 1992) as a "process that evaluates the likelihood that adverse ecological effects may occur or are occurring as a result of exposure to one or more stressors." The primary difference between impact prediction and an ERA is in the estimation of some likelihood (or probability) of an impact occurring in an ERA, rather than an estimation of the actual impact.

The most difficult aspect of either impact or risk assessment is determination of exposure, i.e., an estimate of the number of individuals that are exposed to collisions with turbines. Young et al. (2004) estimated the number of potential bird fatalities that would occur at the proposed Mount Storm Wind Power Project in Grant County, West Virginia. They calculated the potential fatalities using estimates of nocturnal bird-passage rates obtained from X-band marine radar surveys and the dimensions of the proposed wind-energy facility at the Mount Storm site, and estimates of bird fatalities at the nearby Mountaineer wind-energy facility (Kerns and Kerlinger 2004). Rather than estimate the number of fatalities at a site, the ERA approach estimates the probability that an individual bird or bat would be killed. The appeal of the ERA paradigm is that it provides a structure for focusing scientific principles and critical thinking toward the goal of effective environmental management, and integrating the views of diverse stakeholders (EPA 1992). ERA is used by a variety of regulatory agencies, scientists, and industries for environmental decisions (e.g., NRC 1994, 2004; Suter et al. 2000; Efroymson and Suter 2001).

Collecting sufficient data to estimate exposure to wind turbines is problematic for nocturnal migrating passerines and resident and migrating bats. Tucker (1996) and Podolsky (2003) modeled risk of bird collisions with a wind turbine, based on the characteristics of the turbines (e.g., rotor rpm) and the birds (e.g., flight speeds). However, these models do not incorporate behavior (e.g., avoidance or attraction), an important factor in risk assessment, and are of questionable value in estimating actual fatality rates (Chamberlain et al. 2006). More-sophisticated risk models that include turbine and wind-energy-facility characteristics, the environment (e.g., wind, weather), and some surrogate for bird and bat behavior (e.g., flight heights, species presence or absence) are necessary. As with all models, theoretical estimates must be compared to more-deterministic models that are based on empirical data on bird and bat use and fatality rates.

Pre-construction Studies

Pre-construction studies might evaluate a proposed site or sites for potential impacts of developing wind energy, evaluate a selected site to determine the least environmentally damaging development plan, and predict impacts or risk associated with the development of a particular site for wind energy. Both impact and risk assessments are possible with empirical data on exposure and impact, if certain assumptions are made. In addition, either assessment can be used in a comparison of two or more sites. Both approaches for characterizing a site would be improved with additional empirical data on exposure (i.e., abundance of animals at risk of collision) and response (e.g., fatalities, injuries, displacement) from similar sites. Frequently, in impact- and risk-assessment studies, an index of abundance is used, rather than an estimate of absolute abundance. When indices of abundance are used to compare multiple sites, it is essential that the indices be estimated using similar methods and metrics across sites.

Siting a wind-energy facility and individual turbines within a wind-resource area to minimize impacts to wildlife requires knowledge of species presence, relative abundance, behavior, and habitat. Anderson et al. (1999) suggested that pre-permitting studies that result in the collection of empirical data are useful in the following situations:

- A site for a wind-energy facility is selected but the distribution of turbines and turbine strings has not been determined and turbine siting could be influenced by information on potential risk to bird species.
- The decision to construct a wind-energy facility has been made, but development will proceed in phases based on assessment of impacts of construction and operation of the initial phase (i.e., adaptive management).
- Other studies or credible information on bird use and habitat suggest that impacts are likely.

The importance of micro-siting studies is important for a variety of species, and is illustrated by several bird studies at existing wind-energy facilities. Orloff and Flannery (1992) concluded that raptor fatalities at the APWRA were higher for turbine strings near canyons and at turbines that were at the ends of rows. Smallwood and Thelander (2004, 2005) concluded that fatalities were related to turbine-site characteristics and the position of turbines within a turbine string. The implication of both studies is that turbine-siting decisions during the construction of a wind-energy facility could be important. Pre-construction studies identified areas of high raptor use at the Foote Creek Rim site in Wyoming, a flat-topped mesa with a very distinct rim edge. Approximately 85% of the estimated use of this site by raptors occurred within 50 m of the edge of the rim. These high-use areas were avoided by the wind-energy developer when turbines were sited. Anecdotally, the BLM (1995) considered the abundance of golden eagles at the Foote Creek Rim area prior to construction to be similar to that at the APWRA in California. Based on the assumption of similar densities, the BLM predicted fatality rates for the Foote Creek area similar to the AP-WRA, or approximately two golden eagle fatalities per year (BLM 1995). However, over a three-year period, 133 turbines were searched for fatalities, for a total of 202 turbine search years, resulting in the finding of only one dead golden eagle (Young et al. 2003c). Micro-siting of turbines may partially explain why fatalities of golden eagles were lower than predicted at Foote Creek Rim.

Species presence, relative abundance, behavior, and habitat use are determined through sample surveys (Kempthorne 1966), also referred to as observational studies (NRC 1985). The objective of these studies at proposed wind-energy facilities is usually an estimate of parameters necessary to describe the group of animals occurring on the proposed site, such as density and habitat use.

Because these studies are restricted to a single site, their strict statistical inference is restricted to the study site and the protocol used for collecting data, although these studies can provide some valuable insights applicable to other areas. Nonetheless, design principles for best practices, such as randomization of sample-collection locations, replication of sampling, and the use of measures to control or reduce experimental errors are essential to ensure rigorous results (Cox 1958; Cochran 1977; Anderson et al. 1999; Manly 2001). For these types of studies, precision can be increased by refinement of the experimental techniques, including greater sampling precision within experimental units; and improved experimental design, including stratification and measurements of non-treatment factors (covariates) that can potentially influence the outcome of the survey (Cox 1958; Cochran 1977).

Post-construction Studies

Post-construction studies should focus on determination of impacts and evaluation of actual risk versus predicted risk, evaluation of causal mechanisms of impact, evaluation of mitigation and reclamation measures, and evaluation of the ecological or biological significance of the impacts. A relatively small number of post-construction studies of wind-energy facilities have been conducted. With a few exceptions, post-construction studies at new wind-energy facilities have estimated use and fatalities of birds and bats, or fatalities alone (Erickson et al. 2001, 2003a,b, 2004; Howe et al. 2002; Johnson et al. 2002, 2003a,b; Nicholson 2003; Young et al. 2003a, 2005; Kerns and Kerlinger 2004; Koford et al. 2004; Arnett 2005; Arnett et al. in press; Paul Kerlinger, Curry and Kerlinger, LLC, personal communication 2002). Studies estimating raptor use and bird fatalities have also occurred at older wind-energy facilities in California (McCrary et al. 1986; Orloff and Flannery 1992; Howell 1997; Thelander and Rugge 2000; Anderson et al. 2004, 2005; Smallwood and Thelander 2004), primarily focused on raptors. Hunt (2002) conducted a demographic study of the golden eagle population at the APWRA and evaluated its viability in the face of fatalities, primarily from collisions with wind turbines. Responses to wind-energy facilities also have been evaluated for elk (Walter et al. 2004), pronghorns (Johnson et al. 2000a), and grassland birds (Leddy et al. 1999; Johnson et al. 2000b; Erickson et al. 2004).

The approach to estimating fatalities depends on the study objectives (Larkin et al. unpublished material 2007[1]). Fatalities may be estimated so that comparisons can be made to other facilities. For this objective, similar protocols would be important. Alternatively, it may be important to determine the circumstances associated with each fatality (e.g., weather conditions) and thus more-frequent searches would be required (Larkin et al. unpublished material 2007). Frequently, the objective of the study is to estimate the absolute number of fatalities with acceptable precision. In such cases, the protocol should minimize measurement error. The basic components of a fatality-monitoring study are carcass searches of study plots, trials to estimate how effectively searchers detect carcasses, and trials to estimate how quickly scavengers remove carcasses. Potential biases associated with fatality-monitoring studies include use of inappropriate surrogates for bat and bird carcasses, inadequate effort in terms of search interval and search intensity, poorly sized search plots, lack of spatial and temporal replication, and lack of an estimate of background fatalities.

---

[1]Larkin, R.P., B. Cooper, W.P. Erickson, J.P. Hayes, J. Horn, G. Jones, T.H. Kunz, D.S. Reynolds, and M.D. Strickland. Unpublished material. 2007. Ecological impacts of the wind power industry on bats; methods and research protocols.

The estimation procedure for fatalities is important and one approach is described later in this appendix.

Post-construction surveys at wind-energy facilities in the United States have provided relatively little information on the extent of bat fatalities. To date, only 6 of 13 fatality-monitoring efforts have explicitly included bats in their protocols. These efforts vary significantly in landscape conditions, sampling intervals and periods, search methods, and sampling protocols. Potential sources of bias include sample frequency, removal of carcasses by scavengers, and search efficiencies. Sampling intervals in these studies have been relatively infrequent (e.g., 7-30 days), and even fewer studies have been designed to assess expected bias resulting from the removal of carcasses by scavengers. In fact, only five surveys used bat carcasses to correct for observer bias (Arnett 2005). Most studies designed to estimate bat fatalities used small bird carcasses as surrogates of bats to assess searcher efficiency and scavenger removal, largely because dead birds were available (Erickson et al. 2002). In one study, frozen bird carcasses were removed at significantly lower rates than frozen and fresh bat carcasses (Kerns and Kerlinger 2004). Other surveys only covered part of the expected autumn migration period and failed to include the period of summer residency (Kunz et al. 2007). None of the studies have successfully estimated bat abundance at the wind-energy facilities. As a result, estimating risk to individual bats is not possible at present.

To fully interpret bird and bat fatalities it is essential that the number of individuals exposed to collision with turbines be known. To understand the implications of the fatalities it also is important to relate the fatalities to the demographics of the affected populations.

Evaluation of Causal Mechanisms of Impact

Studies to elucidate the causal mechanisms of impact are typically conducted in an effort to identify possible mitigation measures (Kunz et al. 2007). Impact-reduction studies can vary from relatively simple observational studies, such as bat observations using thermal infrared imaging (Horn and Arnett 2005; Horn et al. in press) and bird- and bat-fatality studies (Arnett 2005; Johnson 2005; Arnett et al. in press), to more-complex experiments, such as the work by Hodos (2003), who evaluated bird visual acuity related to potential color schemes for turbine blades.

When studies are designed properly, the ultimate determination of statistical power is sample size. Replication to increase precision can be expensive. Manly (2001) and Morrison et al. (2001) provide a good general discussion of design and analysis when conducting observational and quasi-experimental studies. The study designs discussed above for impact assessment are also the preferred designs for quasi-experiments. Anderson et al.

(1999) suggested modifying the BACI design by applying the treatment and control in the first year to the selected subset of turbines and switching the treatment and control turbines the second year, sometimes referred to as a crossover experiment.

Most statistical texts provide a description of how to design and analyze classical experiments (e.g., Underwood 1997). Krebs (1989) notes that "every manipulative ecological field experiment must have a contemporaneous control..., randomize where possible..., and, because of the need for replication, utilize at least two controls and two experimental areas or units." While classical experiments have limited statistical inference and are practically impossible when studying wind-energy developments, several quasi-experiments have been conducted at wind-energy facilities (e.g., Leddy et al. 1999; Johnson et al. 2000b; Young et al. 2003b). While quasi-experiments improve the confidence in the causal mechanism, the determination of the causal mechanism still requires professional judgment, illustrating the need for a weight-of-evidence approach when investigating the impacts of wind-energy development.

Spatial Component of Impacts

Assessments of impacts of wind-energy facilities typically focus on bird and bat fatalities. Direct habitat loss often is considered relatively minor for wind-energy facilities and is restricted to roads, turbine pads, and construction areas. Habitat loss also is relatively easy to measure using aerial photography, satellite imagery, and GIS. Until recently, impacts on bird habitat were considered the primary impact of wind-energy facilities in Europe (Winkelman 1985, 1990, 1992a,b, 1995). However, impacts on bats also are being documented (UNEP/EUROBATS 2006). Displacement is the greatest concern for most wildlife species and is very difficult to quantify. The cumulative impacts of habitat loss and displacement caused by turbines, roads, and other construction in an area can potentially lead to landscape fragmentation and loss of suitable habitat for wildlife.

Habitat use may be measured by direct observation (e.g., Young et al. 2003b) or with radiotelemetry (Hunt 2002). Habitat use is most meaningful when it is considered in relation to habitat availability. Manly et al. (1993) provided a unified statistical theory for the analysis of use versus availability (resource-selection statistics). The theory and application of resource-selection studies were updated by Manly et al. (2002). Sawyer et al. (2006) used resource-selection study design and analysis to estimate the displacement effect of a gas-field development on mule deer in western Wyoming. This method has considerable potential for addressing concerns regarding displacement and habitat fragmentation associated with wind-energy development (e.g., prairie grouse and large mammals, including black bear).

Impacts and Actual Risk Versus Predicted Risk

Risk is the likelihood (probability) that "adverse ecological effects may occur or are occurring as a result of exposure to one or more stressors" (EPA 1992; Weed 2007). A simple model of risk requires the following (Kaplan and Garrick 1981):

- An existing or planned action leading to the potential of an adverse environmental outcome;
- A qualitative or quantitative statement about the probability of the adverse outcome occurring; and
- A statement about the consequences or advisability of the action.

Pre-siting studies of wind-energy facilities typically incorporate (1) and (3) to estimate potential impacts, but not in the form of a probability statement as in (2) above. Recently, a more-formal risk assessment was attempted at the Chautauqua Wind Power site in Chautauqua County, New York (Chautauqua Windpower, LLC et al. 2004). Impact and risk predictions should be evaluated to allow improved decision making, reduction of adverse ecological effects at existing facilities, and evaluation of the effectiveness of mitigation measures. Predictions of impact and risk require both an estimate of exposure, or the number of organisms that have the potential to suffer impacts, and an estimate of the adverse consequences. In the case of wind-energy production, those adverse consequences for wildlife include mortality, displacement, and habitat loss, and the resulting effect on survival and reproduction. Use of DNA markers for estimating demographic parameters and effective population size of birds and bats promises to provide past and current abundance at local, regional, and continental scales.

Post-construction studies designed to detect impacts and to evaluate pre-project predictions of risk can generally be considered impact-assessment studies as described by Manly (2001). The studies typically are not true experiments, but instead are observational or "mensurative" studies designed to make sure that the data are properly collected to address research questions and hypotheses and to make them amenable to statistical analyses (Anderson et al. 1999; NRC 1994). Additionally, mathematical and statistical models can be important in assessing the significance of estimated impacts. It is common for design/data-based and model-based studies to be conducted in tandem, resulting in inferences based on a number of interrelated arguments. Hunt (2002) illustrates this approach where radiotelemetry data on golden eagle abundance and survival were used to construct a demographic model of the group of birds using the APWRA in an effort to evaluate the significance of estimated eagle fatalities.

Mensurative studies have limited statistical inference because they are not true experiments, and thus must include randomization, replication, and controls (Manly 2001). Most mensurative studies of wildlife lack one or more of these conditions and are referred to as quasi-experiments (Manly 2001). Nonetheless, mensurative studies are essentially the only approach available for impact assessment.

Wind-energy facilities are not scattered randomly over the landscape. They are not even constructed at random within all of the known windy locations. Thus, extrapolation of study results from one site to another is strictly subjective (Gilbert 1987), although confidence can improve in these subjective extrapolations if the studies are conducted using similar methods and metrics in areas with similar ecological conditions.

Social and medical sciences often use meta-analysis (Hedges and Olkin 1985; Hedges 1986) as a statistical approach for analyzing results from several independent studies that are all concerned with the same issue. The purpose of meta-analysis is to provide researchers with a statistical tool to summarize, synthesize, and evaluate independent research studies in order to reach general conclusions (Adams et al. 1997). The troublesome aspect of meta-analysis is that combining different studies requires assumptions about a variety of potentially important issues such as publication bias (the tendency of journals to favor publication of studies with statistically significant results), non-independence among studies, and the quality of studies (Adams et al. 1997). Given the necessary assumptions, a conservative approach to the use of meta-analysis in ecological studies may be prudent. Adams et al. (1997) suggested a non-parametric approach using re-sampling methods (Manly 1991) when combining individual studies that violate necessary assumptions of standard parametric statistics.

Mensurative studies have limitations even when the desire to extrapolate is limited to the specific area of study. For example, a study may indicate that bird fatalities are much higher in one part of a wind-energy facility and the assumption may be that there are specific conditions at that site that may contribute to the difference. Similarly, bird abundance may be declining in the area surrounding a wind-energy facility when compared to a reference area, presumably because the facility is there. However, conclusions on causation are based on assumptions and judgment (Manly 2001).

METHODS AND METRICS FOR BIRD AND BAT STUDIES

This section provides information on the methods for assessing both direct and indirect impacts of wind-energy facilities on bats and birds. The general nature of impacts on birds and bats is similar to that discussed for wildlife earlier in this appendix.

The methods discussed below include observational, remote-sensing (including passive acoustic recording of bird calls, ultrasonic detection of bat calls, radar-imaging, moon-watching, ceilometry, night-vision observations, reflectance-infrared imaging, thermal-infrared imaging, radio-telemetry), and capture protocols.

Comprehensive accounts of every North American bird species are available through the Birds of North America project (Cornell Laboratory of Ornithology 2005). This type of information is also often available through regional field guides for birds and bats that usually provide overviews of what species are known to occur and when they can be expected to occur in an area. Site-specific information can sometimes be obtained from local naturalists, scientists, or published reports, but is rarely at high enough resolution to inform estimations of potential impacts at a specific site. By far the greatest factor of uncertainty, and thus the present limiting factor in any estimate of potential interaction between birds and bats and wind-energy facilities, is natural variation in animal behavior, such as migratory movements, sound production, foraging habits, and mating behavior. Although one could conclude that such variation is so great and so complicated as to make it impossible to enumerate precise numbers of any species at any time, one must be careful not to confuse different spatial and temporal scales of analysis. There are several available methods, both traditional and emerging, for detecting, recognizing, and estimating numbers of birds and bats at given localities. When such methods are integrated with a well-designed sampling strategy, they can allow effective evaluation of the presence and numbers of species present or moving through a particular area.

Methods for Detection, Identification, and Estimating Activity of Flying Animals at Night

Methods of detecting nocturnal flying animals are discussed in Chapter 3. What follows is an overview of the traditional and emerging methods and metrics for detecting, recognizing birds and bats, with particular attention to how they are being used in studies associated with planned and existing wind-energy facilities.

Passive Acoustic Detection of Birds and Bats

Birds, bats, and insects produce sounds for communicating, and a wide variety of bats produce high-pitched sounds in the ultrasonic frequency band (above the range of human hearing) for navigating and finding food.

Most of the sounds, especially from birds, are audible, and there is a long history of detecting and recognizing birds by listening for their species-specific calls and songs. Today, as a result of the intense and dedicated efforts of many amateurs and scientists, the songs of all North American birds and a significant number of their calls are documented and available (Old Bird, Inc. 2005; Macaulay Library 2006), and from Evans and O'Brien (2002). A subset of calls of particular importance in this report is produced by birds at night during migration. These calls are referred to as nocturnal flight calls. Discovery of bat echolocation signals emerged in the early 1960s with the advent of ultrasonic-detection devices (Griffin 2004). Validated libraries of calls for most North American bat species are available online (SonoBat 2005; Batcalls.org 2006).

Passive-acoustic detection can determine presence, not absence, of a species. Once a call is produced and available for detection, there is a secondary level of uncertainty related to the probability of detecting the call once it has been produced. Thus, the overall probability of acoustically detecting a night migrant is the product of the probability of the bird's or bat's producing a call and the probability of detecting that call once it has been produced. By far the greatest source of uncertainty is the calling behavior of individual animals. The rate of calling by birds varies considerably throughout the night and is influenced by a variety of factors such as topography, weather condition, time of night, and time of year. The specific relationships between all the factors and flight-calling behavior are not well understood.

The probability of detecting a call once it is produced depends on its strength, the conditions for sound transmission between the bird or bat and a receiving system (e.g., microphone or microphone array), the bird's distance from the receiving system, and the detection capabilities of the receiving system. There is little information on strengths of sound produced by nocturnal migrants, although there are several valuable models for predicting sound transmission through the atmosphere (typically stated as transmission loss) (Larom et al. 1997; Larom 2002). Detection capabilities vary tremendously, depending on the use of baffles or horns, the quality of the microphone electronics, the application of multiple microphones, and the sophistication of the software analysis. Most passive-acoustic applications have been relatively simple and have not taken advantage of detection gains that can be achieved using microphone arrays combined with advanced signal-processing methods. There is a cost-benefit tradeoff between using a relatively simple method (e.g., for birds a single FET [field-effect transistor] microphone at the base of a flowerpot baffle) that can be installed at modest cost (less than $30 per site) and a more elaborate method (e.g., an array of 16 microphones coupled with mechanical horns) that can cost an order of magnitude more; but the elaborate method has a 30 dB greater gain than the simple method and covers ten times the volume of sky.

Simple methods can provide a determination of species presence for known call types (or species-group presence for call types associated with a species group, for example thrushes), the time of occurrence of each call, the number of occurrences (counts), and estimates of passage rates (in calls per unit time). More complex methods offer additional benefits. For example, with a sparsely distributed array of at least four simple microphones, one can locate the position of a calling bird in 3D space (Fristrup and Dhondt 2001). D.K. Mellinger (Oregon State University, unpublished material) analyzed data on nocturnal flight calls and applied a Doppler algorithm to compute the velocity and direction of nocturnal migrants and to better estimate the number of calls per bird per unit time. The more microphones used in an array and the greater their spatial distribution, the greater the spatial coverage and altitudinal resolution. Estimates of location resolution (range and bearing error) as a function of microphone number and spacing are fairly straightforward. Thus, a metric relating system cost to level of analysis, as measured by spatial coverage and resolution, could be developed.

Recording and Analyzing Audible Bird Calls

More than 200 passerine birds are known to produce audible calls in flight during night migration (Ball 1952; Graber and Cochran 1959). These flight calls often can be used for identification of bird species, and most species produce them in the early morning hours after descending to the ground after migrating throughout the night (Evans and O'Brien 2002). By careful observations and recordings, experts also can visually identify different species while recording their flight calls. The calls of approximately 150 species in the United States have now been validated to either species (e.g., warblers) or a species complex (e.g., thrushes) (Evans 1994; Evans and Mellinger 1999; Evans and Rosenberg 1999; Evans and O'Brien 2002; Farnsworth 2005; Farnsworth and Lovette 2005). With the availability of these acoustic "type specimens," the flight calls of most species of passerine birds can now be identified acoustically. For nocturnally migrating birds, the best available evidence suggests that the number of calls detected per unit time is highly variable and not a good predictor of passage rate (Howe et al. 2002; Farnsworth et al. 2004). Thus, acoustic monitoring is an excellent method for verifying the presence of birds that primarily migrate at night and produce species-specific flight calls, but not for estimating the relative numbers of species, individuals, or passage rates. It will require further research that carefully integrates and compares passive acoustics with other methods (e.g., radar, visual survey) to adequately define the benefits and limitations of acoustic monitoring.

The functions of flight calls are not well understood, but they usually are assumed to serve to maintain flock cohesion (Hamilton 1962) or maintain spacing (Graber 1968). Farnsworth and Lovette (2005) analyzed song and flight-call characteristics of 33 species of wood-warblers to test the hypotheses that the acoustic characteristics produced by these birds were predictable from body size or bill length, but found no statistically significant relationship. They concluded that body size and bill length have not been important factors in the evolution of flight calls, and suggest that different ecological and atmospheric properties, such as sound transmission, might be more important in the selection of these calls. If flight calls have been selected to be optimized for maximum communication range, given the physiological constraints of flight, this should influence the detection range of flight calls and thus be an important factor when designing an acoustic monitoring project. At present, there have been no thorough studies quantifying the strengths and transmission characteristics of flight calls.

Since the mid-1990s, a few studies have used various technologies to detect the occurrence of flight calls, identify migrating species, and record the number of calls per unit time (Evans and Mellinger 1999). Some of these studies have been associated with the development or monitoring of wind-energy facilities. Overall, the type of equipment used to record flight calls of birds and in the sampling strategies used has been highly variable. This lack of consistency makes it difficult to compare, evaluate, and combine the results of various studies to ascertain the effectiveness and efficiency of passive-acoustic monitoring of birds.

Recording and Analyzing Ultrasonic Bat Calls

Ultrasonic-recording devices were first developed to evaluate the structure of echolocation calls of bats that navigate and feed on insects in the laboratory (Griffin 1958; Griffin et al. 1960). These devices first became available on a limited basis for field use in the 1960s (Griffin 2004), but they were not really suitable for field studies. Griffin (1958) provided an early summary of what had been learned from the earliest recordings of echolocation calls in the laboratory (see also Griffin 2004). Enormous strides have been made in the development and use of ultrasonic detectors for recording echolocation calls of bats in both the laboratory and field (Barclay and Brigham 2004; Thomas et al. 2004), and the development of specialized software for analyzing these calls (Britzke 2004; Corben 2004; Jones et al. 2004; Limpens and McCracken 2004; Parsons and Obrist 2004; Swezicak 2004).

From the mid-1970s until the present, several types of bat detector have become available (Ahlén 2004; Limpens and McCracken 2004; Pettersson 2004). In the 1980s the development of several commercial products made it possible for these devices to be deployed for assessing general activity levels, monitoring activity of bats in different habitats, and for educating the public (Fenton 2000, 2004). In the 1990s advances in circuit design, the sensitivity of microphones to ultrasound, and storage and analytical software made it possible to record and store the full range of call structures in the field, and in many situations identify bats to species (Corben 2004; Limpens and McCracken 2004; Pettersson 2004). For some species it may even be possible to discriminate sex and age (Siemers et al. 2005).

Types of Ultrasonic Bat Detectors

The ultrasonic calls of bats are transmitted through the air, captured by a special microphone in a bat detector, and transmitted as an audible sound or as a voltage signal to a built-in speaker, tape recorder, or computer. The most common types of bat detectors are known as heterodyning, frequency division, and time expansion—each based on unique circuitry and types of microphone. Each type of detector has its own advantages and disadvantages. Commercially available detectors may have one or sometimes two or three separate conversion systems (Pettersson 2004).

Heterodyning detectors are designed to detect a narrow range of frequencies, typically a bandwidth of about 10 kHz. A tuning control on the detector is used to center the frequency range for transformation, so that if the detector is set at 40 kHz, it will record frequencies between 35 and 45 kHz. Heterodyne detectors are highly sensitive to bat echolocation calls, making it possible to detect and record relatively weak signals. Echolocation calls of bats that are transformed in a heterodyne detector may sound or appear different depending on the frequency to which the detector is tuned. The narrow bandwidth of a heterodyne detector means that only bats producing sounds within that frequency range will be detected.

Frequency-division detectors transform the entire ultrasonic range of signals (e.g., broadband) to a fixed fraction of the original signal. For example, a given ultrasonic frequency may be transformed or reduced to some fixed fraction of the original frequency. Frequency-division detectors are less sensitive than heterodyne detectors, making it difficult to detect weak signals. However, the transformed signal of a frequency-division detector contains more information than a heterodyne detector. Nonetheless, with frequency-division detectors, the signal is first converted into a square wave, and thus amplitude information is lost. Moreover, if the original call is composed of a fundamental frequency and one or more harmonics, typically only the fundamental frequency will be transformed. Some of the more advanced frequency-division detectors also process amplitude information based on the fundamental frequency of the original signal. If the amplitude is retained, then pulse duration and other temporal parameters can be measured (Ahlén 2004; Pettersson 2004).

A time-expansion detector records the original high-frequency signals and then plays back the original signal at a slower rate, while retaining virtually all of the original characteristics. Thus, the resulting signal is longer and the frequency will be lower than the original signal. With time-expansion detectors, the signal is digitized and a portion of it is stored in digital memory, from which it can be replayed at one-tenth the original speed. Thus, an important advantage of time-expansion signals is that the original signal is stretched out, making it possible to detect details of the call. Most time-expansion detectors also include a built-in heterodyne system (Ahlén 2004; Pettersson 2004).

One of the disadvantages of frequency-division detectors is their lack of sensitivity. By contrast, an important advantage of frequency-division detectors is their inherent stability, which makes them more robust for long-term field use than either heterodyne or time-expansion detectors. In general, the circuitry of heterodyne and time-expansion detectors is more complex, including critical timing elements that can affect the accuracy of the signal detected and recorded. Frequency-division detectors provide the most robust technology and are the most cost-effective for monitoring purposes and for characterizing bat echolocation calls in field conditions. One of the advantages of the time-expansion circuitry is that sounds can be recorded at high speed and played back at slower speeds for detailed analysis. Another advantage of time-expansion detectors is that the storage medium is digital and the recordings can be played back immediately. An obvious disadvantage of time-expansion method is that sounds are not recorded in real time. The greatest advantage of time-expansion systems is that the signals are stored in memory and can be played back for more in-depth analyses, including power spectra, spectrograms, pulse length, interpulse interval, etc. (Pettersson 2004).

A Case Study of Bats

In a recent investigation conducted at the proposed Maple Ridge Wind Project in upstate New York, Reynolds (2006) deployed a spatial array of frequency-division ultrasonic detectors (Anabat Model version 6.2) connected to a CF-SCAIM data-storage unit (Titley Electronics), each housed within a waterproof storage box with a 12 V deep-cycle battery. These detectors were used to sample 35 different sites each for a single night in the period from June 23 through July 5, 2004. Each detector was mounted on a 1.5-m pole, with the microphone pointing toward the ground to prevent the condensation of moisture on the microphone. Echolocation calls produced by the bats were reflected toward the microphone using a 10×10 cm Lexan plate positioned at 45 degrees from the horizontal (Reynolds 2006). The Anabat microphones that were used in this study have the potential to detect approaching bats at a distance up to 11.6 m, with a potential sampling cone of 2,542 m^3 (Larson and Hayes 2000). Actual field tests reported by Reynolds (2006), however, revealed that the microphones that he used could consistently detect ultrasonic signals up to 22 m distant. Nonetheless, this limited detection range makes it difficult to record calls made by bats in the rotor-swept area of industrial-scale wind turbines using single microphones. The application of beam-forming arrays of microphones would significantly increase detection range and provide monitoring within the swept area of turbines.

Spatial data were statistically analyzed by comparing the number of bat passes recorded during three four-hour periods each night. The results, although limited spatially and temporally, show that bats were most frequently recorded in the vicinity of ponds, with different temporal patterns of activity exhibited by the four "species" detected. *Myotis* species (most likely *M. lucifugus* and *M. septentrionalis*), which accounted for 95.7% of all recorded calls, were active throughout the night. By contrast, big brown bat and silver-haired bat calls (*Eptesicus/Lasionycteris*), which could not be distinguished, were recorded most frequently in the early hours of the night. Only 1% of the activity was represented by tree-roosting eastern red and hoary bats (*Lasiurus borealis* and *L. cinereus*).

To further assess possible migratory behavior, Reynolds (2006) used the same bat detectors for spatial analysis from April 10 to June 22, 2004, at two locations at the New York project site. Frequency-division bat detectors were deployed as two vertical arrays on meteorological towers at each of two sites, with one detector at each site positioned at "ground" level (ca. 7 m) and the others positioned at approximately 15 and 50 m above the ground. Analysis of data recorded from the vertical array of ultrasonic detectors revealed that significantly more activity was recorded at 50 m above the ground than at 15 and 7 m (Reynolds 2006). Moreover, the results support the hypothesis that migratory behavior of bats is episodic.

More calls from *Eptesicus/Lasionycteris* were recorded in middle to late spring, although there was no evident seasonal pattern in the calls recorded from *Myotis* during this limited period. From these data, Reynolds (2006) concluded that migratory activity was the highest on the warmest days with low wind speeds. The results are consistent with reports of fatality data at the Mountaineer Wind-Energy Center, Tucker County, West Virginia and Meyersdale Wind-Energy Center, Pennsylvania, where the highest fatalities occurred on warm nights with low wind speed (Arnett 2005). Data from Reynolds (2006) suggest that accurate estimates of migratory behavior will require long-term acoustic monitoring, using both horizontal and vertical arrays of ultrasonic bat detectors. The few recordings made of eastern red and hoary bats (species that have experienced the highest fatality rates at these sites) suggested to Reynolds (2006) that these bats were not common in the study region or, at least that they could not be detected from the deployment heights of the bat detectors. The absence of a call does not imply absence of bats, only that the call may not have been detected. These observations are consistent with Gruver's (2002) study in Wyoming, which reported that hoary bats experienced the highest fatality rate, yet were seldom detected by bat detectors. To adequately assess risks to both resident and migratory bats, ultrasonic bat detectors should be deployed to monitor flight activity within the rotor-swept area of a wind turbine and with the capacity to detect bats flying from different directions. Clearly, additional studies are needed to better detect migratory tree bats, as they seldom appear in acoustic data but are consistently present in fatality samples. In fact, they account for 68% of all bat fatalities recorded at North American wind-energy facilities to date.

Recently, D.S. Reynolds of North East Ecological Services (personal communication 2006) developed a design that consists of a cluster of four ultrasonic detectors attached to meteorological towers with each detector aligned in each of the four cardinal directions (Figure C-1). With this approach, it may be possible to establish the dominant direction of migratory bats that fly within the range of the detectors. More important, few studies to date have used ultrasonic detectors to reliably detect bats flying within the rotor-swept area. Thus, even if bats are active at the height of the rotor-swept area, existing technology cannot detect bats flying in this zone. Moreover, with the increasing size of turbines being developed and installed, deployment of ultrasonic detectors on the turbines cannot provide

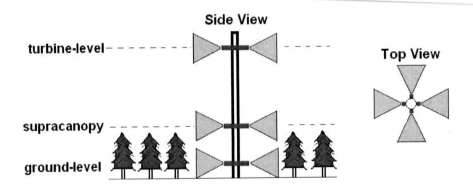

FIGURE C-1 Model of a vertical array of ultrasonic bat detectors for assessing nightly foraging and migratory activity of echolocating bats from ground level to the height of a turbine nacelle. Bat detectors are affixed to meteorological towers. SOURCE: D.S. Reynolds 2006, modified figure printed with permission; copyright 2006, *Journal of Wildlife Management*.

information on the activity of some species within the range of detection, because there is no practical way to install ultrasonic detectors in the rotor-swept area. Instead, they need to be mounted on meteorological towers at comparable heights to the rotor-swept zone to access both migratory and foraging behavior of bats where they are most at risk at wind-energy facilities. However, use of blimps and kites, to which ultrasonic microphones can be attached (Menzel et al. 2005b), holds considerable promise for extending the ranges of detection above the forest canopy and into the range of the rotor-swept area of modern wind turbines.

Strengths and Limitations of Ultrasonic Bat Detectors

Some investigators have wondered whether insectivorous bat species do not echolocate when they migrate, which would render them unable to detect the presence of turbines or moving blades, resulting in collisions. However, a study conducted in Minnesota (Johnson et al. 2004) suggests that migrating bats do echolocate, at least while feeding in the vicinity of wind turbines. Summer resident species in the Mid-Atlantic Highlands and other regions of the United States (e.g., species of *Myotis*, *Eptesicus*, and *Pipistrellus*) as well as insectivorous species (species of *Lasiurus*, *Lasionycteris*, and *Tadarida*) rely on echolocation while foraging. Moreover, Larkin (2006) suggested that because the energetic cost of echolocation is so low (see Speakman and Racey 1991), it is unlikely that bats would forego echolocation during migration. However, even if these bats do echolocate and feed near wind turbines, it is possible that they emit cries too infrequently and detection distances are too limited in range to protect them from encountering a rapidly moving rotor blade (Kunz et al. 2007). Also, given the limited range of detection of most ultrasonic detectors (< 30 m), it is quite likely that some migrating bats would have been missed when detectors were deployed at ground level.

Ultrasonic detectors for recording echolocation calls of bats have several limitations. Different models of detectors may have different microphones and ranges of detection, and even the same models can have different sensitivities, sometimes making it difficult to compare results from one detector to another (Larson and Hayes 2000; Limpens and McCracken 2004). In addition, these devices are not the "silver bullet" that some users have expected or promoted, as they have limited detection ranges (typically less than 25 m) and cannot be used unambiguously to identify all species present in a given area, owing largely to the similarity of calls of some closely related species (e.g., little brown myotis, *Myotis lucifugus*, and Indiana bat, *M. sodalis*), the presence of cryptic species (e.g., Jones and Barlow 2004) and the variation among individual bats associated with both habitat and geographic location (Barclay and Brigham 2004; Gannon and Sherwin 2004).

Bat detectors offer alternatives to more traditional methods, such as mist-netting and harp-trapping,[2] to determine the presence of many insectivorous bat species, especially in environments and in situations where it is not possible to capture them or make direct observations (e.g., open areas near wind turbines and above tree canopies). Bat detectors represent only one of a number of tools available to researchers for investigating the foraging, commuting, and migratory behavior of insectivorous bats (Kunz 2004). Ultrasonic detectors have their inherent biases associated with limits of detection, uncertainty of discriminating certain species based on call signatures, and the inability to discriminate individuals by sex, age, and reproductive condition (Barclay and Brigham 2004). However, a recent study suggests that it may be possible to distinguish sex and age of some bat species based on their echolocation call structures (Siemers et al. 2005). These situations place limits on the applicability of these devices for assessing the full extent of bat activities near wind turbines. Live capture and other methods (see below) are needed to gain a full understanding of which bat species are present in an area for assessing potential impacts in pre-siting surveys and pre-construction monitoring in the vicinity of wind-energy facilities.

The importance of direct observations, in concert with recordings of echolocation calls, is essential for gaining a full understanding of how insectivorous bats are killed by wind turbines. Recordings of echolocation calls synchronized with multiflash, 3D photography or videography, thermal infrared imaging, and tracking radar can also provide additional insight into the behavior and echolocation call structure associated directly with prey capture (Kalko 2004; Horn and Arnett 2005) and responses to novel objects in their environment, such as wind turbines.

Some of the most promising research on migration, commuting, and foraging behavior of insectivorous bats could result from continuous recordings at sites proposed for development of wind-energy facilities. Use of ultrasonic bat detectors for assessing composition of local assemblages in the vicinity of proposed or constructed wind-energy facilities could provide valuable information, but accomplishing these goals requires reliable species identification (Barclay and Brigham 2004; Gannon and Sherwin 2004). Unfortunately, many studies that have used ultrasonic bat detectors have been hampered by inappropriate sample design, and uncertainties regarding species identity (Hayes 1997, 2000, 2003). These uncertainties and protocols must be addressed if ultrasonic detectors are to provide use-

---

[2]A harp trap is a free-standing device consisting of a rectangular frame within which two or more banks of monofilament lines are strung. Bats fly through the first bank of lines but are stopped by the second and fall into a bag attached to the bottom of the frame and can then be removed for examination (Kunz et al. in press a).

ful information on species composition at wind-energy facilities and other areas of interest. Recent developments using time-expanded recordings of search calls produced by echolocating bats hold promise for improving the identification of species (e.g., Preatoni et al. 2005; Siemers et al. 2005).

An exhaustive analysis of empirical data using different bat detectors is probably beyond the scope of most projects. Instead, investigators should try to obtain data that would provide first-order understanding of basic questions, such as:

- Over what volume of air space can reliable detection occur?
- How do characteristics of calls affect detection?
- How does this detection volume change with local conditions?

In cases where multiple sensors are being used to compute direction to or from the location of a calling animal, calibration should determine:

- The directional, 2D, and/or 3D resolution of the system as a function of bearing and range to the source;
- The directional, 2D, and/or 3D resolution of the system as a function of source-center frequency, bandwidth, and time-bandwidth product.

Array performance in the plane of the array can be assessed by projecting sounds with known acoustic characteristics from known x-y positions. Assessing array performance as a function of altitude is more of a challenge since one needs to have an acoustic source at various altitudes above the ground at various declinations from the array. No one, to our knowledge, has performed such a test. It could be conducted by deploying an array on or near a tall tower and placing controlled sources at various elevations on the tower.

Until comprehensive field studies involving acoustic calibrations are completed, it is difficult to develop reliable estimates of the costs and benefits of each method. Ideally, an acoustic calibration in the field would involve deployment of many sensors in a variety of geometries to spatially oversample, to obtain high-resolution location data, and to increase detection probability. During the post-processing phase, one could then subsample to determine the tradeoffs between detection probability, location accuracy, and processing complexity all as a function of call characteristics and environmental variables. A comprehensive field investigation would involve an integrated approach that combines multiple modalities including acoustics, radar, thermal imaging, and visual methods. Ideally one would do this with controlled objects moving in 3D space through the sampling region. Given the difficulty of that task, one could simultaneously use various methods so that at certain times each was operating blind and independently, and coordinate all methods and follow a prescribed protocol for sharing information to optimize detection and recognition within the largest space possible. This type of cross-validation would greatly improve the present state of knowledge by providing information on, for example, types of radar returns coincident with acoustic species-specific identification, acoustic locations associated with radar returns, and the proportion of radar targets that are bats.

The benefits of using different methods similarly could be evaluated using standardized metrics and their performances compared to the optimum as predicted by a model. Such an approach has been used successfully to predict potential impacts from anthropogenic noise and evaluate the efficacy of different animal-detection schemes in the marine environment. In most cases, an integrated approach using a combination of methods is preferred, because no single method can unambiguously assess the behavior of animals in natural populations, let alone measure the effects of wind turbines on a biotic community. Investigators should understand the limitations, applicability, and operational considerations of each method before deploying them in the field. Researchers responsible for evaluating a particular monitoring project should have some way of gauging its likelihood of success, its optimal expected benefits, and certainly whether the proposed effort has the statistical power to address the key issues of concern.

At present, some of the potential benefits from the application of passive-acoustic monitoring techniques for birds and bats have been demonstrated, but the actual benefits have not been rigorously field tested, calibrated, or fully quantified. Birds and bats produce sounds for various reasons, and the assumption is that one can take advantage of their natural behaviors to monitor their presence relative to wind-energy facilities. Technology can be applied to record, detect, and in some cases identify a significant portion of bird and bat species that are resident or occur seasonally as migrants through wind-energy facility areas. However, little is known about how to interpret counts of call detections for actual passage rates. Acoustic technologies exist, and a few pilot projects have demonstrated that they can be applied to characterize the 3D distributions of calling birds as they move through a wind-energy facility (Jacques Whitford Limited 2005). There often is considerable natural variability in the density, spatial distribution, and seasonal occurrence of bird and bat species in the vicinity of wind-energy facilities. However, there is no ideal way to predict or compensate for such variability with much spatial or temporal accuracy other than to conduct site-specific monitoring.

Challenges and Recommendations

Factors that limit the application of passive-acoustic technologies include the cost of the equipment and the time, level of expertise, and signal-processing software required to analyze the data. Significant advances in hardware and software development have occurred in the past few years, and rates of improvements are expected to increase. In the hardware domain, there are several versions of low-powered, autonomous acoustic recorders that can be deployed as stand-alone units, and at least one version where units can be deployed in an array configuration. Such units can operate continuously in remote areas for months at a time and some can collect acoustic data in frequency ranges appropriate for birds and bats. In the software domain, extensible sound-analysis software packages are becoming available that allow automatic detection and 3D localization and batch processing. However, most bioacoustical technologies have not been rigorously calibrated in the field or systematically compared with other methods to help estimate biases in the data.

In cases where single, audible and ultrasonic sensors are being used to detect birds or bats, calibration should determine:

•   The absolute sensitivity of the recording electronics as a function of frequency. This can be accomplished in the laboratory prior to field work and can be used to predict the detection volume as a function of frequency in the ideal context of an acoustic free-field with spherical spreading;
•   The empirical sensitivity of the recording unit as a function of frequency, source location, and environmental conditions. Sensor performance should be verified in the field by projecting sounds of separate frequencies from known positions specified or known environmental conditions (e.g., wind speed, humidity, and temperature). Such data can be used to derive an empirical transmission-loss function that can be used to predict the detection volume under a variety of typical field conditions.

Using Radar to Detect, Monitor, and Quantify the Movements of Bats, Birds, and Insects in the Atmosphere

More than 60 years ago the British discovered that birds were responsible for some of the puzzling radar echoes dubbed "angels" (Lack and Varley 1945; Buss 1946). Subsequently, radar has proven to be a useful tool for the detection, monitoring, and quantification of the movements of organisms in the atmosphere. Radar can be used to investigate the movements of birds (Eastwood 1967; Gauthreaux 1970; Richardson 1979; Kerlinger 1982; Kerlinger and Gauthreaux 1985; Vaughn 1985; Bruderer 1997a,b), bats (Williams et al. 1973), and insects (Riley et al. 1983) in the atmosphere during the day, or at night including small (1-10 km of a tracking or marine radar), intermediate (10-200 km or the surveillance area of a single weather radar), and large spatial scales (continent-wide radar network surveillance).

Radar has been valuable not only for descriptive studies of daily and seasonal movements of flying animals, but technically has also been used to answer important questions related to orientation, aerodynamics, and habitat selection of migrants. During the past two decades, radar has been increasingly used in risk-assessment studies related to projects that could potentially impact species that are migratory, endangered, threatened, or of special concern (Cooper 1996; Gauthreaux and Belser 2003a, 2005; Larkin 2005b).

Tracking Radar

Small tracking radars can detect individual targets within a range of a 100+ m to 4-6 km, and large tracking radars used by the National Aeronautics and Space Administration (NASA), can detect and track individual bird-sized targets for tens of kilometers and a bumblebee out to a range of 10 km. Tracking radar "locks on" a target and follows the subsequent movements of the target in 3D space as long as another target does not enter the beam at the same range. If this happens, the radar may switch targets and track the new target. This is a serious problem when many targets are present. Although tracking radar can provide detailed information on the flight paths of individual targets as well as information on wing-beat patterns (Renevey 1981), the very narrow beam limits simultaneous sampling of several targets.

Small military-surplus tracking radars have been used to monitor the movements of individual birds (Bruderer and Steidinger 1972; Griffin 1972; Able 1977; Larkin and Frase 1988; Bruderer et al. 1995; Bäckman and Alerstam 2003) and insects (Larkin 1991) within a range of a few kilometers, although these units often require regular maintenance and parts are difficult to find. Birds and insects have also been tracked with large tracking radars (SPANDAR) used to track deep-space probes by NASA at the Wallops Island, Virginia facility (Glover et al. 1966, Williams et al. 1972, Demong and Emlen 1978). At present there is only one published study of bat movements recorded by tracking radar (Bruderer and Popa-Lisseanu 2005), although R.P. Larkin (Illinois Natural History Survey, personal communication 2005) has used his tracking radar unit based at the Illinois Natural History Survey to follow bats (*Myotis lucifugus*).

Marine Radar

Most of the small, mobile radars used in wildlife and entomological studies to date have been 5 to 60 kW marine-surveillance radars of 3- or 10-cm wavelengths. Many are commercial off-the-shelf units that are used without modification. In typical horizontal surveillance mode, the marine-radar antenna samples a large airspace (20-25°) in the vertical and much less (1.0-2.3°) in the horizontal. Because of the broad vertical coverage of the radar beam, measuring the altitude of targets is impossible in horizontal surveillance mode. To address this limitation, the radar can be tilted 90° so that the sweep of the antenna is vertical. In this mode accurate measurements of target altitude are possible without echoes from ground clutter. Some investigators have used a single unit for both horizontal and vertical surveillance (Harmata et al. 2003), whereas others have used two radars, one for horizontal and one for vertical surveillance (Harmata et al. 1999).

Another approach uses a second radar with either a non-rotating parabolic dish that can be positioned at any elevation angle between horizontal and vertical (Gauthreaux 1985a,b) or a non-rotating parabolic dish that is directed vertically and is mounted on top of the transmitter/receiver unit (Cooper et al. 1991). It also is possible to replace the open-array antenna with a rotating, parabolic antenna that projects a narrow, conical (e.g., 2.5-4.0°) beam (Gauthreaux and Belser 2003a). When the conical beam is elevated in the horizontal surveillance mode, the altitude of an echo is a trigonometric function of the range of the echo and the angle of antenna tilt.

Each of the above configurations has advantages and disadvantages. The open-array antenna samples a greater air space, but this reduces the range of detection and the altitude of a target in the vertical scan cannot be linked to the track of a target in the horizontal scan. The parabolic antenna samples a smaller volume of atmosphere but has a greater detection range and three-dimensional information on each target can be measured.

When tuned properly, high-resolution marine radars with parabolic antennas can readily detect small flocks of birds out to 6 km from the radar (Gauthreaux and Belser 2003a) and insects out to 1.5-2.5 km (Riley 1989). Low-powered marine radar can detect individual birds within a range of 2-3 km and flocks of birds out to 10 km, and a 25 kW marine radar with an open-array antenna can detect small passerines out to a range of 800-1,000 m, and European thrushes (*Turdus* spp.) can be detected by 10 and 12 kW units at the same range (Desholm et al. 2004). With 25 kW 3-cm-wavelength radar in clear weather, the range at which an 800 g duck can be detected has been computed to be 2.19 km for short-pulse and 3.2 km for long-pulse radars; with 60 kW 10-cm-wavelength radar in clear weather, a 500 g pigeon-like target can be detected at 3.97 km for short-pulse and 5.5 km for long-pulse radar (Desholm et al. 2004). Cooper et al. (1991, 2004a) noted that 12 kW radar with an open-array antenna can routinely detect flocks of waterfowl out to 5.6 km, individual hawks out to 2.3 km, and single, small passerines out to 1.2 km.

Based on simultaneous visual and vertical radar observations during the day, when birds are flying high and near the visual limits of observers, observers noted that fewer than half of the birds were detected by radar (Harmata et al. 1999). In a European study, about 40% of the birds flying below 50 m were missed by marine radar, but when birds were flying above 50 m, only 8% went undetected by the radar (Knust et al. 2003). Moreover, low-flying birds often were obscured by the ground-clutter pattern of the radar display. On short pulse lengths, minimum detectable range can be as close as 20-30 m, and range discrimination depends on the pulse length used.

Several technical limitations affect the quality of quantitative data gathered by marine radar. The aspect of the target relative to the radar beam affects the amount of energy reflected back to the radar receiver. Head-on and tail-on detections have significantly smaller radar cross-sections than broadside detections. In addition, a radar's beam width is defined as the angle where the energy of the beam is reduced by one-half (or -3 dB) of the maximum at the center of the beam. If one of two identical targets at the same range occurs at the very edge of the radar beam and the other is positioned at the center of the beam, then the target at the edge of the beam will produce a weaker echo than the one at the center of the beam. Similarly, a strong target outside the "beam" can be detected as a weak target. The problem is amplified when using wide-beam (20-25°) radar, because of the rapid power loss with range.

The following factors are known to affect the performance of marine-surveillance radars and influence the results obtained from different radar studies of bird, bat, and insect movements:

- Pulse length and corresponding pulse repetition frequencies.
- Transmitter power (e.g., 5, 10, 25, 50, or 60 kW).
- Sea clutter and rain clutter settings.
- Tuning of the receiver.
- Antenna rotation speed.
- Antenna beam characteristics.
- Range setting.
- Beam-brilliance setting.
- Gain setting.
- Frequency or wavelength.

When impact assessment studies use different brands of surveillance radar, and the above factors are not the same, the results are not likely to be quantitatively comparable. Without some form of operational standards this problem will persist. Moreover, not all marine radars detect biological targets equally (J. Kube, Institut für Angewandte Ökologie, personal communication 2005). It is essential that calibration of the unit be performed before studies are begun, and the calibrations be conducted periodically during the study. This can be done with a radar reflector of standard cross-section. The reflector can be lifted by balloon or kite at a fixed range from the radar, and the characteristics of the echo produced by the reflector target compared to some reference standard.

The transmitter power of the marine radar should be as high as possible (25 kW or greater), because long pulse lengths enhance detectability but suffer a loss of resolution, and short pulse lengths increase resolution with a loss of detectability. The greater the transmitter power the greater the cost, but 50 kW radar operating on short pulse will produce superior results in an assessment study over a 10kW unit operating on short pulse. Marine radars can be purchased in either of two wavelengths—3 cm (X-band) or 10 cm (S-band)—and there is considerable debate among users of these two radar types regarding which one is best. Both have been used to study bird movements aloft, but no published study has compared them at the same location and under similar weather conditions. It is well known that precipitation attenuates 3 cm wavelengths considerably more than it does 10 cm signals, and as a result, precipitation will greatly decrease the chances of a 3 cm radar's detecting targets beyond a shower. Irrespective of wavelength, small target detection within an area of heavy precipitation is not likely.

A comparison of small radar systems that currently are being used to study bird movements can be found in Desholm et al. (2004) and in MacKinnon (2006). The former reference examines the performance of marine-surveillance radars used to study bird and bat movement in the vicinity of wind turbines and includes a discussion of the characteristics of an ideal bird/bat detecting radar. The latter is a compilation of information on small radars used to detect, monitor, and quantify bird movements that pose a threat to aircraft and includes information on recent developments in digital processing of marine-radar detections of biological targets in the atmosphere (e.g., Nohara et al. 2005).

The methods of collecting and processing marine-radar data differ among studies. In some cases, investigators extract the echo data from the radar display (or a digital image of the display) manually (Figure C-2A) and then perform analyses to compute descriptive statistics of migration direction and passage rate. In other cases commercial radars with digital processors gather raw radar data from the receiver and then use proprietary algorithms to process the data. The algorithms reduce ground clutter and the processed data from targets are reported out either in a spreadsheet format with information on target track (direction and velocity), reflectivity, and size or as plots (Figure C-2B). Hundreds of targets can be tracked at once, but as the number of targets increases, so does the possibility of tracking errors, because the tracking algorithms may switch between nearby targets. Manual data extraction is labor-intensive and time-consuming and there is always the possibility of some bias in manual extraction of the data. Automatic digital processing is extremely fast and eliminates the possibility

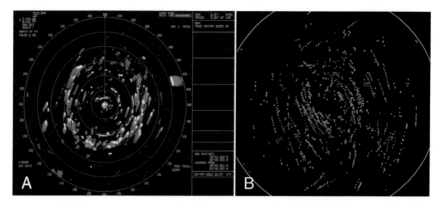

FIGURE C-2 (A) A frame from a Furuno 2155-BB radar with a parabolic dish (4° beam width) elevated 30° above the horizontal. Most of the echoes are from migrating birds and bats flying toward NNE. The yellow echoes show the position when the image was generated and the blue trails show past positions of the echoes. The large echo coming in from ENE is a helicopter. (B) An image generated from digitally processed data from same radar system showing tracks (green) and current position and heading (red symbols).
SOURCE: Sidney Gauthreaux, Clemson University.

of such bias, but the processing algorithms must be carefully evaluated to exclude the possibility that systematic biases are not introduced into the algorithms by the developer (e.g., algorithms requiring a certain number of detections before tracking begins and consequently excluding fast targets that produce less than the required number of detections). There is clearly a need to carefully ground-truth the reports of data from digitally processed radar return, but no published studies have done so.

The identification of birds, bats, and insects and determination of the number of targets per echo on marine radars can be problematical. The echo information returned to a radar receiver from birds and bats in the atmosphere makes it nearly impossible to discriminate echoes from bats and echoes from similarly sized birds, and this is particularly true for migrating bats and birds engaged in linear flight. Foraging bats may produce irregular echo movement patterns, but similar echo movement patterns may be produced by nocturnally foraging birds (nighthawks). Because of the inability to discriminate between radar reflections from birds and bats, investigators typically refer to the sources of echoes in radar studies as "targets." One generally cannot discriminate an individual target from a tight cluster of targets, because a single large target may produce the same echo as a tight group of smaller targets. If one knows the speed and direction of the wind at the altitude where a target is detected, the airspeed of the target can be calculated, and this information can be used to assign targets to categories based on flight speed (Harmata et al. 1999). Similarly, the flight behavior of a target may offer clues to its identity (e.g., very irregular flight path of a foraging bat, circling of a raptor in a thermal). Claims of target identification based on size of target (number of pixels) are likely incorrect. When echo characteristics of hundreds of known targets have been analyzed statistically, the relationship is not significant (O. Hüppop, Institut für Vogelforschung Vogelwarte Helgoland, personal communication 2006). Thus far, the best studies have employed simultaneous visual observations (Newman and Lowery 1959) and thermal imaging (Gauthreaux and Livingston 2006) during the daylight hours and moon-watching (Lowery 1951), ceilometry, image intensification (Gauthreaux 1969; Able and Gauthreaux 1975; Plissner et al. 2006), and thermal-imaging (Gauthreaux and Livingston 2006) observations at night (Horn et al. in press).

Marine Radar Studies at Eastern Wind-Energy Projects: Case Studies

Cooper et al. (2004a) studied spring bird-migration concurrently with radar and direct visual observations for about 4 hours per day and with radar observations alone for about 5 hours per night from April 15 through May 15, 2003, at the site of the proposed Chautauqua Wind-Energy Facility in western New York. The radar was an X-band Furuno Model FR-1510 MKIII with a peak power of 12 kW and a 2-m-long open array (T-bar) antenna that produces a horizontal-beam width of 1.23° and a vertical-beam width of 25° (± ~10° side lobe). Pulse lengths used were 0.07 and 0.5 sec.

The mean flight direction of targets observed with radar at night was 029° ± 40° and mean passage rates were 395 ± 69 targets $km^{-1}h^{-1}$, significantly higher than the daytime rate of 79 ± 13 targets $km^{-1}h^{-1}$. The mean flight altitude during the daytime was 372 ± 6 m above ground level (agl); the mean nocturnal flight altitude was 528 ± 3 m agl. The mean percentage of all targets flying below 125 m agl, a height that could bring them into contact with wind-turbine blades, was 17.2% during the day and 3.8% at night.

Cooper et al. (2004b) also conducted a radar study of nocturnal migration at the proposed Chautauqua facility site from September 2 through October 10, 2003. Sampling occurred for about 6 hours per night during 30 nights within the 40-day study period. The radar equipment and surveillance protocols were the same as described in Cooper et al. (2004a). The mean flight direction of birds and bats was toward the SSW (199° ± 58°). The mean (± SE) fall passage rate was 238 ± 48 targets $km^{-1}h^{-1}$ and ranged from nightly rates of 10 to 905 targets $km^{-1}h^{-1}$ with higher rates after mid-September. The mean altitude of flights in fall was 532 ± 3 m agl.

Mabee et al. (2004) presented the results of a radar study of nocturnal bird and bat migration at the site of the proposed Mt. Storm wind-power facility in northeastern West Virginia from September 3 through October 17, 2003 for approximately 6 hours per night on 45 nights. Vehicle-mounted radars (X-band Furuno Model FR-1510 MKIII) monitored five locations within the proposed site. Horizontal surveillance covered a 1.5-km radius and vertical surveillance covered altitudes up to about 1.5 km. The mean flight direction of targets detected by radar was toward $184° \pm 1°$, and the night-to-night passage rates varied from 8 to 852 targets $km^{-1}h^{-1}$ with a mean rate of 241 ± 33 targets $km^{-1}h^{-1}$ at the central study site and a mean of 199 targets $km^{-1}h^{-1}$ for all the fall sample sites in the project area. The authors found variation in passage rates among some ridge sites (e.g., central and southern sites) and between ridge and off-ridge sites (e.g., central and western sites) but did not find that nocturnal migrants concentrated along the Allegheny Front. The general direction of migration over the site was toward the southwest. According to the authors, "of 1,733 targets at the central station that could be tracked as they approached the primary ridgeline long enough to determine a response to the ridgeline, 5.3% of targets approached and turned greater than 10° before crossing or turned and did not cross the ridge, 49.7% approached and crossed the ridge, and 45% did not approach the ridge." The mean altitude of flight measured with vertically scanning radar was 410 ± 2 m agl, but night-to-night mean altitudes were highly variable as were the hour-to-hour flight altitudes during the night. Migration at lower altitudes occurred later in the evening. The mean passage rate for the 13% of the targets that flew below 125 m agl was 36.3 targets $km^{-1}h^{-1}$.

Marine-radar and night-vision studies of bird and bat migration between August 16 and October 14, 2005, at two sites in the proposed Highland New Wind Development area in the Allegheny Mountains of western Virginia found a mean nocturnal passage rate of 385 ± 55 targets $km^{-1}h^{-1}$ with a range among nights between 9 and 2,762 targets $km^{-1}h^{-1}$ (Plissner et al. 2006). The study found no differences in passage rates, flight altitudes, or observed proportions of targets between the two survey sites. These passage rates are among the highest measured at eastern U.S. wind-energy development sites where similar equipment and methods have been used. The mean flight direction of targets observed on the marine radar was SSW (204°). For the entire fall season, the mean nocturnal flight altitude measured with vertical radar was 442 ± 3 m agl. The mean altitude of flight varied from 211 to 721 m agl, and during autumn 11.5% of targets flew at an altitude of 125 m agl or below.

In a recent review of radar studies at proposed and existing wind-energy projects in the United States, Young and Erickson (2006) indicated that the mean passage rates of targets (targets $km^{-1}hr^{-1}$) in the east (21 studies) was similar in spring and fall (258 vs. 247, respectively). Their analysis also showed that the mean height of flights was 409 m agl in spring and

470 m agl in fall, and 14% of the targets were below 125 m agl in the spring and 6.5% were below 125 m agl in the fall. Mean flight directions were toward the NNE (31°) in spring and toward the SSW (193°) in the fall. The passage rate of birds and bats during spring and fall migration periods is greater in the east than that measured in the central and western United States, and in general the passage rates are greater in autumn than in the spring, with Chautauqua, New York being a notable exception (Table C-3). The altitudinal distribution of targets detected by radar covers several hundreds of meters agl, and relatively few of these targets fly below 125-127 m agl in the potential strike zones of wind turbines (Table C-4). The strike zone is typically defined as the area swept by the turbine, but it is possible that some birds and bats collide with the supporting monopole below the rotor-swept area due to turbulence. It is also possible that birds flying higher than the zone could descend if attracted by warning lights on the turbines, but Kerlinger and Kerns (2003) found that birds were not attracted to FAA-recommended lighting on wind turbines (Johnson et al. 2002). Marine-surveillance radar does not detect low-flying birds and bats

TABLE C-3 Mean Nocturnal Passage Rates (targets $km^{-1}\,hr^{-1}$) in Spring and Fall for Different Regions of the United States

Location	Spring (year)	Fall (year)	Reference
Eastern U.S.			
Chautauqua, NY	395 (2003)	238 (2003)	Cooper et al. (2004a,b, 2005b)
Copenhagen, NY	170	242	Cooper and Mabee (2000)
Wethersfield, NY	62	180	Cooper and Mabee (2000)
Carthage, NY		225	Cooper et al. (1995a,b)
Mt. Storm, WV		54-241 (2003)	Mabee et al. (2004)
Nantucket Sound, MA	53 (2000)	135 (2000)	Curry and Kerlinger LLC and ESS Group (2004)
New England, ME	71 (1994)	478 (1994)	Northrop, Devine, and Tarbell, Inc. (1995a,b)
Highland, VA		385 (2005)	Plissner et al. (2006)
Central U.S.			
Thief River-W, MN	63	83	Day and Byrne (1989, 1990)
Thief River-E, MN	43	108	Day and Byrne (1989, 1990)
Buffalo Ridge, MN	93 (1997)	98 (1996)	Hawrot and Hanowski (1997)
Amhurst-S, SD	83	40	Cooper et al. (2004c)
Amhurst-N, SD	23	27	Cooper et al. (2004c)
Western U.S.			
Vansycle, OR	48 (2001)	26 (2001)	Mabee and Cooper (2002, 2004)
Hatch Grade, OR	45 (2001)	22 (2001)	Mabee and Cooper (2004)

TABLE C-4 Percentage of Radar Echoes Flying Between Ground Clutter and 125-127 m Above Ground Level

Location	Spring	Fall	Reference
Nantucket Sound, MA	21.8	7.7	Curry and Kerlinger LLC and ESS Group (2004)
New York	3.8	4	Cooper et al. (2004a,b,c)
West Virginia		16	Mabee et al. (2004)
Virginia		11.5	Plissner et al. (2006)
Oregon	15	6	Mabee & Cooper (2002, 2004)

SOURCES: Cooper et al. (2004c); Curry & Kerlinger LLC and ESS Group (2004).

(less than 30 m agl), because of the recovery time of the radar and because ground clutter echoes obscure targets flying at very low altitudes. The tuning of the sensitivity-time control (STC) curve also is critically important, because high STC settings for short ranges will make detection of close-in targets more difficult.

Weather-Surveillance Radar (WSR-88D or NEXRAD)

The Weather-Surveillance Radar-1988, Doppler (WSR-88D) is the backbone of the national network of weather radars in the United States operated by the National Weather Service (NWS) in the National Oceanic and Atmospheric Administration (NOAA) of the Department of Commerce, the Department of Defense (units at military bases), and non-continental-United States (CONUS) Department of Transportation sites (Crum and Alberty 1993; Crum et al. 1993; Klazura and Imy 1993). There are 155 WSR-88D radars in the United States, including the Territory of Guam and the Commonwealth of Puerto Rico.

These powerful and sensitive S-band Doppler weather-surveillance radars (also known as NEXRAD for next generation radar) have a 1.0° beam, and when the beam is tilted 0.5° above the horizontal, the radar can detect concentrations of biological targets out to a range of 240 km and intense precipitation at a maximum range of 460 km. The WSR-88D detects the intensity of reflected energy from objects in the atmosphere as well as the Doppler shifts of returned frequencies from moving targets, and the latter information can be used to calculate radial velocities of the moving objects relative to the radar station. NEXRAD radar is extremely sensitive and can detect birds, bats, and concentrations of insects in precipitation mode. When there is no precipitation detected, the radar operates in "clear-air mode" and samples the same volume of air space more slowly, making it possible to detect the reflected energy from very small objects such as insects and even dust and smoke particles.

NOAA and the National Climatic Data Center (NCDC) in Asheville, North Carolina, have archived data from each WSR-88D station since 1995, although there are several gaps in the archived records. Level II archived data contain the three basic moments from the radar: reflectivity, radial velocity, and spectrum width (Figures C-3A-3C). Level III archived data contain derived products such as base reflectivity, base velocity, vertical-wind profile, echo tops—products that are very useful for biological studies. These archived data are freely accessible from the NCDC. Data from one or more stations can be integrated to support analysis of nightly behavior, seasonal trends, and population shifts of bats and birds at local, regional, and national scales.

The antenna of the WSR-88D is computer-controlled and repetitively scans the atmosphere through a sequence of predefined elevation angles, antenna rotation rates, and pulse characteristics (volume coverage patterns or VCP) depending on the radar's mode of operation. Two operational modes exist—a precipitation mode and a clear-air mode—and selection of an operational mode is closely related to the detected coverage of precipitation. Precipitation mode has four VCPs (11, 12, 21, and 121) and clear-air mode has two VCPs (31 and 32) (Table 4-1 in the Federal Meteorological Handbook No. 11, 2006). VCP 21 is the standard precipitation mode and is typically used when precipitation is first detected. VCP 11 provides better vertical sampling of weather echoes close to the antenna than VCP 21 and is usually preferred in situations where convective precipitation is within 120 km of the antenna. VCP 12 has the same number of elevation angles as VCP 11 but denser vertical sampling at lower-elevation angles. VCP 121 has the same elevation angles as VCP 21, but it has more scans and lessens the range and velocity aliasing (producing a false frequency with the correct one—the Doppler Dilemma). In the clear-air mode of operation, VCP 31 uses a 4.7 µsec pulse length and VCP 32 uses a 1.57 µsec pulse length. The longer pulse length of VCP 31 provides greater sensitivity, and the display threshold is reduced to detect minute amounts of energy returning from weak reflectors in the atmosphere (e.g., refractive index gradients, cloud droplets, dust and smoke particles, and pollen grains). Each operational mode is also associated with a product generation list (Table 4-2 in the Federal Meteorological Handbook No. 11, 2006), and additional characteristics of the WSR-88D can be found in that handbook (OFCM 2006).

WSR-88D digital-base data (Level II) from the signal processor of the Radar Product Generator (RPG) are recorded at all NWS and several select CONUS DOD WSR-88D sites on 8-mm magnetic tape and sent to the NCDC for archiving and dissemination. Other data include information on synchronization, calibration, date, time, antenna position, and operational mode. From these base data, additional computer processing generates a set of pre-determined products known as Level III data as defined in Federal

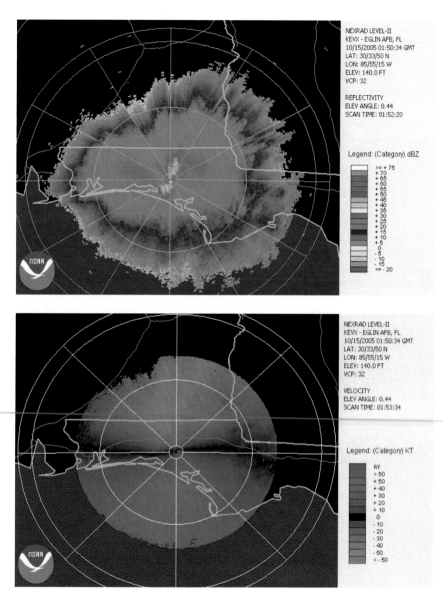

FIGURE C-3 (A) Base reflectivity image from WSR-88D Level II data showing the night during the fall of 2005 with the highest density of bird migration (ca. 1,800 birds km^{-3}) over northwestern Florida. North is at the top and east is to the right in this and following figures. (B) Base-velocity image from WSR-88D Level II data showing the radial velocity of targets displayed in Figure C-3A. Greens indicate movement toward the radar and reds are moving away.
SOURCE: Sidney Gauthreaux, Clemson University.

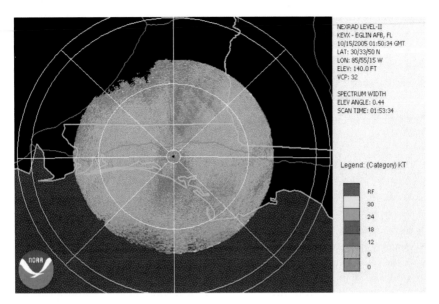

FIGURE C-3 (C) Spectrum width display from WSR-88D Level II data that corresponds to displays in Figures C-3A and B.
SOURCE: Sidney Gauthreaux, Clemson University.

Meteorological Handbook No. 11, Part A. Level III products are recorded at 155 of the 159 worldwide sites, and sent to NCDC for permanent storage.

The archive of digital data from the WSR-88D is stored on the NCDC Robotic Mass Storage System, commonly known as the Hierarchical Data Storage System (HDSS). WSR-88D data may be downloaded at no cost from NCDC, and they are available from 1991 to one day from present and are easily accessible with the NEXRAD Inventory Search tool. This tool can be used to view the data for completeness and to download data. The requested data are ready for use with the NCDC Java NEXRAD Viewer and Data Exporter. Each order may contain up to 24 hours of data at a time for a single site. The data may also be ordered from the NCDC HDSS Access System (HAS) web page. This is a better option for users needing large amounts of data in a compressed-archive format. Up to a week of data for multiple sites may be ordered. Downloaded radar data from NCDC are in a unique digital binary format. Special software must be used to visualize the data. Several free visualization and analysis products are available for download at the NCDC web site (NCDC 2006).

Detection of Biological Targets on the WSR-88D

Aerial biological targets are readily detected by the WSR-88D, and several investigators have detailed its use for studying bird migration (Gauthreaux and Belser 1998, 1999, 2003b; Diehl and Larkin 2005), bird roosts (Russell and Gauthreaux 1998; Russell et al. 1998), bat colonies (McCracken 1996; McCracken and Westbrook 2002), and concentrations of insects aloft (Westbrook and Wolf 1998). The WSR-88D can be used to quantify the amount of bird migration aloft (Gauthreaux and Belser 1998, 1999; Black and Donaldson 1999) and it has been used to study regional patterns of migration (e.g., Great Lake Region [Diehl et al. 2003]). Detailed methods of analyzing Level III data from the WSR-88D can be found in Gauthreaux and Belser (2003b).

WSR-88D technology has made it possible to characterize and quantify the nightly behavior of some bat species as they disperse nightly from their roosts to forage at high altitudes. For example, in the southwestern United States, Brazilian free-tailed bats (*Tadarida brasiliensis*) form enormous colonies that disperse nightly, as far as 70 km from their daily refuges. The first observation of expanding ring-like formations at WSR-57 radar installations in Texas was confirmed as bats flying at high altitudes (Williams et al. 1973). To quantify both direction and speed of nightly dispersal from the selected colonies, the density of bat reflectivity can be quantified. WSR-88D radar images taken at weather-monitoring stations in south-central Texas have been used by Horn (2007) to estimate colony size and patterns of nightly dispersal of Brazilian free-tailed bats from cave and bridge roosts remotely (Figure C-4). To accomplish this, a mosaic radar image is produced from different weather stations by translating reflectivity from radial to a geo-referenced grid format, and then filtered to remove weather activity and enhance the visibility of the biological activity. Spatiotemporal statistics are used to describe rates of dispersion, duration of foraging bouts, direction vectors for movements of foraging groups, activity centers where foraging may occur, and summary measures of overall activity from the reflectance data. Computer-vision algorithms provide the framework for spatial dependence and for calculating the probability of transition from one reflectivity level to another (Horn 2007).

While WSR-88D technology may have limited applicability for assessing impacts of wind-energy facilities on birds and bats at many locations because it cannot detect low-flying targets, it offers a powerful tool for quantifying nightly dispersal and migratory activity of bat species that form large cave- and bridge-roosting colonies and disperse over the landscape within the range of radar coverage, especially where wind-energy facilities have been and are being developed in Oklahoma, Texas, and New Mexico, where there are large colonies of Brazilian free-tailed bats (Horn 2007). Similarly WSR-88D analysis of bird-migration patterns and bird-roost

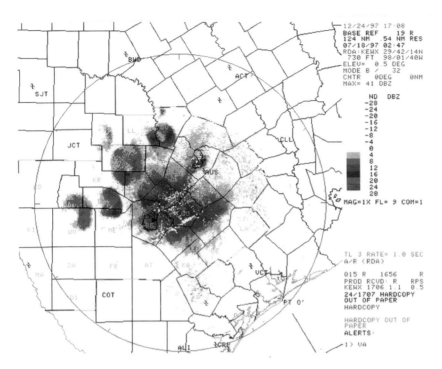

FIGURE C-4 NEXRAD, WSR-88D Doppler radar images of Brazilian free-tailed bats dispersing nightly from selected cave and bridge roosts in south-central Texas.
SOURCE: Kunz 2004; National Oceanic and Atmospheric Administration, National Weather Service.

locations in the general area of a planned wind-energy facility can help in the initial assessment of aerial bird activity at alternate development sites. Data from WSR-88D sites around the country have been archived for more than a decade and therefore could provide opportunities to assess annual variations in bird migration over large areas.

Within 60 km of the radar, WSR-88D can be used to delimit important migration stopover areas by measuring the density of birds (birds km^{-3}) in the beam as they begin a migratory movement (exodus). Within minutes of the onset of nocturnal migration, the distribution and density of echoes in the radar beam can provide information on geographical ground sources of the migrants (migration stopover areas), and satellite imagery can be used to identify the topography and habitat type that characterizes these areas (Gauthreaux and Belser 2003a). At a larger spatial scale (that of the surveillance area of a single Doppler weather radar, out to 240 km), this approach can also be used to quantify the density of birds and to delimit locations of post-breeding, nocturnal roost sites of birds such as purple martins (*Progne subis*) and other species. Martins flying toward the roost late in the day generally fly low, often under radar coverage. However, when they depart the roost near dawn (Figure C-5) they fly higher and can be easily detected by Doppler radar (Russell et al. 1998; Russell and Gauthreaux 1998). At a continental scale, the national network of WSR-88D radars can be used to measure the direction and quantity of bird migration over the United States on an hourly basis at different altitudes dependent on distance from the radar (Gauthreaux et al. 2003). This ability is significant because it provides a means of monitoring the seasonal and annual variation in the patterns of migration at different altitudes for different geographical regions and the nation as a whole.

Although the WSR-88D is highly beneficial for the study of bird migration, it does have limitations with respect to its use in environmental

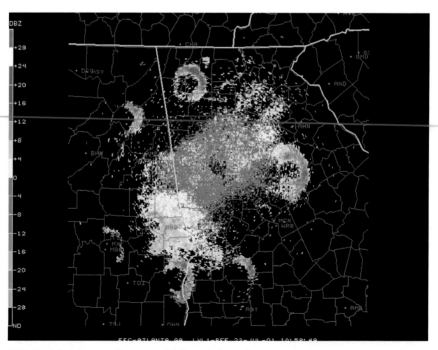

FIGURE C-5 Base reflectivity display of the Atlanta, GA (KFFC) WSR-88D showing the exodus of purple martins from roost sites near sunrise at 10:58 UTC on July 23, 2001.
SOURCE: National Oceanic and Atmospheric Administration.

risk assessment studies of wind-energy development. The pulse lengths of the radar are relatively long and the resolution of the data is rather coarse. The maximum resolution of the WSR-88D for reflectivity measurements and for Level III radial velocity measurements at 240 km range is 1° × 1 km. For Level III velocity data at 60 km range and for Level II velocity and spectrum-width data, the maximum resolution is 1° × 250 m. The lowest tilt of the WSR-88D antenna averages 0.5° above the horizontal, and over most of the surveillance coverage the base of the beam is too high to detect low-flying birds and bats. The beam width of the WSR-88D is 1°, and at a distance of 30 km the beam is 262 m wide. This eliminates the possibility of precise altitudinal measurements of targets.

Thus far, relatively few pre-construction studies at proposed wind-energy development sites have used WSR-88D data for a general assessment of migration volume over the site, but an analysis of these data can be valuable for assessing the temporal and spatial variability of daily (bird) and nightly (bats) dispersal and seasonal migration over the general area where wind-energy facilities are being considered proposed. WSR-88D data have been used to assess migration density during pre-construction studies at the proposed Mt. Storm Wind Power Site in northeastern West Virginia (Mabee et al. 2004). This study found no strong correlations between radar migration-passage rates measured with marine radar and NEXRAD reflectivity values (representing bird densities) during 25 nights of comparable data. The authors acknowledge that the WSR-88D station near Pittsburgh, Pennsylvania—170 km distant—is the closest to Mt. Storm. At such a distance, the effective coverage area of the lowest scan (0.5° above horizontal) of the WSR-88D overshoots most of the migration layer and does not allow for a direct comparison of the migration density recorded by both radar systems. Despite this shortcoming, mean flight directions of migrants recorded by both systems were correlated.

Because the radar pulse volumes of the WSR-88D are large, a given pulse volume often includes birds, bats, and insects, and one must use the mean airspeeds of targets to discriminate between slow-flying insects, foraging bats, and faster-flying migrating birds and bats (Gauthreaux and Belser 2003b). There is a need to compare migration-passage rates measured with marine radars and migration-traffic rates measured with thermal imagers and vertically pointing, fixed-beam radars and the WSR-88D. Larkin et al. (2002) related the number of flight calls from a nocturnally migrating bird, the dickcissel (*Spiza americana*), to displays of bird-migration density on the WSR-88D and concluded that the two measures were highly correlated. In contrast, Farnsworth et al. (2004) found a weak correlation between the number of flight calls per night and the maximum density of bird migration displayed on the WSR-88D, but the hour-to-hour pattern of flight calls during the night was not correlated with the hour-to-hour changes in the density of migration displayed on the radar. This is not surprising because the frequency of flight calling by birds increases when the sky is overcast and the cloud ceiling low. Likewise, flight calls of birds increase dramatically during foggy, misty conditions. The WSR-88D was designed as a weather radar and was not intended to be used by biologists to study biological targets in the atmosphere. Despite its limitations, the WSR-88D has proven to be of great value for studying the distribution and abundance of biological targets in the atmosphere and different aspects of their flight behavior (e.g., foraging, roosting, daily [birds], and nightly [bats] dispersal and migration) over large geographical areas.

Moon-Watching, Ceilometer, Thermal Imaging, and Chemiluminescent Tags

Moon-Watching

One of the first techniques developed to observe bird migration at night was moon-watching (Lowery 1951; Lowery and Newman 1955). By directing a telescope of sufficient power (20-30×) toward the disc within two days of the full moon during periods of migration, it is possible to observe silhouettes of birds and bats (and an occasional insect) as they pass before the illuminated disc of the moon. By following clearly defined methods for making observations and analyzing data (Nisbet 1959, 1963a,b), one can quantify the magnitude of migration (migration-traffic rate) passing over an area, and an experienced observer can easily distinguish between birds and bats and even identify some birds to family. The migration-traffic rate (as originally defined by Lowery 1951) is the number of birds crossing a mile of front per hour up to 1 mile in altitude. A shortcoming of the technique is its limitation to the five-day periods from two days before to two days after full moon without an obscuring cloud cover.

Ceilometry

The need to visually investigate the overhead passage of nocturnal migration when no moon is visible prompted the development of additional techniques. Howell et al. (1954) and Graber and Hassler (1962) demonstrated that migrating birds could be observed with powerful light beams, and in 1965 and 1966, S. Gauthreaux made observations with 10× binoculars and a 20× telescope of nocturnal migration at fixed-beam ceilometers operated by the NWS at Lake Charles and at New Orleans, Louisiana. Fixed-beam ceilometers are intense, narrow light beams of almost a million candlepower. Although these instruments killed many birds during inclement weather (Gauthreaux and Belser 2006), observations of Gauthreaux showed that during fair weather very few birds were affected by the light; most flew through the beam without hesitation or deviation in direction. No kills of migrants were recorded at either of the ceilometers during a two-year study.

Because fixed-beam ceilometers at airport weather stations were rapidly being replaced by rotating-beam devices that were unsuitable for migration studies, Gauthreaux (1969) constructed and tested a small, portable ceilometer for visual studies of nocturnal migration. The apparatus was inexpensive and easily constructed. Gauthreaux's (1969) portable ceilometer is best suited for studies of bird migration below 760 m, and the technique can be used to compute quantitative estimates of nocturnal migration-traffic rates (Able and Gauthreaux 1975; Gauthreaux 1980). Ceilometers have been used to quantify nocturnal bird migration in many different locations within the United States (e.g., Maine [Northrop et al. 1995a,b], New Hampshire [Williams et al. 2001], Vermont [Kerlinger 2002], North Dakota [Avery et al. 1976], California [McCrary et al. 1983]). The use of an image intensifier instead of binoculars or a telescope greatly enhances the detection of targets as they pass through the vertical light beam. An approach similar to this was used by Plissner et al. (2006), who made visual observations with Generation III night-vision goggles with a 1× eyepiece (Model ATN-PVS7, American Technologies Network Corporation) during nighttime radar sampling to determine relative numbers and proportions of birds and bats flying at altitudes at or below 150 m agl, the approximate maximal distance that passerines and bats could be distinguished. They used two 3-million-candlepower spotlights with infrared lens filters to illuminate targets flying overhead. The filters help prevent insects, birds, and bats from being attracted to the lights. They mounted a "fixed" spotlight on a tripod with the beam oriented vertically, and used a second, handheld light to track and identify targets flying through the fixed spotlight's beam. During the study between August 16 and September 29, 2005, they found the proportions of birds and bats below maximal turbine height to be 88% and 12%, respectively.

Thermal Infrared Imaging

Although the ceilometer's vertically pointing narrow beam of light did not appear to influence migrating birds or bats during fair weather, there was still concern that the light could possibly influence flight behavior. To address this problem the use of thermal infrared imaging cameras to monitor bird and bat movements aloft was explored. One of the advantages of thermal imaging over natural or artificial illumination is that thermal infrared cameras can detect warm objects independent of any visible or infrared light source (Hill and Clayton 1985). Thermal infrared cameras are designed to detect heat emitted by objects as long as they are warmer than the background, and thus these devices can be used at night and during the day (Figure C-6). A drawback to the widespread use of thermal imaging

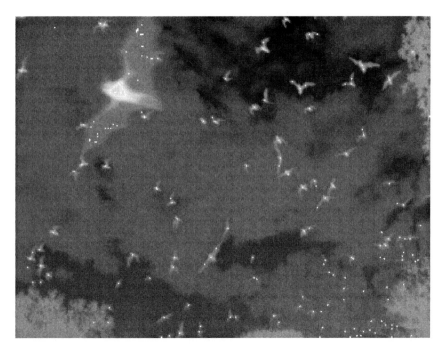

FIGURE C-6 Thermal infrared image of foraging Brazilian free-tailed bats in south-central Texas.
SOURCE: Thomas Kunz, Boston University.

cameras for monitoring movements of birds, bats, and insects is the cost of the units. A single camera can range from $60,000-$120,000 depending on specifications (e.g., resolution and sensitivity) and lenses.

Thermal infrared imaging has been compared with other techniques (moon-watching and tracking radar) for detecting and monitoring aerial bird movements (Liechti et al. 1995), and the results suggest that a long-range thermal imaging unit (LORIS, IRTV-445L, Inframetrics) with a 1.45° telephoto lens could detect nearly 100% of the small passerines within 3000 m. The same unit has been used in southern Sweden to study fall bird migration (Zehnder and Karlsson 2001; Zehnder et al. 2001) and in Africa on the edge of the Sahara to study nocturnal bird migration across the desert (Liechti et al. 2003).

Thermal imaging has also been used to monitor bird movements near wind turbines (Winkelman 1992b; Desholm 2003). A thermal-imaging camera (UA 9053, Philips Usfa) with three different lenses was used to observe the flight behavior of nocturnal songbird migrants as they flew through land-based turbines in Holland (Winkelman 1992c). A 15° lens could detect passerines out to a distance of 50-250 m and a pigeon out to 250-300 m. A 5° lens could detect a pigeon out to 600 m and a 3° lens could detect a duck out to 3 km. A long-wave (7-15 μm) thermal-imaging camera (Thermovision IRMV 320V, FLIR Systems 2000) was used for automatic detection of avian collisions at offshore wind-energy facilities by Desholm (2003). This device can be triggered automatically when a target is detected and can be aimed remotely (Desholm et al. 2004). Vertically pointing thermal-imaging cameras and fixed-beam radars were combined to monitor aerial bird, bat, and insect movements (Gauthreaux and Livingston 2006). The thermal imager and radar data are combined into a single video image (Figure C-7) and stored on digital video tape for analysis. This approach produced quantitative measures of migration-traffic rate for any combination of altitudinal bands, and the technique is useful for distinguishing birds from insects and foraging bats (Gauthreaux and Livingston 2006).

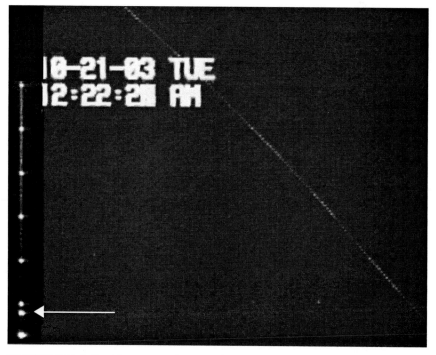

FIGURE C-7 Split video screen showing the vertical radar beam (left) and the field of view of the thermal imager (right). An arrow indicates the echo of a bird in the radar beam. The other marks in the radar beam are range marks over 463 m. The track of the bird as a thermal image is brighter when wings are open and duller when wings are closed. The target in this image is flying toward the southwest. SOURCE: Sidney Gauthreaux, Clemson University.

Thermal infrared imaging has proven valuable for censusing roost sites for the presence and seasonal activity of bats near proposed or developed wind-energy facilities, as well as for observing their flight activity (commuting, foraging, and migratory activity) in the vicinity of wind turbines (Desholm 2003; Horn and Arnett 2005; Horn et al. in press). Coupled with portable computers, thermal images of flying bats, birds, and insects can be recorded and analyzed, making it possible to quantify flight trajectories as well as the relative densities of these animals in different landscapes.

Thermal infrared cameras were used at the Mountaineer Wind Energy Center in 2004 (Horn and Arnett 2005; Horn et al. in press) to investigate whether, when, where, and how bats were killed by the wind turbines. Three FLIR S-60 infrared thermal cameras were positioned at one of 44 wind turbines where the highest fatalities of bats were observed in 2003. Each camera was positioned to record images at overlapping locations that spanned the upper region of the monopole, the nacelle, and most of the rotor-swept zone (Figure C-8). The thermal images recorded in this study clearly indicated that bats were killed by direct contact with moving turbine blades. Additionally, the images showed some bats flying in the vicinity of

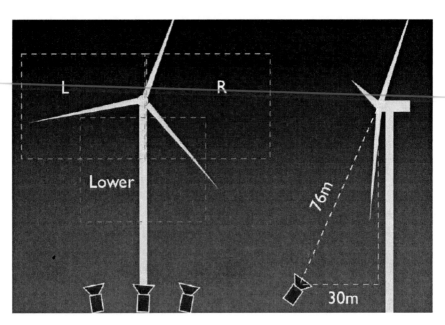

FIGURE C-8 Configuration of thermal infrared cameras (FLIR S-60) used to record the flight behavior of bats in the rotor-swept zone at a wind turbine at the Mountaineer Wind-Energy Center, Tucker County, West Virginia.
SOURCE: Horn et al. in press.

moving rotor blades, as if the bats were "inspecting" or being attracted to the blades, possibly by insects that also were in the vicinity of the rotors. Other images suggest that some bats may have followed the tips of the turbine blades, or were possibly caught in the blade-tip vortices (Figure C-9).

Because large numbers of insects accumulate on the surfaces of turbine blades at some localities (Corten and Veldkamp 2001), it is possible that insects may be attracted to the turbines. If so, wind turbines may be creating patches of aerial insects that bats feed on, a topic that needs additional research. Observations made with thermal-imaging cameras have recorded bats interacting with moving and stationary turbine rotors in various ways, and thus, images derived from thermal infrared cameras can provide valuable information on how bats interact with wind turbines and insects at these sites. Additional information of this type will be critical for identifying possible ways to mitigate the high fatality rates that have been observed and reported at various wind-energy facilities.

Thermal infrared-imaging cameras also can provide accurate and reliable census information about bats that roost in caves and similar structures, from which they emerge at dusk (Sabol and Hudson 1995; Frank

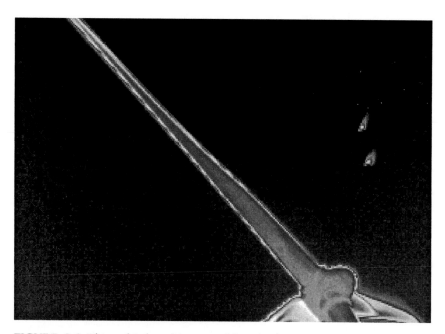

FIGURE C-9 Thermal infrared images of bats in the rotor-swept zone of a wind turbine at the Mountaineer Wind-Energy Center, Tucker County, West Virginia. SOURCE: Adapted from Horn et al. in press.

et al. 2003; Betke et al. in press; Kunz et al. in press a). This type of information could be critical for assessing the long-term and cumulative impacts on bat colonies in the vicinity of proposed and developed wind-energy facilities. Recent developments of computer vision and tracking algorithms have advanced the ability to automatically census bats that roost in large colonies (Figure C-10) and also to record flight trajectories of foraging and migrating bats (Kunz 2004; Horn and Arnett 2005; Betke et al. in press; Horn et al. in press; Kunz et al. in press a).

Chemiluminescent Light Tags

Chemiluminescent light tags (Cyalume®, Cyalume Light Technologies 2006) also offer potential for observing the flight behavior of bats, including those flying in the vicinity of proposed and operational wind-energy facilities. Buchler (1976) and Buchler and Childs (1981) used chemiluminescent light tags to observe the dispersal, commuting, and foraging behavior of selected North American insectivorous bats. Other investigators (e.g., LaVal and LaVal 1980; Aldridge and Rautenbach 1987) also have used this technique with the greatest success rates when observations were made in open areas, in flyways, and along forest edges. Buchler and Childs (1981)

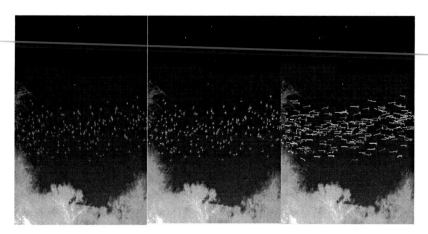

FIGURE C-10 Sequence showing thermal infrared image of Brazilian free-tailed bats emerging from Davis Blowout Cave in south-central Texas. "Left": image as detected with Merlin mid thermal infrared camera; "Middle": same image as in "Left" to which an automated detection algorithm was applied—identifying centroids; and "Right": same image as "Left" and "Middle", but with the automatic tracking algorithm applied.
SOURCE: Kunz et al. in press a.

observed the dispersal of light-tagged *E. fuscus* and postulated that individuals navigated to feeding grounds by following acoustic cues produced by calling frogs and stridulating insects. Use of chemiluminescent light tags may offer opportunities to observe the behavior of bats in response to sounds produced by moving wind-turbine blades or to insects that may be attracted to these structures. Light tags also have been used successfully to follow known individuals while their echolocation calls are monitored using ultrasonic detectors, and thus can be used to validate species-specific calls and therefore used to identity calling bats. The greatest limiting factor of using light tags to investigate flight behavior of bats is that they often quickly disappear from view, especially in heavily forested areas, and in some instances are difficult to distinguish from fireflies.

Radiotelemetry

Radiotelemetry has primarily been used to assess the roosting habits, foraging behavior, and home ranges of bats and birds, and less for assessing migratory behavior. The most comprehensive radiotelemetry study of birds associated with a wind-energy facility was Hunt's (2002) four-year study of golden eagles at the APWRA. Hunt used telemetry to evaluate spatial and temporal use and to estimate reproduction and fatalities of a local population of golden eagles in northern California, and to relate these population parameters to impacts from the wind-energy facilities within the APWRA. Radiotelemetry generally has not been used to investigate migratory behavior of small bats and birds, largely because the detection ranges of most small transmitters are limited to a few kilometers. Nonetheless, some investigators have successfully used fixed-wing aircraft to follow dispersal movements of the endangered Indiana bats (*Myotis sodalis*) over considerable distances (C.W. Butchkoski, Pennsylvania Game Commission, personal communication 2006; J. Chenger, Bat Conservation and Management Inc., personal communication 2006; A. Hicks, New York Department of Environmental Conservation, personal communication 2006). The latter studies have shown that Indiana bats do fly along and actually cross over ridge tops in the eastern United States. Radiotracking by satellite telemetry currently is feasible only for birds and bats that are able to carry radiotransmitters weighing in excess of 10 g (Spencer et al. 1991; Eby 1991; Millspaugh and Marzliff 2001; Fleming and Eby 2003; Tidemann and Nelson 2004). In contrast, because of their small size, bat species from the Mid-Atlantic Highlands generally cannot carry transmitters heavier than 0.5 to 1.0 g and maintain normal flight behavior. Recent developments using satellite tracking of small vertebrates hold considerable promise for tracking small (~30 g) bats and birds (M. Wikelski, Princeton University, personal communication 2006).

Stable Isotopes and Genetic Markers

Stable isotopes used to assess geographic variation in patterns of precipitation and the unique stable-isotope signatures that are transferred from precipitation to biological primary producers (plants) and ultimately to consumers (herbivores and carnivores) have provided new tools for understanding migration of birds and bats (e.g., Chamberlain et al. 1997; Kelly and Finch 1998; Marra et al. 1998; Hobson 1999; Hobson and Wassenaar 2001; Bowen and Wilkinson 2002; Rubenstein et al. 2002; Rubenstein and Hobson 2004), and genetic data (Berthold 1991; Clegg et al. 2003; Royle and Rubenstein 2004; Kelly et al. 2005). Stable-isotope techniques have been used mostly to associate breeding areas (where molt to new plumage or hair growth or replacement occurs) to migratory stopover areas and wintering areas. The resolution of the signatures is rather crude with respect to latitude and longitude so that it may not be possible to precisely discriminate source areas within a small geographical region. However, as geographical distance increases so does the reliability of the isotope signature. Stable-isotope signatures are also sensitive to elevation; thus altitudinal migration may be confounded with latitudinal migration.

Analysis of stable isotopes shows promise in differentiating migratory status. Kelly et al. (2002) used stable isotopes of hydrogen contained in feathers to estimate hydrogen stable-isotope ratios (dD) of feathers from breeding, migrating, and wintering Wilson's warblers (*Wilsonia pusilla*). They found that feathers from museum specimens collected throughout the western portion of the breeding range indicate that dD values were significantly and negatively related to latitude of collection, which is an indication that dD values provide a good descriptor of the latitude at which breeding occurs. They also found by analyzing feathers collected on the wintering grounds that the hydrogen isotope ratio was significantly positively related to wintering latitude. Gannes et al. (1997) pointed out the importance of identifying assumptions inherent in stable-isotope analysis and called for laboratory experiments to validate the method. Advances in genetic analysis also show promise in determining the population and geographical origin of individual birds (Webster et al. 2002), and may assist in identifying the origin of bird and bat fatalities at wind-energy facilities.

Mitochondrial and nuclear DNA markers have provided valuable for determining source populations of animals that move long distances. They could make it possible to identify geographic origin of bats and birds killed at wind-energy facilities if investigators collected hair samples from bats and feather samples from birds, and compared the isotope signatures and genetic markers on a geographic scale. This kind of information could aid in determining whether bats and birds killed at wind-energy facilities were residents or migrants. Moreover, established DNA sequences for different species can aid in the identification of birds and bats (or parts thereof) from the carcass remains of individuals found during fatality searches.

CAPTURE TECHNIQUES

Capture methods are invaluable for assessing and confirming the presence of a species, although it may not be possible or practical to capture all species in an area. Some forage and migrate well above the practical limits of capture, although many species fly closer to ground level or forage over water or within the subcanopy of forests. While capture may be challenging for many nocturnal species, captures of migrating passerines are more likely during stopovers. Correct identification of species present in the area of interest is essential for assessing potential risks of wind-energy facilities to different species. For bats and migratory birds, this usually requires that live or dead animals be available for study.

Methods and equipment used to capture live bats have been thoroughly described (Greenhall and Paradiso 1968; Tuttle 1976; Kunz and Kurta 1988; Kunz et al. 1996a; Kunz et al. in press a). Methods for assessing colony size, demographics, and population status of bat species are in O'Shea and Bogan (2003) and Kunz and Parsons (in press) and methods for landbirds are in Sutherland et al. (2004). Many of the methods used to capture birds and bats are similar, albeit with some differences. If bats are to be captured at roost sites to assess the species present in the vicinity of wind-energy facilities, or to monitor changes in colony size, harp traps are preferable to mist nets (Kunz et al. in press a). Most important, efforts should be made to minimize disturbance to bat colonies. No single capture method is suitable for all bat species, although mist nets and harp traps are the most commonly used devices for capturing these animals while in flight because they are relatively easily deployed and can be used in a variety of situations.

A mist net consists of a nylon mesh supported by a variable number of taut, horizontal threads, or shelf strings. Bats and birds are captured after they become entangled in the mesh of the nets. Mist nets are available from manufacturers in different colors and sizes. For nighttime netting black is the preferred color. They may be set as single net at ground level or stacked on top of one another to form a canopy net (Figure C-11). Simple canopy nets can be modified by restringing horizontal nets (Munn 1991; Rinehart and Kunz 2001). Ground-level nets are generally most practical to deploy, but are biased against species or individuals that do not fly close to the ground. Use of elevated canopy nets can provide researchers access to the aerial space where some bats and birds may commute and forage, although even with canopy nets erected into or suspended in the subcanopy, elevated nets are not suitable for capturing species that typically fly above

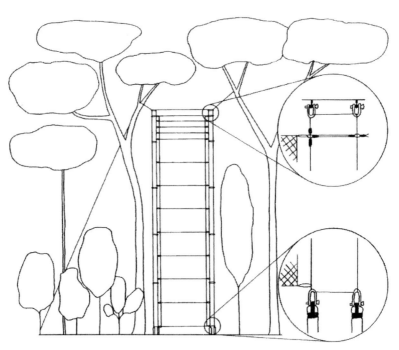

FIGURE C-11 Multiple stacked horizontal mist nets used for capturing bats and birds from ground level into the forest subcanopy.
SOURCE: Hodgkison et al. 2002. Reprinted with permission; copyright 2002, Global Canopy Programme.

the canopy. In these situations, other tools such as ultrasonic detectors and audible sound recordings may be more appropriate.

During pre-construction surveys where the local bat fauna and possible colony sizes are unknown, harp-trapping may be used successfully at expected or potential commuting, foraging, drinking, and roosting sites. Prior assessment of local topography, habitat structure (foliage density), and visual or acoustic surveys often can facilitate the selection of a potential capture site and the appropriate deployment of mist nets and harp traps.

VISUAL ESTIMATES OF OCCURRENCE AND USE

The most common approach to estimate species occurrence and relative abundance of diurnally active bird species is through visual observation (Ralph et al. 1993; Bibby et al. 2000). Quantification of abundance is achieved by sampling an area of interest, usually using line-transect

(Burnham et al. 1980) or point-count (Reynolds et al. 1980) sampling. The opportunity to estimate species-specific abundance and behavior is a valuable asset of visual-estimation methods. While this discussion is focused on visual surveys, the theory and application of line-transect and point-count sampling is well suited to nocturnal surveys using radar and other survey methods.

Line-transect sampling is typically applied with a line randomly or systematically located on a baseline as the basic sampling unit, and is extended across the study region (Morrison et al. 2001). Objects on either side of the line are recorded based on some rule of inclusion. Line-transect sampling, where an effort is made to count all organisms within a certain distance, is equivalent to a belt transect (or rectangular plot). When surveys are completed according to a standard protocol, without correction for detection bias, the counts can be considered an index of abundance (e.g., Conroy et al. 1988). Line-transect counts are most often considered incomplete when used to estimate absolute abundance, because objects are always missed and the probability of detection must be estimated. The theory and application of this sampling method have received much attention in the scientific literature (e.g., Burnham et al. 1980; Buckland et al. 1993; Manly et al. 1996; Quang and Becker 1996, 1997; Beavers and Ramsey 1998). Line transects are commonly used in bird surveys and are best suited to grassland and shrub-steppe landscapes.

Counts from a variable circular plot often are applied as a variation of the line-transect sampling method for estimating the number of birds in an area (Reynolds et al. 1980). The variable circular plot is more useful than the line transect in dense vegetation and rough terrain, where attention may be diverted from the survey and toward simply negotiating the transect line (Morrison et al. 2001). One major advantage of the circular plot is that the observer can allow the subjects of the counts to become accustomed to the observer. In breeding-bird surveys (Reynolds et al. 1980), observers wait several minutes after their arrival at a point before counts begin. Stationary surveys also allow the observer to use both visual and auditory senses to detect birds. Program DISTANCE (Laake et al. 1993) can be used to estimate densities from circular-plot data (see also Rosenstock et al. 2002). Johnson et al. (2000b) described the use of circular plots in the estimate of relative abundance of songbirds from small plots (i.e., 100-m radius) and large birds from larger plots (i.e., 0.8 km) in pre-project studies at the proposed Buffalo Ridge wind-energy facility in southwestern Minnesota.

Estimates of Fatalities

Fatalities are typically estimated from carcasses located on standardized search plots at turbines, turbine strings, meteorological towers, and reference areas. Search plots at wind-energy facilities may take many shapes from circular to rectangular and typically contain one or more turbines, depending on the spacing of individual turbines. Plot boundaries are delineated at a minimum distance from the turbines, usually based on the size of the turbine. Plots most often are circular or elliptical and are centered on the turbine or turbine string, with the edge of the plot from 30 to 100 m from the nearest turbine. Studies conducted at wind-energy facilities in Oregon (Erickson et al. 2000, 2003b), Minnesota (Johnson et al. 2003a), Wyoming (Young et al. 2003b), and Washington (Erickson et al. 2003a) found most dead bats (more than 80%) within one half the maximum distance from the tip height to the ground from the monopole of the turbine. Arnett (2005) found that 93% of all fatalities at the Mountaineer site and 84% of all the fatalities at Myersdale were found less than or at 40 m from the nearest turbine. At both sites, fewer than 3% of fatalities were found more than 50 m from the nearest turbine. Preliminary evaluation of the distribution of bird carcasses within search plots at the Stateline wind-energy facility in Oregon and Washington (Erickson et al. 2004) suggests that bird carcasses occur further from turbines than bats. However, few birds were located on the periphery of the 63-m-radius search plots, and thus may not represent what actually occurs. At the Mountaineer wind-energy Center in West Virginia, Kerns and Kerlinger (2004) searched out to 60 m from the base of each tower and found birds and bats out to 60 m, although the majority of the carcasses were between 16 and 30 m of the base of turbine towers. The size of search plots should increase with turbine height and diameter of the rotor, using a minimum plot radius approximately equal to diameter of the rotor. Turbine plots to be searched should be selected through a probabilistic sampling process allowing extrapolation to the entire wind-energy facility, after considering variation in topography and type of vegetation present at each site. A systematic selection process with a random start is the most effective method for most sites (Morrison et al. 2001).

Personnel trained in proper search techniques typically conduct standardized carcass searches by walking parallel transects within the search plot at a predetermined speed. The cause of death of each carcass should be determined so that fatalities determined not to be related to the wind-energy facility could be discounted. Suggested criteria for identifying bird and bat remains as a bird or bat carcass are:

- *Intact:* A carcass that is completely intact, is not badly decomposed and shows no sign of being fed upon by a predator or scavenger.
- *Scavenged:* An entire carcass that shows signs of being eaten by a predator or scavenger, or portions of a carcass in one location (e.g., wings, skeletal remains, legs, pieces of skin, etc.).

And for birds only:

• *Feather Spot:* 10 or more feathers at one location indicating predation or scavenging.

Some bat and bird fatalities that are discovered and used in fatality-rate estimation may not be related to wind-energy projects, even though the cause of death cannot be determined. Natural mortality and predation may be responsible, but the level of this background mortality in project areas typically has not been studied. However, background fatalities can be significant. For example, of the 86 avian fatalities found during a four-year study at the Buffalo Ridge wind-energy facility in Minnesota, Johnson et al. (2002) found 31 fatalities (36%) at reference plots. Thus, including background fatalities in calculations of fatality estimates may contribute to overestimation of project-related fatality rates, particularly for smaller species. By contrast, failure to detect bird and bat fatalities outside a designated search area may underestimate fatality rates. Care should be taken to insure that fatality counts at reference and turbine plots are independent.

Carcass Survey Biases

Carcass searchers no doubt fail to locate some carcasses in search plots. Carcass detection is affected by topography, vegetation within the plot, the size of the search plot, size of the remains of the bird or bat, climate, weather, and observer skill. Observer-detection bias or searcher-efficiency studies are necessary to estimate the percentage of actual bird and bat fatalities that searchers are able to find (Anderson et al. 1999). Typically, these studies are conducted in the same area in which standardized searches occur and thus should include all habitat types. Trials should be conducted in each season in each monitoring year. Search efficiency can be improved when trained dogs are used to find carcasses (Arnett 2006). Estimates of observer-detection rates are used to adjust the number of carcasses found for detection bias.

Carcasses also may be removed from search plots before they are searched. This removal is most often by scavengers, but carcasses could be removed by other causes (e.g., human activity, wind). Carcass-removal bias is estimated by conducting experimental studies that estimate the length of time bird and bat carcasses remain in the search area before being removed by scavengers or other means. Carcass-removal studies should be conducted during each season of each monitoring year in the vicinity of, but not on the search plots. Estimates of carcass removal are used to adjust carcass counts for removal bias. Daily searches are essential when evaluating carcass removal of bats. Arnett (2005) estimated that 35% of randomly placed test carcasses were removed with the first 24 hours, 48% were removed within 48 hours, and by the 18th day more than 90% of the test carcasses were removed. The rate of removal for small birds appears to be less than for bats, and weekly or biweekly searches for birds may be adequate. The mean duration for small-bird test carcasses have ranged from 4.69 days at Buffalo Ridge (Johnson et al. 2003a) to 16.7 days at Stateline (Erickson et al. 2004). Scavenging rates may differ seasonally and from year to year following construction of wind-energy facilities. It is possible that scavenging may actually increase because scavengers develop search images and return to sites more frequently once carcasses have been discovered.

Carcass-removal and searcher-detection trials use carcasses placed in areas either in plots used in standardized searches (searcher detection trials) or in nearby areas of similar characteristics (carcass removal trials). Carcasses of varying sizes should be placed in most of the habitats being searched. Carcasses of native bats and birds found within the wind-energy facility are ideal for use. The experimental placement of frozen instead of fresh carcasses and using birds as surrogates for bats may contribute to biases in estimated removal rates. Ideally, fresh carcasses should be used in these experiments, because they more closely mimic what occurs near a wind turbine. However, adequate supplies of native bats and birds are seldom available and surrogate carcasses may be used, even though this approach may yield biased results. The efficacy of using surrogate carcasses and fresh versus frozen specimens needs further investigation.

STATISTICAL METHODS FOR FATALITY ESTIMATES

Methods for estimation of the total number of wind-facility-related fatalities are taken from Erickson et al. (2004) and are based on:

- Observed number of bat and bird carcasses found during standardized searches for which the cause of death is either unknown or is probably facility-related;
- Searcher efficiency expressed as the proportion of planted carcasses found by searchers during the entire survey period; and,
- Non-removal rates expressed as the estimated average probability that a carcass will remain in the study area and be available for detection by the searchers during the entire survey period.

Definition of Variables

The following variables are used in equations (1-3) below:

c_i number of carcasses detected at plot I for the study period of interest (e.g., one year) for which the cause of death is either unknown or is attributed to the facility

n number of search plots

k number of turbines searched (includes the turbines centered within each search plot and a proportion of the number of turbines adjacent to search plots to account for the effect of adjacent turbines within the arch plot buffer area)

$\bar{c}$ average number of carcasses observed per turbine per year

s number of carcasses used in removal trials

s_c number of carcasses in removal trials that remain in the study area after 40 days

se standard error (square of the sample variance of the mean)

t_i time (days) a carcass remains in the study area before it is removed

$\bar{t}$ average time (days) a carcass remains in the study area before it is removed

d total number of carcasses placed in searcher efficiency trials

p estimated proportion of detectable carcasses found by searchers

I average interval between searches in days

$\hat{\pi}$ estimated probability that a carcass is both available to be found during a search and is found

m estimated annual average number of fatalities per turbine per year, adjusted for removal and observer-detection bias

Observed Number of Carcasses

The estimated average number of carcasses ($\bar{c}$) observed per turbine per year is:

$$\bar{c} = \frac{\sum_{i=1}^{n} c_i}{k} . \tag{1}$$

Estimation of Carcass Removal

Estimates of carcass removal are used to adjust carcass counts for removal bias. Mean carcass-removal time ($\bar{t}$) is the average length of time a carcass remains at the site before it is removed:

$$\bar{t} = \frac{\sum_{i=1}^{s} t_i}{s - s_c} . \tag{2}$$

This estimator is the maximum-likelihood estimator assuming the removal times follow an exponential distribution. When the estimate is that no carcasses will be left, the collection of data ends, or is censored. The probability of finding a carcass decreases with time, by convention to the right from the origin, and thus the data are said to be "right-censored." Erickson et al. (2004) collected trial bird carcasses still remaining at 40 days, yielding censored observations at 40 days. If all trial bird carcasses are removed before the end of the trial, then sc is 0, and $\bar{t}$ is simply the arithmetic average of the removal times. For bats, carcasses were monitored every day for 20 days. Removal rates are estimated by carcass size (small and large) and season.

Estimation of Observer-Detection Rates

Observer-detection rates (i.e., searcher-efficiency rates) are expressed as p, the proportion of trial carcasses that are detected by searchers. Observer-detection rates are estimated by carcass size and season.

Estimation of Facility-Related Fatality Rates

The estimated per-turbine annual fatality rate (m) is calculated by:

$$m = \frac{\bar{c}}{\hat{\pi}},$$ (3)

where $\hat{\pi}$ includes adjustments for both carcass removal (from scavenging and other means) and observer-detection bias assuming that the carcass removal times $\bar{t}_i$ follow an exponential distribution. Data for carcass removal and observer-detection bias are pooled across the study to estimate $\hat{\pi}$. Under these assumptions, this detection probability is estimated by

$$\hat{\pi} = \frac{\bar{t} \cdot p}{I} \cdot \left[\frac{\exp\left(\frac{I}{\bar{t}} \right) - 1}{\exp\left(\frac{I}{\bar{t}} \right) - 1 + p} \right].$$ (4)

This equation has been independently verified by Shoenfeld (2004).

Fatality estimates should be calculated for the species and size class of interest. Erickson et al. (2004) used 7 groups including all birds, small birds, large birds, raptors, grassland birds, nocturnal migrants, and bats. The final reported estimates of fatalities and associated standard errors and 90% confidence intervals can be calculated using bootstrapping (Manly 1997), a computer-simulation technique that is useful for calculating point estimates, variances, and confidence intervals for complex test statistics.

TABLE C-5 Neotropical Migrant Species that Have Shown a Negative Population Trend During the Time Period 1978-1987

Common Name	Scientific Name	Trend, %/year
Broad-winged Hawk	*Buteo platypterus*	−2.3
Black-billed Cuckoo	*Coccyzus erythrophthalmus*	−5.9
Yellow-billed Cuckoo	*Coccyzus americanus*	−5.0
Chuck-will's Widow	*Caprimulgus carolinensis*	−2.0
Whip-poor-will	*Caprimulgus vociferus*	−0.8
Olive-sided Flycatcher[a]	*Contopus borealis*	−5.7
Eastern Wood-pewee[a]	*Contopus virens*	−0.7
Acadian Flycatcher	*Empidonax virescens*	−1.3
Least Flycatcher[a]	*Empidonax minimus*	−0.2
Great crested Flycatcher[a]	*Myiarchus crinitus*	−0.3
Veery[a]	*Catharus fuscescens*	−2.4
Swainson's Thrush[a]	*Catharus ustulatus*	−0.2
Wood Thrush[a]	*Hylocichla mustelina*	−4.0
Gray Catbird	*Dumetella carolinensis*	−1.4
White-eyed Vireo	*Vireo griseus*	−1.2
Solitary Vireo[a]	*Vireo solitarius*	−0.1
Yellow-throated Vireo	*Vireo flavifrons*	−0.9
Blue-winged Warbler	*Vermivora pinus*	−1.0
Golden-winged Warbler[a]	*Vermivora chrysoptera*	−1.9
Tennessee Warbler	*Vermivora peregrina*	−11.6
Northern Parula	*Parula americana*	−2.1
Chestnut-sided Warbler[a]	*Dendroica pensylvanica*	−3.8
Cape May Warbler	*Dendroica tigrina*	−2.3
Black-throated Green Warbler[a]	*Dendroica virens*	−3.1
Blackburnian Warbler[a]	*Dendroica fusca*	−1.1
Yellow-throated Warbler	*Dendroica dominica*	−0.4
Prairie Warbler	*Dendroica discolor*	−0.4
Bay-breasted Warbler	*Dendroica castanea*	−15.8
Blackpoll Warbler	*Dendroica striata*	−6.3
Cerulean Warbler[a]	*Dendroica cerulea*	−0.9
American Redstart[a]	*Setophaga ruticilla*	−1.2
Worm-eating Warbler[a]	*Helmitheros vermivorus*	−2.0
Ovenbird[a]	*Seiurus aurocapillus*	−1.0
Louisiana Waterthrush	*Seiurus motacilla*	−0.4
Kentucky Warbler[a]	*Oporornis formosus*	−1.6
Mourning Warbler[a]	*Oporornis philadelphia*	−1.6
Common Yellowthroat	*Geothlypis trichas*	−1.9
Wilson's Warbler	*Wilsonia pusilla*	−6.5
Canada Warbler[a]	*Wilsonia canadensis*	−2.7
Summer Tanager	*Piranga rubra*	−0.8
Scarlet Tanager[a]	*Piranga olivacea*	−1.2
Rose-breasted Grosbeak[a]	*Pheucticus ludovicianus*	−1.4
Indigo Bunting	*Passerina cyanea*	−0.7
Baltimore Oriole	*Icterus galbula*	−2.9

[a]Denotes species that breed in the Mid-Atlantic Highlands. All others are known from migration records only.
SOURCES: Data from Robbins et al. (1989); breeding status follows Hall (1983) and Buckelew and Hall (1994).

TABLE C-6 Bird Species of Conservation Concern that Potentially Occupy Ridge-Top Habitats in the Mid-Atlantic Highlands[a]

Common Name	Scientific Name	Status[b]
Bald Eagle	*Haliaeetus leucocephalus*	MD-T
Peregrine Falcon	*Falco peregrinus*	MD-I, VA-T, PA-E
Northern Goshawk	*Accipter gentilis*	MD-E
Olive-sided Flycatcher	*Contopus cooperi*	MD-E
Alder Flycatcher	*Empidonx alnorum*	MD-I, VA-SC
Sedge Wren	*Cistothorus platensis*	MD-E, VA-SC, PA-E
Winter Wren	*Troglodytes troglodytes*	VA-SC
Appalachian Bewick's Wren	*Thyromanes bewickii altus*	VA-E
Golden-crowned Kinglet	*Regulus satrapa*	VA-SC
Red-breasted Nuthatch	*Sitta canadensis*	VA-SC
Blackburnian Warbler	*Dendroica fusca*	MD-T
Blackpoll Warbker	*Dendroica striata*	PA-E
Magnolia Warbler	*Dendroica magnolia*	VA-SC
Swainson's Warbler	*Limnothlypis swainsonii*	MD-E, VA-SC
Mourning Warbler	*Oporornis philadelphia*	MD-E, VA-SC
Nashville Warbler	*Vermivora ruficapilla*	MD-I
Hermit Thrush	*Catharus guttatus*	VA-SC
Red Crossbill	*Loxia curvirostra*	VA-SC

[a]WV has no statutes requiring the development of State Endangered and Threatened species lists.
[b]MD = Maryland, PA = Pennsylvania, VA = Virginia, E = Endangered, T = Threatened, SC = "Species of Special Concern", I = "In Need of Conservation."
SOURCES: VA (Roble 2006), MD (MDDNR 2003), PA (PAGC 2006).

BIRD SPECIES OF CONCERN FOR THE
MID-ATLANTIC HIGHLANDS

Concern exists regarding the status of a number of bird species potentially occurring in the Mid-Atlantic Highlands. Table C-5 contains a list of neotropical-migrant bird species and Table C-6 contains a list of bird species of conservation concern. Because such species should receive careful attention when considering the impacts of a proposed wind-energy facility for the Mid-Atlantic Highlands. These lists should be updated as the status of these and other species changes.

APPENDIX
D

A Visual Impact Assessment Process for Evaluating Wind-Energy Projects

Evaluating aesthetic impacts requires a process of information-gathering, analysis, and evaluation. This appendix provides a more detailed outline than is in Chapter 4 of the steps involved and some of the underlying visual principles that form the basis of aesthetic impact assessment.

The steps are as follows:

- Project Description.
- Project Visibility, Appearance, and Landscape Context.
- Scenic-Resource Values and Sensitivity Levels.
- Assessment of Aesthetic Impacts.
- Mitigation Techniques.
- Determination of Acceptability or Undue Aesthetic Impacts.

PROJECT DESCRIPTION

A detailed description of all elements of a proposed project is an essential first step. All site alternatives that will have potential visual impacts should be identified by the developer in detail. These should include the characteristics of the turbines (e.g., height, rotor diameter, color, rated noise levels), the number planned, their locations; information about meteorological towers; roads; collector, distribution, and transmission lines; temporary or permanent storage ("laydown") areas; substations; and any structures associated with the project. In addition all site clearing should be identified, including clearing for turbines, roads, power lines, substations, and laydown areas. Information also is needed on all site regrading that will be engineered, including the amount of cut and fill, locations, and clearing required. This information forms the basis for all aesthetic review.

PROJECT VISIBILITY AND LANDSCAPE CONTEXT

A number of tools and techniques are available for determining visibility and for describing relevant landscape and project characteristics. The key techniques outlined below often are required as part of a permit application.

Computer Viewshed Analysis

Computer-generated maps based on digital-elevation models (DEMs) illustrate where any hypothetical point (such as the tip of a turbine blade) could potentially be visible within a given area, such as a 10-mile radius around the proposed project (Figure D-1). They also can indicate approximately how many turbines are likely to be visible from a given point. They are based on digital-terrain modeling and may not account for surface elements like vegetation or buildings that might block views. Field analysis is essential to verify actual visibility. It also is possible to do a "partial viewshed analysis," which examines the visibility of particular turbines, or to look at a particularly sensitive viewing point on the ground to examine an area of potential visibility.

Line-of-Sight Visual Analysis

When complex topography makes it difficult to determine whether a particular turbine or other object will be visible from a particular point, a line-of-sight analysis can provide a useful check (Figure D-2).

Simulations (Visualizations)

Several types of simulations can be used to help predict how the project will appear. Photographic simulations or photomontages based on still photographs taken from selected viewpoints are the most common (Figures D-3 and D-4). Some professionals prefer 3D visualization models, which create a digital image from selected viewpoints. These images eliminate the variability and lack of clarity in some photographs and can depict conditions ranging from clear blue skies to nighttime lighting conditions, but they are not as realistic in appearance and details as a photographic simulation. Animated simulations illustrate the rotation of the blades on the turbines at accurate speeds. Photographic simulations generally show only a narrow window of a particular view (wide-angle lenses result in inaccurate perspec-

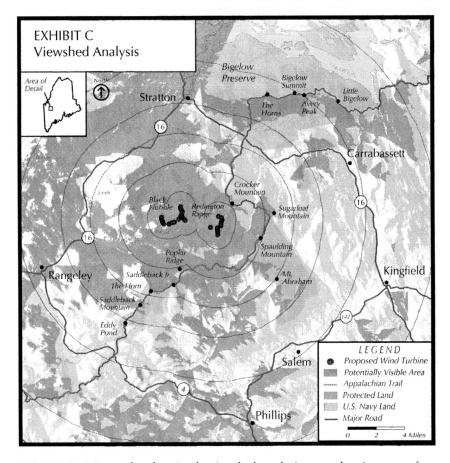

FIGURE D-1 Example of a simple viewshed analysis map showing areas from which a proposed wind-energy project would potentially be visible (shaded areas). Field assessment is necessary to determine actual visibility and the characteristics of the views. Source: Appalachian Trail Conservancy 2007. Reprinted with permission; copyright 2007, Appalachian Trail Conservancy.

tives). In understanding visual impacts it is useful to understand the broader context of the view. Whether the broader panorama will contain turbines as well, or whether it will remain undeveloped, will be an equally important part of the analysis. Several 3D visualization programs allow "fly-through" simulations, and are based on a virtual landscape.

Creating technically accurate simulations is critically important. Simulations can be manipulated to produce images that either exaggerate or minimize the visual impacts of a proposed project. Accuracy should be

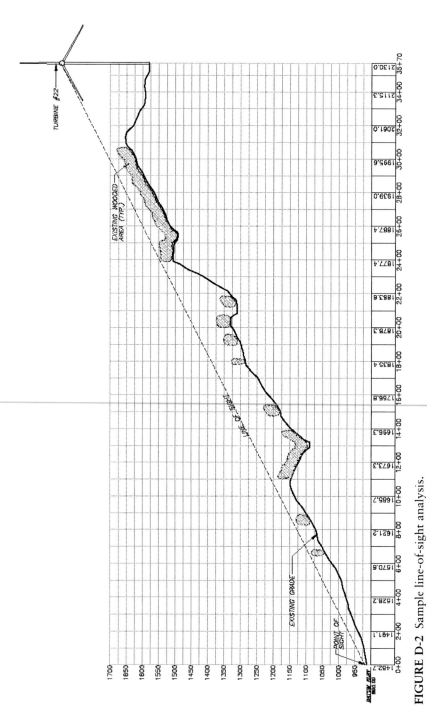

FIGURE D-2 Sample line-of-sight analysis.
SOURCE: NYSERDA 2005b. Reprinted with permission; copyright 2005, prepared by Saratoga Associates under contract for the New York State Energy Research and Development Authority.

Exhibit "B"
Simulation of Redington Wind Energy Project

Viewpoint #1, Appalachian Trail at Saddleback Mt.
Distance to nearest turbine: 5.6 miles
Number of turbines visible: 30
Overall turbine height: 410 feet
New road construction visible: Yes
Horizontal field of view: 38.6 degrees
Equivalent focal length: 50mm

Note: To replicate "real-world" scale, simulation
should be viewed 21 inches from eye when
displayed in 10" x 15" format.

Simulation Produced by Erik Crews
USDA Forest Service
July 12, 2006

FIGURE D-3 Example of a "virtual" simulation using 3-D Nature Studio software. Whether they are based on photographs or created entirely from a Digital Terrain Model and 3-D software, the image must rely on accurate terrain modeling and GPS (Global Positioning System) point recording, and on an image that represents the equivalent of a 50-mm lens or a field of view of 38.6 degrees.
SOURCE: Erik Crews, Department of Agriculture, National Forest Service.

checked by experts in the field of digital images. Another check is to have at least two independent parties provide simulations from the same point. The following description provides an overview of good practice, but consulting technical experts and developing standards will be important.

Photographs should be taken with a 50-mm lens or digital equivalent that creates a 38.6° angle of view, which most closely matches human visual perception. Shorter focal lengths tend to flatten out topography and the vertical impression of the turbines, while longer focal lengths tend to exaggerate these features. However, the human eye is much sharper than any camera lens, and so photographs should be taken at high resolution, whether a film or a digital camera is used. Clear weather provides the best clarity of the scene as well as "worst-case conditions," which should be represented in all simulations to allow a complete evaluation. Foreground clutter such as power poles should be avoided if possible in the photograph. Global-positioning system (GPS) location points should be recorded for each simulation viewpoint, preferably using a GPS unit with submeter ac-

FIGURE D-4 Photographic simulation (photomontage) showing proposed 1.5 MW turbines with existing 0.55 MW turbines (right), Wilmington, Vermont.
SOURCE: Photograph by Jean Vissering, Jean Vissering Landscape Architecture; Simulation by AWS-Truewind for Vermont Environmental Research Associates.

curacy, but at least 3 m accuracy, to ensure repeatability. Some landscape architects fly weather balloons to mark locations of the nacelle in the field, but on windy sites it may be difficult to get a vertical position.

Using a DEM, various 3D programs create accurate digital images of the terrain from a particular point that has GPS coordinates recorded, along with the angle of view. Exact turbine locations as well as roads, meteorological towers, and other project infrastructure can be inserted into the model. Available Geographic Information System (GIS) data may vary from 10- to 30-m digital elevation (DE). For example, 30-m DE is accurate to within 15 m vertically and 12 m horizontally, while 10-m DE can be accurate to within several meters. Once the DEM is created, the photograph that contains important detail information such as structures and vegetative patterns can be superimposed on the DEM. Images of the turbine and other structure can be created on the DEM using programs such as Visual Nature Studio and merged with a photograph using a digital photo-editing program. The color, brightness, shadows, and sharpness of the turbines can be adjusted to appear consistent with the photograph. Depending on lighting conditions, the turbines may appear white, or black if they are silhouetted against the sky. Illustrating various lighting conditions can be helpful.

The relationship between the size of the photograph and the distance of the observer is important for creating a realistic image. A minimum image size of "10×12" can be viewed at a comfortable arm's length, and it is preferable to smaller simulations. Poster-size simulations that can be viewed from about 4-5 feet away are suitable for public display. The formula for determining the correct size of the image in relation to the distance viewed is as follows:

Distance from viewer = Width of image $/(2 \bullet \tan (HFOV^1 / 2))$

HFOV should equal 38.6 when using a 50-mm lens or equivalent. Animated images illustrating the rotation of the blades can be projected using PowerPoint and are particularly useful.

Field Assessment and Inventory of Views

A field inventory of views of all public viewpoints within a 10-mile radius of the project provides the basis for evaluating the extent of visibility as well as the visual characteristics of views in the study area. In addition to photographically documenting and mapping viewing locations, the following information should be recorded: distance from project, duration of view,[2] characteristics of the view (intermittent, panoramic, and foreground, middleground and background elements in the view) (Table D-1). Views should be recorded from parks and recreations areas, hiking trails, natural areas, wilderness areas, designated scenic areas or roads, areas with panoramic views, village or town centers, water bodies, state and federal highways, designated scenic roads, other roads receiving heavy traffic (the U.S. Forest Service defines this as an average of 150 vehicles/day), areas with concentrations of residences, and historic sites. Any sites noted in local, regional, and state planning documents as having scenic, recreational, cultural, or natural values can be considered to be potentially sensitive sites.[3] Some viewpoints are more sensitive than others because of differences in viewer expectations, the duration of view, proximity to the project ridges, or the scenic quality of the viewpoint.

---

[1]Horizontal field of view.

[2]Duration of view refers to how long an object remains visible while traveling past it. The term applies to mechanized transport as well as non-mechanized activities such as hiking or canoeing.

[3]It is not a problem for wind-energy projects to be visible from these areas; rather how they are seen and the extent to which they degrade the views or the experience of these landscapes by visitors or residents is critical and is discussed below.

TABLE D-1 Sample Summary of the Characteristics of Inventoried Viewpoints

VP#	Location	Distance from Turbines (miles)	Extent of View Duration or Area
1	Rt. 9 East of Wilmington SIMULATION POINT	4.8-8	0.6 mile; intermittent views for 2.5 miles into Wilmington
2	Fire Tower Molly Stark State Park	6.5	Point
3	Stowe Hill Road	6-6.5	0.4 mile; plus 0.4 mile intermittently

Visual assessment is particularly important in sensitive areas. Residential areas generally cannot be inventoried in detail, but information can be provided about the number of residences that may be affected. In addition to views of the project ridges, other scenic features within the study area need to be documented.

Public Participation in Identifying Viewpoints

For people who live, work, and recreate in a region, the landscape consists of layers of meaning that may not be understood by an outside professional conducting a visual assessment. If local residents and other interested parties can participate in the selection of sites to be inventoried

Description of Existing View	Relationship of New Turbines to Existing Context
For travelers heading west on Route 9, views begin near the top of the ridge just east of Molly Stark State Park. Views focus on a sequence of hills to the west including rounded foreground hills and the flat ridge with the existing turbines near the center of this view. This view is relatively narrow. West of Lake Raponda Road, views become difficult to see due to foreground trees, hills, and buildings interfering with the view.	The turbines along the eastern string will be visible along the background ridge, and will be in the center of the view. Several of the western expansion turbines will be visible behind the existing turbines but will be farther away. As travelers descend into the Wilmington valley, closer hills and ridges will increasingly interfere with the view of the turbines. Rt. 9 is the gateway into the Wilmington valley.
The fire tower offers a 360° view of the myriad hills, mountains and ridges in the area. Since it is close to the ridge dividing Wilmington from Marlboro and Brattleboro, it offers views much farther to the east and west than anywhere else in the area. A communications tower can be seen in the foreground. The fire tower is a popular hike especially during the summer and fall.	The proposed towers will be easily visible from this vantage point. They will occupy a small portion of the overall view, and will be seen in the background of the view.
Broad views open up around White Road. Due to trees along the road the views alternate between the southern hills and; or to the northern mountains, Haystack and Mt. Snow. The existing turbines are easily visible but appear as a small part of the overall view. Several houses are in the foreground view.	The larger size of the new turbines will make them more visually dominant. A foreground hill will partially obscure some of the eastern string of turbines. The turbines will not be visible in the northwestern views of Haystack and Mt. Snow.

and the simulations to be produced, the result of the process usually is more widely accepted. Pre-construction surveys of residents, business owners, and tourists can provide a useful complement to public hearings to the degree that they reflect expertise in survey design and are free from bias. Other public-participation techniques are discussed in Chapter 5.

SCENIC RESOURCE VALUES AND SENSITIVITY LEVELS

Evaluating the aesthetic impacts of wind-energy projects ideally begins with an understanding of the elements and locations of the proposed project, as well as particular visual characteristics of the surrounding area that contribute to or detract from scenic or visual quality.

FIGURE D-5 Examples of landscapes of increasing visual diversity. (A) Landscape with no topographic and little vegetative diversity. (B) Increasing topographic diversity, some vegetative diversity (meadow, deciduous, and evergreen) and foreground, middleground, and background distance zones.
SOURCE: Photographs by Jean Vissering, Jean Vissering Landscape Architecture.

FIGURE D-5 (C) The contrast between high, irregular mountains and the flat lake create a dramatic setting. (D) The combination of highly diverse topography, exposed ledges, water, and vegetation in this scene make it highly scenic.
SOURCE: Photographs by Jean Vissering, Jean Vissering Landscape Architecture.

Regional Landscape Character and Distinctive Features

Landscape character depends on a combination of the natural and human or built landscapes. All landscapes are composed of unique combinations of topography (land form), vegetative patterns, and water features (lakes, rivers, streams, wetlands) that contribute to visual character. Superimposed on the natural landscape is the human or built landscape, also characterized by distinct patterns. For example, patterns of towns or villages may contrast with patterns of farms, fields, and forests. Some regions are characterized by numerous hills and ridges, while others have only a few distinct and prominent ridges or mountains. In some landscapes, certain natural or cultural features become focal points. Forestry practices, mining, suburban development, and recreational structures also are superimposed on the landscape and become part of its overall visual character.

Identifying Important Scenic Resources, Focal Points, and Unique Areas

Processes for determining relative scenic quality are well documented (USFS 1974, 1995; MADEM 1982; RIDEM 1990) (Box D-1). As noted above, however, these processes need to be combined with public review since landscape features that are locally or regionally valued may not be obvious to outside professionals. Identifying areas of high, medium, and low scenic quality is not difficult, although scenic quality is relative. A highly scenic area in upstate New York, for example, looks different from a highly scenic area in the Rocky Mountains. Scenic resources may be of local, regional, statewide, or even national significance. The underlying visual principles, however, are the same. Scenic quality alone is not necessarily sufficient reason to exclude a wind-energy project.

ASSESSMENT OF AESTHETIC IMPACTS

Factors affecting the visual impacts of a wind-energy project are listed below. The first set of factors concerns the particular landscape characteristics of the *site* and its surrounding context that may affect the sensitivity of views and the degree of aesthetic impact. The second set of factors relates to the characteristics of the *project* itself, how it is seen in these views, and how these may affect the overall experience of the landscape context. Visual impact assessments consider the combined effects of a proposed project throughout a region or on a locality as it is seen from all views, and particularly from sensitive viewpoints. No single view is likely to create serious impacts. Wind-energy projects inevitably are visible, but how they are seen within views, their relative prominence as seen throughout the region, and the degree to which they interfere with regional focal points or degrade unique or highly sensitive landscapes are important factors.

BOX D-1
Principles for Determining Scenic Quality

• *Visual Diversity (Variety Type):* The USFS uses the term "variety class" to describe a fundamental principle of landscape aesthetics: the greater the variety or diversity in the landscape, the more scenic it is likely to be. For example, landscapes with greater diversity in vegetation and topography are more likely to be scenic than a flat landscape with uniform vegetation. Water features such as rivers or ponds tend to add diversity, as do natural rock outcroppings. High scenic quality often results from the contrast among landscape features such as field and forest, steep and flat or rolling terrain, village and countryside. Particularly dramatic landscape features often stand out because of their contrast in form, line, color, or pattern (texture) (Figure D-5A-D).

• *Intactness (Order):* The principle of visual diversity relating to scenic quality generally holds for both natural and built landscapes. But in the human landscape too much diversity can lead to visual chaos or clutter (strip development being a good example, where every business vies for attention). Landscapes with a clear underlying order or logic tend to be more visually appealing (Lynch 1960, 1971). Undeveloped landscapes or those that retain 19th- or early 20th-century landscape patterns are becoming increasingly rare, and provide further examples of intact landscapes that may be of value. In some respects, wind-energy projects can provide a sense of order in the landscape because of their logical connections with very windy sites. The repetition of similar elements in many wind-energy projects can result in less visual clutter than the combined effect of other types of development.

• *Focal Points:* Focal points are elements in the landscape that stand out because of their contrasting shape (form), line, color, or pattern. They may also be elements of cultural importance. Often distinct focal points enhance scenic quality. They can be natural elements such as a lake, river, or mountain; or they can be built elements such as an important public building or central green. Some focal points may be locally important, others are regionally important and become landmarks that are visible from many vantage points. Occasionally, built elements that are viewed negatively become focal points, such as large clearcuts, mining operations, or power plants. Appropriate siting and design often can prevent developments from being viewed negatively by preventing them from conflicting with or degrading important regional focal points.

• *Unique Visual Resources:* Some visual resources may not meet the threshold of being highly scenic or sensitive, but may have visual value because of their uniqueness. Examples might include large tracts of wild or undeveloped land, some of which might even appear bleak and desolate.

Factors Affecting the Landscape Context

• *Distance from the Project:* In general, visual impacts are greater when objects are seen at close range (Figure D-6A-B; compare Figure 3-3 for a close view of the Mountaineer facility in West Virginia). In foreground

FIGURE D-6 View of wind-energy projects at various distances. (A) Madison Wind Project, Madison, New York from approximately 1 mile. (B) Simulation of proposed Equinox Wind Project in Manchester, Vermont, at 2.4 miles.
SOURCE: Photographs by Jean Vissering, Jean Vissering Landscape Architecture; simulation by EDR for Bennington County Regional Commission.

areas (up to a half-mile away) details can be seen and objects appear large and often occupy a large part of one's overall view.[4] Middle-ground views extend up to 5 miles away.[5] At this distance landscape patterns can be perceived, as can individual wind turbines, although they will appear smaller and part of a larger context than turbines in a foreground view. Background views are those greater than 5 miles where larger landforms tend to dominate the view. Wind turbines may be seen from 15 miles away, and even farther under optimal atmospheric conditions, but they appear very small at such distances, and appear as small portions of a larger panorama. Noise also diminishes with distance, and is of greatest concern within a half-mile (Chapter 4). Shadow flicker is also experienced only within close range (Chapter 4).

• *View Duration:* View duration refers to how long the project is visible as one drives along a road or paddles along a lake, for example. In many cases views of the project may be intermittent and seen through groupings of trees or buildings as one moves through the landscape. As with all considerations, view duration is evaluated along with other factors such as the distance of the project, sensitivity of the viewing area, and prominence of the land feature involved.

• *Angle of View:* Whether the project is seen directly ahead in views or to one side may influence the degree to which it is likely to be a focal point in views. Viewing a project from above usually makes roads and site clearing more visible than if seen from below.

• *Panoramic versus Narrow View:* When one sees a project as part of a wide panorama, it may appear to occupy a relatively small part of the view unless a particular landscape features make it a focal point.

• *Scenic Quality of View:* Highly scenic views are generally those with a high degree of landscape diversity, and with little or no landscape degradation (Figure D-7). Landscape degradation results from development that erodes existing scenic landscape patterns, or land uses that become unintended focal points due to their contrast in form, color or pattern with their surroundings. Panoramic views of high scenic quality are considered to be visually sensitive.

---

[4]Because of the larger scale—both vertical and horizontal—of more recent wind-energy projects, distance zones may need to be extended, with 2-3 miles considered a "foreground" area of greater potential visual effects.

[5]The original Forest Service Visual Management System used 5 miles to define the outer limits of the middleground zone. The more recent Scenery Management System changed this for purely clerical reasons rather than for reasons of visual perception (E. Crews, USFS, personal communication 2006). In fact the boundary is not sharp and particular topographic and air-quality conditions can affect the level of detail and significance of these distances. Nevertheless 5 miles is an appropriate distance, because land-use patterns are clearly visible within 5 miles.

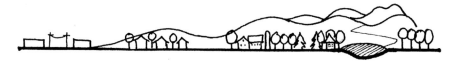

FIGURE D-7 Diagram of increasing scenic quality.
SOURCE: Jean Vissering, Jean Vissering Landscape Architecture.

• *Focal Point within a View:* Distinct cultural or natural focal points often enhance scenic quality (Figure D-8). When a focal point exists, new development will generally be more adversely perceived if it conflicts with or degrades the visual quality and prominence of a focal point.

• *Number of Observers:* Heavily used public areas, such as a heavily traveled road or a popular recreation area, are sometimes considered to be more sensitive than other areas. This criterion needs to be compared with other factors such as viewer expectations (below).

• *Viewer Expectations:* For certain uses there may be expectations for a primitive setting (wilderness camping) or for a natural setting (natural area) (Figure D-9A,B). Recreational areas restricted to non-motorized uses may be more sensitive to changes involving built elements than other settings.

• *Documented Scenic Resources:* Local, regional, or state planning documents that have been publicly adopted and that identify a particular site or area as having particular values merit serious attention. National and state recognition may carry greater weight than local recognition, but the latter still is worthy of attention.

FIGURE D-8 Haystack Mountain is a regional focal point due to its pyramidal shape (right). The proposed wind project would be located quite far away and along a less visually distinct ridgeline.
SOURCE: Photographs by Jean Vissering, Jean Vissering Landscape Architecture.

FIGURE D-9 Viewer expectations. (A) Water bodies used exclusively for non-motorized boats may be more visually sensitive than those used predominantly by motorized craft. (B) Wilderness areas can be considered highly visually sensitive, but are often predominantly wooded. Nevertheless, there may be views during leaf-off conditions that should be inventoried.

SOURCE: Photographs by Jean Vissering, Jean Vissering Landscape Architecture.

• *Visibility:* Projects that would be seen with great frequency within the study area may have higher impacts than projects that would be seen infrequently. Visibility must be studied along with the sensitivity, resource values, and prominence of the project within the views for an adequate assessment.

• *Weather Conditions:* Generally, projects are evaluated using "worst-case conditions," e.g., leaf-off visibility and clear skies. An abundance of clear skies makes aesthetic impacts in that area no worse or better than visual impacts in a region that has more cloudy skies. Indeed, a scenic view that is only rarely visible may be even more highly valued than one that usually can be seen.

Project Characteristics That May Affect Scenic Resources

• *Scale:* We perceive the size of an object in relation to its surroundings. The actual size of a wind turbine is less relevant than its perceived size in relation to its surroundings. Vertical scale (apparent height) in relation to the associated landmass, horizontal scale, and the overall project size are relevant. Despite the height of modern wind turbines, it is difficult for most people to distinguish between a 200-foot turbine and a 400-foot turbine unless they are side by side. Both appear much larger than surrounding trees and buildings, but the size becomes relevant in most cases only when it begins to appear to diminish the size and importance of a nearby natural feature such as a ridgeline.[6]

Horizontal scale contributes to the relative prominence of the project throughout the region. Certain western landscapes can accommodate larger projects than eastern landscapes of smaller scale. Projects may be too large when turbines become a constant occurrence within a landscape and when it is difficult to enjoy any views or ridgelines without wind turbines. Overall project size appears to be a significant issue in public acceptance of wind-energy projects in the United States (Figure D-10) (Pasqualetti et al. 2002).

• *Number of Turbines in the View:* The number of turbines visible at any one time may affect the prominence or relative scale of the project (Figure D-11). When wind turbines would be seen looking in all directions, or entirely covering the major landforms within a locality, the project may be viewed negatively, and further study probably will be needed.

• *Visual Clutter:* The accumulation of diverse built elements on a site, especially elements that contrast with their surroundings in form, color, and texture, can result in visual clutter (Figure D-12A,B). While it may seem

---

[6]Often the larger turbines appear less visually intrusive due to their greater spacing and the smaller numbers required for an equivalent power output.

FIGURE D-10 This project in Fenner, New York, generally works well in this high-elevation rolling agricultural landscape. The vertical relationship of turbines to distinct hills or ridgelines needs to be examined in simulations. The ridge above does not appear as prominent from most vantage points, but the issue could arise in other situations.
SOURCE: Photographs by Jean Vissering, Jean Vissering Landscape Architecture.

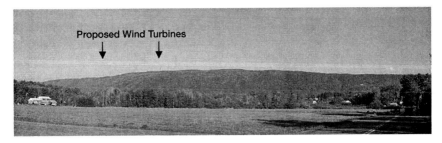

FIGURE D-11 Simulation of a proposed project in the Berkshire Mountains in Massachusetts. The proposed project would occupy only a portion of this longer ridge.
SOURCE: Photographs by Jean Vissering, Jean Vissering Landscape Architecture.

FIGURE D-12 (A) The repetition of identical elements that is characteristic of wind-energy projects helps to create a sense of order. (B) The valley location at San Gorgonio (Palm Springs, CA) diminishes the scale of this large project, but the overall accumulation of different turbine types results in a much more cluttered appearance than is likely in future project planning and maintenance.
SOURCE: (A) Photograph by Sandy Wobeck, East Montpelier Gully Jumper; (B) Photograph by David Policansky.

logical to place wind-energy projects in already-built landscapes, too much development can result in an increasingly chaotic or cluttered landscape. Because wind-energy projects involve the repetition of like elements, they often result in greater unity and less clutter than some other types of development. Even combining wind turbines with cell towers may increase visual clutter and therefore, visual impact. The introduction of different sizes and types of wind turbines over the life of a project can potentially severely degrade a landscape (Gipe 2003).

- *Visibility of Project Infrastructure:* Visibility of project roads, power lines, substations, and other infrastructure can substantially increase visual clutter (see above) and therefore visual impacts. These also increase the perceived scale of a project. In wooded landscapes, clearing resulting from installation of roads, power lines, and grade changes can visually alter a forested landscape.

- *Noise:* To the extent that noise degrades the character and experience of a particular landscape, it is an aesthetic concern. Most modern turbines are relatively quiet, but noise can be an aesthetic concern primarily for residents living within half a mile of a wind-energy project. Careful siting of individual wind turbines as well as selection of turbines rated for low noise can help to reduce these impacts.

- *Lighting:* Night lighting can be one of the most difficult aspects of a wind-energy project to evaluate, and may result in some of the greatest concerns. The importance of changes in landscape depend on where it occurs on the continuum of urban to wild landscape, as well as the project's overall visibility and proximity. In many landscapes where projects have been built or proposed, there currently is little night lighting. Red lights have less contrast than white lights with the night sky in terms of value, but they differ markedly from colors typically observed in the night landscape (except where other objects occur with obstruction lighting).

Other Issues Affecting Visual Impacts

- *Cumulative Impacts:* This issue relates both to the expansion of existing projects and to the addition of new projects within a geographic area. The first possibility raises concerns of the overall project scale and its appropriateness for the particular landscape. The second raises concerns of both scale and overburdening a particular locality with development impacts. Developing state-wide or region-wide siting guidelines can help prevent the undue impacts that may result from numerous projects being proposed over time within certain areas.

- *Meaningful Benefits:* Perceptions of aesthetic attractiveness are often linked to real or tangible benefits. For many people, however, the benefits of "cleaner air" or "less dependence on foreign fuels" may seem too intangible, and usually they occur at least in part away from the areas subject to aesthetic impacts. Linking wind-energy development to both economic benefits at the local level and a meaningful program of pollution reduction at the state, local, and federal levels can enhance public perception of the benefits of wind energy. Developing direct community participation and links to the wind-energy projects they are hosting also can help these projects become a meaningful part of "place" (Pasqualetti et al. 2002).

Other Methods for Identifying Aesthetic Impacts

Public Participation and Surveys

Communities around the country have used a range of techniques for eliciting public opinions, and the effectiveness of these approaches needs further study. When a specific project is proposed in a particular area, the focus must be on understanding the site and the perceptions of the community members who live and work in the area. Aesthetic effects are site-specific and individual communities react differently. There is considerable evidence that public acceptance increases with a sense of involvement in the project. Involvement includes active efforts to inform neighbors, providing thorough analyses, responding to expressed concerns with alterations in project design, and providing material or monetary benefits to affected individuals or to the community at large.

Much of what we know about public reactions is anecdotal. Statistically valid and independently conducted pre- and post-construction surveys provide useful information about public perceptions of wind-energy projects and help determine what factors are important in public perceptions. Such surveys are commonly conducted in Europe, but much less often in the United States. To permit generalization of information gathered from public perceptions, surveys need to be carefully designed to factor in particular project attributes, site features, and the public processes followed in presenting the project to the public (Priestly 2006). Attitudes of nearby residents and recreational users from elsewhere may be quite different.

Independent and Peer Review

Experts in aesthetics hired by developers may be perceived as biased in favor of the developer. Two approaches have been used for obtaining independent reviews of proposed wind-energy projects. Some state or local governments hire independent experts to conduct visual impact assessments. In other states a process of peer review is used. Two or more independent experts in aesthetics review the work of the developer's consultant. Usually they are presented with project information including visibility maps, simulations, and photographs of landscape character. They are asked to evaluate a number of sensitive viewpoints for which simulations have been prepared and to score the degree of contrast resulting from the proposed project. This process could easily be institutionalized by reviewing bodies. In both cases, the developer generally pays for this independent review process.

MITIGATION TECHNIQUES

Some visual impacts will be inevitable with any wind-energy project. Reducing or minimizing negative impacts can be achieved in a number of ways. A well-sited and designed project will have incorporated some of the techniques into the original application. If there appear to be significant visual impacts resulting from the project, additional mitigation approaches can be used. If none can adequately reduce the visual impacts, the project may be found to be unsuited for the particular site. Mitigation techniques include the following:

• *Appropriate Siting:* This critical mitigation technique involves avoiding a site that is located on valued regional scenic resources, or that appears very prominent throughout a region. Selecting a site that can comfortably accommodate the number of turbines desired without visually overwhelming sensitive scenic resources on or near the site and the region as a whole also is important. Appropriate siting may also need to address potential issues of cumulative impacts (see below) so that a particular area or landscape type is not overburdened with wind-energy development.

• *Downsizing:* Reducing the scale of the project (numbers of turbines or height of turbines)[7] can help the project fit more comfortably into its surroundings. In some cases one or more turbines may be particularly prominent from sensitive viewpoints, or the overall scale of the project may overwhelm the particular land form or surrounding landscape. In most settings the difference in overall turbine height are difficult to distinguish. The difference between a 200-foot turbine and a 360-foot turbine (hub or nacelle height) can be difficult to perceive, especially when the turbines are seen against the sky. Size may make a difference if the height of the land-form begins to be overwhelmed by the height of the turbine. Generally, fewer larger turbines can result in a better visual outcome than a larger number of smaller turbines.

• *Relocation:* Moving turbines from one location to another can help, but it may not be possible in all cases. Relocation can be used to

---

[7]Turbine heights also have effects on project productivity and on avian and bat mortality, which must be balanced with aesthetic issues.

avoid proximity to residences or visual prominence from sensitive viewing areas.[8]

- *Lighting:* The revised Federal Aviation Administration (FAA) lighting guidelines reduce lighting impacts. Lighting impacts often are of greatest concern to residents and recreationists, and should be minimized to the greatest extent possible. Any new technologies or modification of FAA lighting requirements that can further reduce lighting for wind turbines ideally should be incorporated into design standards.

- *Turbine Pattern:* In most cases turbines are located to take advantage of small rises in the land, flatter terrain, or other site features that determine their pattern or organization on the ground. Some studies suggest that turbine configurations can be designed to respond in meaningful or visually pleasing ways to their surroundings. In rolling landscapes a less rigid arrangement that reflects topography may be preferable, while in flatter landscapes, especially with patterns of rectangular fields or roads, a more geometric or linear pattern may work better. Simulations provide a useful way to study the effects of different turbine patterns from sensitive viewing areas.

- *Infrastructure Design:* Paying attention to project infrastructure such as meteorological towers, substations, power poles, and project buildings in addition to the turbines themselves is important. Generally, it is advisable to screen all project infrastructure from view to the greatest extent possible.

- *Color:* A recent FAA study showed that daytime lighting could be eliminated provided that turbines are white. White often is regarded as more cheerful and less industrial than other colors, which may be part of the reason some people find wind turbines more visually appealing than, for example, cell towers. Bright patterns and obvious logos can be avoided. Unobtrusive colors are important in other project infrastructure such as operations buildings, transmission support poles, and road surface materials. In general, darker colors are less noticeable, especially against a background of vegetation.

- *Maintenance:* People find wind turbines more visually appealing when the blades are rotating than when they are still (Pasqualetti et al. 2002). Requirements for immediate repairs of wind turbines can be part of permit requirements. Also the replacement of wind turbines with visually different wind turbines can result in visual clutter, so replacing wind turbines with the same or a visually similar model over the lifetime of the project may be an important requirement. Sufficient funds need to be assured for this purpose.

---

[8]Moving turbines away from a high point of land often results in minimal aesthetic benefits in contrast to a fairly significant reduction in electrical production.

• *Decommissioning:* Once a project or individual turbine can no longer function, requirements for removing the project infrastructure and reclaiming the site are important. A plan for decommissioning may be required as part of the permit application. In some cases, money is reserved in escrow for this purpose.

• *Non-reflective Materials:* Use of materials that will not result in light reflection may be required, for all project components.[9]

• *Minimizing Vegetation Removal:* Ideally, existing vegetation should be retained to the greatest extent possible. Clearcuts generally have negative visual impacts (Brush 1979). Screening areas of cleared forest may be advisable, as well as maintaining vegetation along roadsides and around turbines.

• *Screening:* While turbines cannot be screened from view, other project infrastructure (roads, power lines, substations, and buildings) can be. Existing vegetation is usually preferable, but plantings may be needed and should incorporate typical indigenous vegetation.

• *Noise:* Noise and siting standards can help reduce impact on residents near the project (generally within half a mile). Noise standards can be set at firm levels such as 40 dB(a)h (decibels corrected or A-weighted for sensitivity of the human ear) nighttime and 50 dB(a) daytime at the property line or at residential structures; or can be set as an increment above ambient noise levels (e.g., a maximum of 5 dB(a) above ambient noise levels). Post-construction monitoring is important here as in many aspects of the impacts of wind-energy facilities.

• *Burial and Sensitive Siting of Power Lines:* Collector lines often are buried between turbines. In very sensitive viewing locations other collector and transmission lines may also need to be buried (see Figures 3-2A and 3-2B).

• *Offsets:* In some cases protecting an offsite visual resource can help to offset the impacts of the project if mitigation cannot be accomplished on site.

DETERMINATION OF ACCEPTABLE OR UNDUE AESTHETIC IMPACTS

Decision makers usually need guidance to evaluate under what circumstances the degradation of aesthetic resources may outweigh the benefits of a proposed project. The immediate question may be: *would this particular project result in undue harm to valuable aesthetic resources in this particular setting?* At a policy level, the question is broader: how can wind-energy projects be accommodated while retaining the valued scenic resources of

---

[9]Color and reflectivity may also be a consideration for avian and bat mortality.

the state and of individual communities? These questions can be addressed systematically using the process described above and relying on well-established aesthetic principles. Many sites are likely to be suited to wind-energy development, and where these occur, the question becomes: *does this project as designed work on this site or will mitigation be required?* Mitigation possibilities are discussed above, but there will be circumstances when mitigation techniques fail to address critical problems with the site itself. Visibility alone generally does not result in a wind-energy project's being perceived as unacceptable. If the project appears to result in many issues, to involve important regional scenic resources, and to significantly affect the ability of people to enjoy these resources, then the project may be perceived as or judged to be unacceptable. Some questions to consider in revieweing wind-energy projects are listed below. Assuming that a high-quality wind site is involved, decision-making agencies may feel more comfortable in concluding that the aesthetic impacts are undue if more than one of the following concerns is involved. Ideally, the criteria will be weighed against the overall public benefits of the project and along with the general suitability of the site in other respects (see Box D-2 and Chapter 5 of this report).

Questions to Consider in Determining Acceptability of Visual Impacts

• Is the project located within an area of identified scenic or cultural significance?[10]

• Would the project significantly degrade views or scenic resources of regional or statewide significance?

• Is the project on or close to a natural or cultural landscape feature that is a regional focal point?

• Is the project in a landscape area that is visually distinct and rare or unique?

• Is the project unreasonably close (usually less than a half-mile) to many residences that would be severely affected, especially as a result of noise, shadow flicker, or by being completely surrounded by wind turbines?

• Will the project occupy an area valued for its wildness and remoteness? If these values have been specifically documented, then consideration of the appropriateness of a wind-energy project becomes even more important.

• Would the project's scale in terms of turbine height or numbers

---

[10]Preferably the scenic values have been identified in public documents rather than merely identified through the aesthetic impacts assessment process. However, few states or localities have taken steps to document scenic resources, so a careful visual impact assessment process may be the only available tool.

BOX D-2
Maine's Department of Environmental Protection Visual Impact Assessment Criteria (MEDEP 2003)

• *Landscape Compatibility:* Which is a function of the subelements of color, form, line, and texture. Compatibility is determined by whether the proposed activity differs significantly from its existing surroundings and the context from which they are viewed such that it becomes an unreasonable adverse impact on the visual quality of protected natural resources as viewed from a scenic resource.

• *Scale Contrast:* Which is determined by the size and scope of the proposed activity given its specific location within the viewshed of a scenic resource.

• *Spatial Dominance:* Which is the degree to which an activity dominates the whole landscape composition or dominates landform, water, or sky backdrop as viewed from a scenic resource.

---

of turbines overwhelm the landscape in which it occurs? (For example, would scenic views that are free of turbine remain throughout the region, or would wind turbines occupy all or most notable ridgelines within view of the area?)

• Will the project result in unreasonable visual clutter due to its combination with existing built features that already degrade landscape features? This is an issue of cumulative impacts.

• Has the applicant used reasonable and available mitigating techniques that would reduce the project's impacts?

• Does the project violate a clear, written community standard intended to protect the aesthetics or scenic beauty of the area? Such a standard ideally will be legally adopted by a community or state, and provide clear guidance to developers and be based on sound principles of aesthetic resource assessment.

SAMPLE PEER REVIEW EVALUATION SHEET[11]

Panel Member: _____

Date: _____

Viewpoint #: _____

Viewpoint Description: _____

_____

_____

_____

Visual Impact:

Rate the Project's contrast with existing conditions on a scale of 1 (completely compatible) to 5 (strong contrast). Under comments, explain the reason for rating focusing on the elements of line, scale, color, texture, and form. Then provide your overall assessment of the project's aesthetic impact from this viewpoint.

Landscape Component	Contrast	Comments
Vegetation		
Land Use		
Land Form		
Viewer Activity		
Water		
Total		
Average Score		

Overall Aesthetic Impact:

_____

_____

[11]This form was adapted from one used by Michael Buscher ASLA of T. J. Boyle and Associates.